Lecture Notes in Artificial Intelligence 4048

Edited by J. G. Carbonell and J. Siekmann

Subseries of Lecture Notes in Computer Science

Lou Goble John-Jules Ch. Meyer (Eds.)

Deontic Logic and Artificial Normative Systems

8th International Workshop on Deontic Logic in Computer Science, DEON 2006
Utrecht, The Netherlands, July 12-14, 2006
Proceedings

Springer

Series Editors

Jaime G. Carbonell, Carnegie Mellon University, Pittsburgh, PA, USA
Jörg Siekmann, University of Saarland, Saarbrücken, Germany

Volume Editors

Lou Goble
Willamette University, Department of Philosophy
Salem, OR 97301, USA
E-mail: lgoble@willamette.edu

John-Jules Ch. Meyer
Utrecht University, Institute of Information and Computing Sciences
Intelligent Systems Group, 3508 TB Utrecht, The Netherlands
E-mail: jj@cs.uu.nl

Library of Congress Control Number: Applied for

CR Subject Classification (1998): I.2.3, I.2.11, I.2, F.4.1

LNCS Sublibrary: SL 7 – Artificial Intelligence

ISSN 0302-9743
ISBN-10 3-540-35842-0 Springer Berlin Heidelberg New York
ISBN-13 978-3-540-35842-8 Springer Berlin Heidelberg New York

Springer is a part of Springer Science+Business Media

springer.com

Printed in Germany

Typesetting: Camera-ready by author, data conversion by Scientific Publishing Services, Chennai, India
Printed on acid-free paper SPIN: 11786849 06/3142 5 4 3 2 1 0

Preface

This volume presents the papers contributed to DEON 2006, the 8th International Workshop on Deontic Logic in Computer Science, held in Utrecht, The Netherlands, July 12–14, 2006. These biennial DEON (more properly, ΔEON) workshops are designed to promote international cooperation among scholars across disciplines who are interested in deontic logic and its use in computer science. They support research that links the formal-logical study of normative concepts and normative systems with computer science, artificial intelligence, philosophy, organization theory, and law.

Papers for these workshops might address such general themes as the development of formal systems of deontic logic and related areas of logic, such as logics of action and agency, or the formal analysis of all sorts of normative concepts, such as the notions of rule, role, regulation, authority, power, rights, responsibility, etc., or the formal representation of legal knowledge. They might also be more concerned with applications, such as the formal specification of systems for the management of bureaucratic processes in public or private administration, or the specification of database integrity constraints or computer security protocols, and more. Of particular interest is the interaction between computer systems and their users. (The DEON 2006 website, `http://www.cs.uu.nl/deon2006/`, contains links to previous workshops and their papers. This history reveals a vibrant interdisciplinary research program.)

In addition to those general themes, the 2006 iteration of the workshop focused also on the special topic of artificial normative systems, their theory, specification and implementation, such as electronic institutions, norm-regulated multi-agent systems and artificial agent societies generally. Here too the concern is both with theoretical work, such as the design of formal models and representations, and also work more oriented toward implementation, such as architectures, programming languages, design models, simulations, etc.

The 18 papers printed here were selected for presentation at the workshop after a thorough process of review and revision. All are original and presented here for the first time. They range from studies in the pure logic of deontic operators to investigation of the normative extension of the computer language C+ to examination of the structure of normative systems and institutions. The titles themselves demonstrate commitment to the themes of the workshop. In addition to these full papers, we present abstracts of the talks of our three invited speakers, José Carmo (University of Madeira), Frank Dignum (University of Utrecht), and Paola Petta (University of Vienna).

We are grateful to all who contributed to the success of the workshop, to our invited speakers, to all the authors of the presented papers, to all who participated in discussion. Special thanks go to the members of the Program Committee for their service in reviewing papers and advising us on the program and to the members of the Organization Committee for taking care of all the countless details that a workshop like this requires, especially Jan Broersen for setting up and maintaining the DEON 2006 website and Henry Prakken for financial arrangements, sponsorships, and more. Thanks too to Richard van de Stadt whose CyberChairPRO system was a very great help to us in organizing the papers from their initial submission to their final publication in this volume. We are also very grateful to the several sponsoring organizations for their essential support. Finally, we wish to express our appreciation to Springer for publishing these proceedings in their LNCS/LNAI series. This is the second such volume in this series; the first was from DEON 2004, *Deontic Logic in Computer Science*, LNAI 3065, edited by Alessio Lomuscio and Donald Nute. We hope these volumes may continue into the future to provide a record of research in this rich and growing field.

April 2006

Lou Goble
John-Jules Ch. Meyer

Workshop Organization

Organization Committee

General Co-chairs	Lou Goble, Willamette University
	John-Jules Ch. Meyer, University of Utrecht
Local Organization Co-chairs	Henry Prakken, University of Utrecht
	Jan Broersen, University of Utrecht
Local Committee	Gerard Vreeswijk, University of Utrecht
	Davide Grossi, University of Utrecht
	Susan van den Braak, University of Utrecht

Program Committee

Co-chairs

Lou Goble	Willamette University
John-Jules Ch. Meyer	University of Utrecht

Members

Paul Bartha	University of British Colombia
Jan Broersen	University of Utrecht
Mark Brown	Syracuse University
José Carmo	University of Madeira
Frédéric Cuppens	ENST-Bretagne Rennes
Robert Demolombe	ONERA Toulouse
Frank Dignum	University of Utrecht
Risto Hilpinen	University of Miami
John Horty	University of Maryland
Andrew Jones	King's College London
Lars Lindahl	University of Lund
Alessio Lomuscio	University College London
David Makinson	King's College London
Paul McNamara	University of New Hampshire
Ron van der Meyden	University of New South Wales / NICTA
Donald Nute	University of Georgia
Rohit Parikh	City University of New York
Henry Prakken	University of Utrecht / University of Groningen
Filipe Santos	ISCTE Portugal
Giovanni Sartor	University of Bologna

Krister Segerberg	Uppsala University
Marek Sergot	Imperial College London
Carles Sierra	IIIA-CSIC Barcelona
Leon van der Torre	University of Luxembourg

with the assistance of:

Guido Boella	University of Turin
Robert Craven	Imperial College London
Dorian Gaertner	Imperial College London
Davide Grossi	University of Utrecht
Olga Pacheco	University of Minho

Sponsors

SIKS (Dutch Research School of Information and Knowledge Systems)
JURIX (Dutch Foundation for Legal Knowledge and Information Systems)
BNVKI (Belgian-Dutch Association for AI)
NWO (Netherlands Organisation for Scientific Research)
KNAW (The Royal Netherlands Academy of Arts and Sciences)
Faculty of Science, University of Utrecht
Department of Information and Computing Sciences, University of Utrecht

Table of Contents

Abstracts of Invited Papers

Contributed Papers

Roles, Counts-as and Deontic and Action Logics

José Carmo

Engineering and Mathematics Department
University of Madeira
Campus Universitário da Penteada
9000-390 Funchal, Madeira, Portugal
jcc@uma.pt

An organization may be the subject of obligations and be responsible for not fulfilling its obligations. And in order for an organization to fulfill its obligations, it must act. But an organization cannot act directly, so someone must act on its behalf (usually some member of the organization), and this must be known by the "external world" (by the agents that interact with the organization).

In order to account for this, the organization is usually structured in terms of what we may call *posts*, or *roles* within the organization, and the statute of the organization distributes the duties of the organization among the different posts, specifying the norms that apply to those that occupy such positions (that hold such roles), and describing who has the power to act in the name of the organization. But this description is abstract, in the sense that it does not say which particular person can act in the name of the organization; it attributes such power to the holders of some roles. Depending on the type of actions, the power to act in the name of an organization may be distributed through different posts, and the holders of such posts may (or may not) have the permission or the power to delegate such power. On the other hand, those that can act in the name of an organization can establish new obligations for the organization through their acts, for instance by establishing contracts with other agents (persons, organizations, etc.). And in this way we have a dynamic of obligations, where the obligations flow from the organization to the holders of some roles, and these, through their acts, create new obligations in the organization.

On the other hand, a person (or, more generally, an agent) can be the holder of different roles within the same organization or in different organizations (being the subject of potentially conflicting obligations), and can *act by playing different roles*. And in order to know the effects of his acts we must know in which role they were played. Thus, it is fundamental to know which acts *count as* acts in a particular role.

If we want a logical formalism to abstractly specify and reason about all these issues, we need to consider and combine deontic, action and counts-as operators. Particularly critical is to decide which kind of action logic we consider. For some aspects, like that of describing how the obligations flow from the organization to the holders of some posts and how some of the acts of the latter count as acts of the organization, it seems it is better to consider a "static" approach based on the "brings it about" action operators. On the other hand, if we want to be able to describe the dynamics of the obligations deriving, for instance, from the contracts that are made in the name of the organization, it seems that a dynamic logic is necessary, or at least very useful. However, the combination of the two kinds of logic of actions has proven to be not an easy task. This paper addresses these issues.

L. Goble and J.-J.C. Meyer (Eds.): DEON 2006, LNAI 4048, p. 1, 2006.

Norms and Electronic Institutions

F. Dignum

Institute of Information and Computing Sciences
Utrecht University, The Netherlands

Abstract. The term *Electronic Institution* seems to be well accepted in the agent community and to a certain extent also in the e-commerce research community. However, a search for a definition of an electronic institution does not yield any results on the Internet. This is different for the term *institution*. North [9] defines institutions (more or less) to be a set of norms that govern the interactions of a group of people. Examples are *family* and *government*. Here we are not so much interested in giving a very precise definition of an institution, but just want to note that the concept refers to a very abstract notion of a set of norms or social structure.

It is not immediately clear how such an abstract set of norms can be "electronic". The term electronic institution is therefore a bit misleading. It actually refers to a description of a set of electronic interaction patterns that might be an instantiation of an institution. E.g. an electronic auction house (which can be seen as an instantiation of the auction institution). So, it is not referring (directly) to a set of norms or a social structure. However, because the term is widely used by now, although it is not entirely appropriate, we will stick with using the term "electronic institution" to refer to such a kind of specification.

In contrast to the situation in human society, where these interaction patterns might emerge over a long period of time and the institutionalization follows after the stabilizing of these patterns, the electronic institutions are specifically designed by humans to fit with existing institutions. E.g. an electronic market can be designed to instantiate the auction institution. So, the mechanism and interactions to be used in the electronic market can be designed such that they comply to the norms specified in the auction institution (e.g. following a bidding protocol to ensure a fair trade).

If the electronic institution is specified and implemented using a tool like AMELI [5] then the agents interacting in the institution can only follow precisely the pre-specified interaction patterns. Any attempt to perform a deviating action is caught by the so-called governors and has no effect. Thus if the interaction patterns are such that agents always interact in a way that keeps the system in a non-violation state according to the norms of the institution then by definition the agents will never (be able to) violate any of the norms of the institution.

However, this is not the only way to "instantiate" the set of norms that define the institution. One of the main characteristics of norms is that they can be violated. So, how does this relate to the design of electronic institutions? Should they also allow for violations? If they allow for violations, what should happen in these violation states?

L. Goble and J.-J.C. Meyer (Eds.): DEON 2006, LNAI 4048, pp. 2–5, 2006.

What we are actually looking for is what it means for an electronic institution to *instantiate* an (existing) institution (seen as a set of norms) in this context. And subsequently what are the concepts necessary to describe all elements of the electronic institution such that one could "prove" that it actually instantiates the abstract institution.

One obvious way to go ahead is to use (a) deontic logic to specify the norms and use the same formalism to specify the electronic institution. This specification can then be used to describe ideal vs. real behavior. It can also be used to verify compliance to norms and/or to reason about the combination of norms. We can even use it to check whether the system has means to return from any possible violation state to a permitted state.

However, although this approach is a good first step, it does not capture all the subtleties involved. We will briefly touch upon a few of these issues in the following.

One immediate problem is the connection of an abstract set of norms with a concrete specification of interaction patterns. Almost by definition, the terms used in the norm specification are more abstract than the terms used in the specification of the electronic institution. E.g. a norm might mention a prohibition to "reveal" certain information. Because agents will not have a specific action "reveal", they will by definition comply to this norm. However, there is of course a connection between (sets of) actions that an agent can perform and the abstract action of revealing. This relation is usually given using the *counts-as* relation. Some important groundwork on this relation has already been done in [8] but much work still needs to be done to capture all aspects of it (see e.g. [7]).

Another issue that is also related to levels of abstraction are the temporal aspects of norms. Often norms are abstracting away from the use of temporal aspects. E.g. the winning bidder in an auction has to pay for the item she has won. However, in order to compare a norm with a concrete specification of interactions the temporal aspects are of prime importance. Does the winning bidder have to pay right away, before some deadline, at some time in the future,...? So, it seems to be important to specify the norms at least with some kind of temporal logic in order to establish this relation. Some first steps in this direction are taken in [2, 1], but no complete formal analysis is as yet given of a temporal deontic logic.

A third issue that arises is that some norms seem to relate to the behavior of a complete organization. E.g. the auction house should ensure the payment of auctioned items. The question is which agents of the auction house should take care of this? Should it be only one agent or more than one? Should there be backups for if an agent fails? In general this is the question on how a norm for an organization dissipates over the members of that organization. This depends, of course, on the organizational structure(s), power relations, knowledge distribution, etc. Some preliminary work is described in [6, 10].

A fourth issue is that of norm enforcement within the electronic institution. Basically this can be done in two ways: Preventing a norm from being violated or checking for the violation of a norm and reacting on the violation. A decision on which method to choose depends on aspects such as efficiency and safety of the electronic institution. It also reveals another important aspect, if an agent has to enforce a certain norm it should be able to "know" whether the norm is violated or not. Often this aspect leads to certain additional interactions which have as only purpose to gather information necessary to check a violation of a norm [11]. E.g. if a participant of an auction should be at least 18 years old, the auction house might institute a registration protocol in which a participant has to prove he is over 18.

A last issue I would like to mention here is the influence of the existence of the norms themselves on the behavior of the agents. In human situations the existence of the norms alone influences the decision process of the persons. In an electronic institution one might have a more complicated situation where some agents are software agents, while others are human agents. How does this influence the interactions? Can we effectively build norm aware agents? Some theory does exist (e.g. [4, 3]), but no practical implementations yet. Does this function better, more efficient, or not?

In the above I have risen more questions than given answers. However, I think they are interesting questions and very relevant if one considers the more general relation between deontic logic and computer science. The relation between deontic logic and computer science is also a relation between the abstract (philosophical logic) and the concrete (engineered processes). So, besides giving an idea of the place of norms in electronic institutions, I hope this presentation also encourages some people to (continue to) perform research in the areas mentioned above.

References

1. H. Aldewereld, F. Dignum, J-J. Ch. Meyer, and J. Vázquez-Salceda. Proving norm compliance of protocols in electronic institutions. Technical Report UU-CS-2005-010, Institute of Information & Computing Sciences, Universiteit Utrecht, The Netherlands, 2005.
2. J. Broersen, F. Dignum, V. Dignum, and J-J. Ch. Meyer. Designing a deontic logic of deadlines. In A. Lomuscio and D. Nute, editors, *Proceedings of DEON'04*, LNAI 3065, pages 43–56. Springer, 2004.
3. C. Castelfranchi, F. Dignum, C. Jonker, and J. Treur. Deliberate normative agents: Principles and architectures. In N. Jennings and Y. Lesperance, editors, *Intelligent Agents VI*, LNAI 1757, pages 364–378. Springer-Verlag, 2000.
4. F. Dignum. Autonomous agents with norms. *Artificial Intelligenc and law*, 7:69–79, 1999.
5. M. Esteva, J.A. Rodríguez-Aguilar, B. Rosell, and J.L. Arcos. Ameli: An agent-based middleware for electronic institutions. In *Third International Joint Conference on Autonomous Agents and Multi-agent Systems*, pages 236–243, New York, US, 2004. IEEE Computer Society.
6. D. Grossi, F. Dignum, L. Royakkers, and J-J. Ch. Meyer. Collective obligations and agents: Who gets the blame. In A. Lomuscio and D. Nute, editors, *Proceedings of DEON'04*, LNAI 3065, pages 129–145. Springer, 2004.
7. D. Grossi, J-J. Ch. Meyer, and F. Dignum. Counts-as: Classification or constitution? an answer using modal logic. In J-J. Ch. Meyer and L. Goble, editors, *Proceedings of DEON'06*, LNAI this volume. Springer, 2006.

8. A. J. I. Jones and M. Sergot. A formal characterization of institutionalised power. *Journal of the IGPL*, 3:429–445, 1996.
9. D. C. North. *Institutions, Institutional Change and Economic Performance*. Cambridge University Press, Cambridge, 1990.
10. L. Royakkers, D. Grossi, and F. Dignum. Responsibilities in organizations. In J. Lehman, M. A. Biasiotti, E. Francesconi, and M. T. Sagri, editors, *LOAIT - Legal Ontologies and Artificial Intelligence Techniques*, volume 4 of *IAAIL Workshop Series*, pages 1–11, Bologna, June 2005. Wolf Legal Publishers.
11. J. Vázquez-Salceda, H. Aldewereld, and F. Dignum. Norms in multiagent systems: from theory to practice. *International Journal of Computer Systems Science & Engineering*, 20(4):95–114, 2004.

Emotion Models for Situated Normative Systems?

Paolo Petta[1,2]

[1] Institute of Medical Cybernetics and Artificial Intelligence
Center for Brain Research, Medical University of Vienna
Freyung 6/2, A-1010 Vienna, Austria (EU)
[2] Austrian Research Institute for Artificial Intelligence*
Freyung 6/6, A-1010 Vienna, Austria (EU)
Paolo.Petta@ofai.at

Research in logic-based multi-agent modelling has been pushing steadily the boundaries of the domain models adopted, while associated enquiries on the relations between constituent entities contribute in turn to an improved understanding of the underlying domain as well as pave the way for moving beyond static scenarios of analysis (see e.g., Munroe et al. 2003, Boella and van der Torre 2004, Dastani and van der Torre 2005, as well as theoretical work on dynamic semantics in logics).

The present talk results from a thread of activities including an ongoing investigation into the relation between the Emotional and computational models of situated normative systems (Staller and Petta 2001, Petta 2003) and work towards the realisation of dynamical representations in multi-agent systems (e.g, Jung and Petta 2005). In it, we will draw a picture of today's status in emotion theorising from the perspective of the ongoing dialogue between computational and psychological research. We will develop a view of the domain of human emotions as informed in particular by cognitive appraisal theories and situated cognition research that illustrates the role of emotions within the coordination of action and (different kinds of) cognition in social scenarios and tries to clarify the nature of processes and concepts involved.

References

Boella, G., Torre, L. van der: Δ: The social delegation cycle, in Lomuscio A., Nute D. (eds.), *Deontic Logic: 7th International Workshop on Deontic Logic in Computer Science, DEON 2004*, Madeira, Portugal, May 26-28, 2004. Proceedings, Lecture Notes in Computer Science 3065, Springer-Verlag Berlin Heidelberg New York, pp.29–28, 2004.

* The Austrian Research Institute for Artificial Intelligence is supported by the Austrian Federal Ministry for Education, Science and Culture and by the Austrian Federal Ministry for Transport, Innovation and Technology. This work is supported by the European Union's FP6 Network of Excellence Humaine (Contract No. 507422). This publication reflects only the authors' views: The European Union is not liable for any use that may be made of the information contained herein.

L. Goble and J.-J.C. Meyer (Eds.): DEON 2006, LNAI 4048, pp. 6–7, 2006.

Dastani, M., Torre, L. van der: Decisions, deliberation, and agent types: CDT, QDT, BDI, 3APL, BOID. *Artificial Intelligence and Computer Science*, Nova Science, 2005.

Jung, B., Petta, P.: Agent encapsulation in a cognitive vision MAS, in Pechoucek M. et al. (eds.), *Multi-Agent Systems and Applications IV, 4th International Central and Eastern European Conference on Multi-Agent Systems, CEEMAS 2005*, Budapest, Hungary, September 2005, Proceedings, Springer-Verlag Berlin Heidelberg New York, pp.51–61, 2005.

Munroe, S.J., Luck, M., d'Inverno, M.: Towards motivation-based decisions for worth goals, in Marik V. et al. (eds.), *Multi-Agent Systems and Applications III, 3rd International Central and Eastern European Conference on Multi-Agent Systems, CEEMAS 2003*, Prague, Czech Republic, June 2003, Proceedings, Lecture Notes in Artificial Intelligence 2691, Springer-Verlag Berlin Heidelberg New York, pp.17–28, 2003.

Petta P.: The role of emotions in a tractable architecture for situated cognizers, in Trappl R. et al. (eds.), *Emotions in Humans and Artifacts*, MIT Press Cambridge, Massachusetts London, England, 251-288, 2003.

Staller A., Petta P.: Introducing emotions into the computational study of social norms: A first evaluation, *Journal of Artificial Societies and Social Simulation*, **4**(1), 2001.

Addressing Moral Problems Through Practical Reasoning

Katie Atkinson and Trevor Bench-Capon

Department of Computer Science
University of Liverpool
Liverpool L69 7ZF UK
{katie, tbc}@csc.liv.ac.uk

Abstract. In this paper, following the work of Hare, we consider moral reasoning not as the application of moral norms and principles, but as reasoning about what ought to be done in a particular situation, with moral norms perhaps emerging from this reasoning. We model this situated reasoning drawing on our previous work on argumentation schemes, here set in the context of Action-Based Alternating Transition Systems. We distinguish what prudentially ought to be done from what morally ought to be done, consider what legislation might be appropriate and characterise the differences between morally correct, morally praiseworthy and morally excusable actions.

1 Introduction

In Freedom and Reason [7], R.M. Hare, perhaps the leading British moral philosopher of the twentieth century, notes that:

> "There is a great difference between people in respect of their readiness to qualify their moral principles in new circumstances. One man may be very hidebound: he may feel that he knows what he ought to do in a certain situation as soon as he has acquainted himself with its most general features ... Another man may be more cautious ... he will never make up his mind what he ought to do, even in a quite familiar situation, until he has scrutinized every detail." (p.41)

Hare regards both these extreme positions as incorrect:

> "What the wiser among us do is to think deeply about the crucial moral questions, especially those that face us in our own lives, but when we have arrived at an answer to a particular problem, to crystallize it into a not too specific or detailed form, so that its salient features may stand out and serve us again in a like situation without so much thought." (p.41–2)

Thus, for Hare, while everyday moral decisions may be made by applying principles and norms, serious moral decisions require reasoning about the particular situation,

L. Goble and J.-J.C. Meyer (Eds.): DEON 2006, LNAI 4048, pp. 8–23, 2006.

and it is such reasoning that gives rise to moral principles. Moral norms are an output from, not an input to, serious moral reasoning. In this paper we will try to model such reasoning, with a view to enabling autonomous software agents to engage in this form of reasoning. In doing so we will distinguish at least three things that might be intended by "agent A should ϕ". We might mean something like "it is prudent to ϕ", as when we say "you should wear a coat when the weather is cold". Here the obligation is determined only by reference to the interests of the agent doing the reasoning. Alternatively, we might mean "it is morally right to ϕ", as when we say "you should tell the truth". Here the obligation is required to reflect the interests not only of the reasoning agent, but also of other agents affected by the action. Thirdly, we might mean "it is legally obligated to ϕ" as in "you should pay your taxes", where the obligation derives from a legal system with jurisdiction over the agent. We will explore the differences between these three senses of "should": in particular we will explain the difference between prudential "should" and moral "should" in terms of the practical reasoning involved, and consider the reasoning that might be used in devising appropriate legislation.

We will base our considerations on the representation and discussion of a specific example, a well known problem intended to explore a particular ethical dilemma discussed by Coleman [5] and Christie [4], amongst others. The situation involves two agents, called Hal and Carla, both of whom are diabetic. Hal, *through no fault of his own*, has lost his supply of insulin and urgently needs to take some to stay alive. Hal is aware that Carla has some insulin kept in her house, but Hal does not have permission to enter Carla's house. The question is whether Hal is justified in breaking into Carla's house and taking her insulin in order to save his life. It also needs to be considered that by taking Carla's insulin, Hal may be putting her life in jeopardy. One possible response is that if Hal has money, he can compensate Carla so that her insulin can be replaced. Alternatively if Hal has no money but Carla does, she can replace her insulin herself, since her need is not immediately life threatening. There is, however, a serious problem if neither have money, since in that case Carla's life is really under threat. Coleman argued that Hal may take the insulin to save his life, but should compensate Carla. Christie's argument against this was that even if Hal had no money and was unable to compensate Carla he would still be justified in taking the insulin by his immediate necessity, since no one should die because of poverty. Thus, argues Christie, he cannot be *obliged* to compensate Carla even when he is able to.

In section 2, we model our agents as simple automata and describe Action-Based Alternating Transition Systems (AATS) [10], which we use as the semantic basis of our representation, and instantiate an AATS relevant to the problem scenario. In any particular situation, the agents will need to choose how to act. In section 3 we model this choice as the proposal, critique and defence of arguments justifying their available strategies in the manner of [2]. In section 4 we show how reasoning about the resulting arguments can be represented as an Argumentation Framework [6, 3] to enable the agents to identify strategies that are prudentially and morally justified. In section 5 we consider how this framework can also be used to answer the question of what would be appropriate legislation for the situation, and what could be appropriate moral principles to take from the reasoning. Section 6 concludes the paper.

2 Representing the Problem

For the purposes of our representation three attributes of agents are important: whether they have insulin (I), whether they have money (M) and whether they are alive (A). The state of an agent may thus be represented as a vector of three digits, IMA, with I, M and A equal to 1 if the agent has insulin, has money and is alive, and 0 if these things are false. Since I cannot be true and A false (the agent will live if it has insulin), an agent may be in any one of six possible states. We may now represent the actions available to the agents by depicting them as automata, as shown in Figure 1. An agent with insulin may lose its insulin; an agent with money and insulin may compensate another agent; an agent with no insulin may take another's insulin, or, with money, buy insulin. In any situation when it is alive, an agent may choose to do nothing; if dead it can only do nothing.

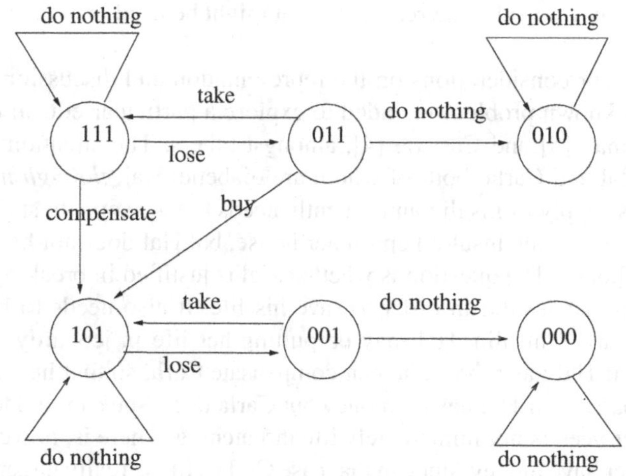

Fig. 1. State transition diagram for our agents

Next we draw upon the approach of Wooldridge and van der Hoek [10] which formally describes a normative system in terms of constraints on actions that may be performed by agents in any given state. We will now briefly summarise their approach.

In [10] Wooldridge and van der Hoek present an extension to Alur et al's Alternating-time Temporal Logic (ATL) [1] and they call this extension Normative ATL* (NATL*). As Wooldridge and van der Hoek explain, ATL is a logic of cooperative ability. Its purpose is to support reasoning about the powers of agents and coalitions of agents in game-like multi-agent systems. ATL contains an explicit notion of agency, which gives it the flavour of an action logic. NATL* is intended to provide a link between ATL and deontic logic and the work presented in [10] provides a formal model to represent the relationship between agents' ability and obligations. The semantic structures which underpin ATL are known as *Action-based Alternating Transition Systems* (AATSs) and they are used for modelling game-like, dynamic, multi-agent systems. Such systems comprise multiple agents which can perform actions in order to modify and attempt to control the system in some way. In Wooldridge and van der Hoek's approach they use an AATS to model the physical properties of the system in question - the actions that agents can perform in the empty normative system, unfettered by any considerations of their legality or usefulness. They define an AATS as follows.

Firstly the systems of interest may be in any of a finite set Q of possible *states*, with some $q_0 \in Q$ designated as the *initial state*. Systems are populated by a set Ag of *agents*; a *coalition* of agents is simply a set $C \subseteq Ag$, and the set of all agents is known as the *grand coalition*. Note, Wooldridge and van der Hoek's usage of the term 'coalition' does not imply any common purpose or shared goal: a coalition is simply taken to be a set of agents.

Each agent $i \in Ag$ is associated with a set Ac_i of possible actions, and it is assumed that these sets of actions are pairwise disjoint (i.e., actions are unique to agents). The set of actions associated with a coalition $C \subseteq Ag$ is denoted by Ac_C, so $Ac_C = \bigcup_{i \in C} Ac_i$.

A joint action j_C for a coalition C is a tuple $\langle \alpha_1,...,\alpha_k \rangle$, where for each α_j (where $j \leq k$) there is some $i \in C$ such that $\alpha_j \in Ac_i$. Moreover, there are no two different actions α_j and $\alpha_{j'}$ in J_C that belong to the same Ac_i. The set of all joint actions for coalition C is denoted by J_C, so $J_C = \prod_{i \in C} Ac_i$. Given an element j of J_C and an agent $i \in C$, i's complement of j is denoted by j_i.

An *Action-based Alternating Transition System* (AATS) is an $(n + 7)$-tuple $S = \langle Q, q_0, Ag, Ac_1, \ldots, Ac_n, \rho, \tau, \Phi, \pi \rangle$, where:

- Q is a finite, non-empty set of *states*;
- $q_0 \in Q$ is the *initial state*;
- $Ag = \{1,...,n\}$ is a finite, non-empty set of *agents*;
- Ac_i is a finite, non-empty set of actions, for each $i \in Ag$ where $Ac_i \cap Ac_j = \emptyset$ for all $i \neq j \in Ag$;
- $\rho : Ac_{Ag} \rightarrow 2^Q$ is an *action precondition function*, which for each action $\alpha \in Ac_{Ag}$ defines the set of states $\rho(\alpha)$ from which α may be executed;
- $\tau : Q \times J_{Ag} \rightarrow Q$ is a partial *system transition function*, which defines the state $\tau(q, j)$ that would result by the performance of j from state q - note that, as this function is partial, not all joint actions are possible in all states (cf. the precondition function above);
- Φ is a finite, non-empty set of *atomic propositions*; and
- $\pi : Q \rightarrow 2^\Phi$ is an interpretation function, which gives the set of primitive propositions satisfied in each state: if $p \in \pi(q)$, then this means that the propositional variable p is satisfied (equivalently, true) in state q.

We now turn to representing the Hal and Carla scenario as an AATS. Recall from section 2 that each agent may independently be in one of six states, giving 36 possible states for the two agents, q_0 .. q_{35}. Normally both agents will have insulin, but we are specifically interested in the situations that arise when one of them (Hal) loses his insulin. The initial state therefore may be any of the four states in which $I_H = 0$. Moreover, since Hal is supposed to have no time to buy insulin, his only available actions in these states, whether or not $M_H = 1$, are to take Carla's insulin or do nothing. If Hal does nothing, neither agent can act further. If Hal takes Carla's insulin and if $M_H = 1$, then Hal can compensate Carla or do nothing. Similarly, after Hal takes the insulin, Carla, if

$M_C = 1$, can buy insulin or do nothing. The possible developments from the four initial states are shown in Figure 2. States are labelled with the two vectors $I_H M_H A_H$ (on the top row) and $I_C M_C A_C$ (on the bottom row), and the arcs are labelled with the joint actions (with the other labels on the arcs to be explained in section 4).

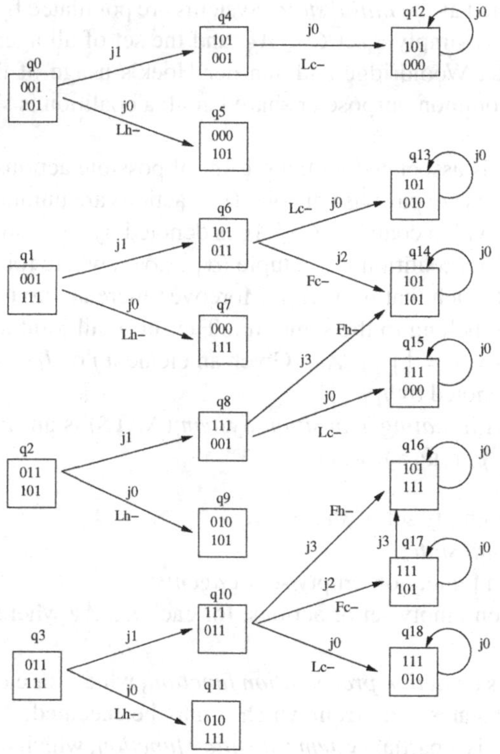

Fig. 2. Developments from the four possible initial states

The instantiation of the problem as an AATS is summarised below. We give only the joint actions and the transitions relevant to this particular scenario.

<u>States and Initial States:</u>
$Q = \{q_0, ..., q_{35}\}$. The initial state is one of four, as shown in the diagram in Figure 2.

<u>Agents, Actions and Joint Actions:</u>
$Ag = \{H, C\}$ $Ac_H = \{take_H, compensate_H, do\_nothing_H\}$ $Ac_C = \{buy_C, do\_nothing_C\}$

$J_{AG} = \{j_0, j_1, j_2, j_3,\}$, where $j_0 = \langle do\_nothing_H, do\_nothing_C \rangle$, $j_1 = \langle take_H, do\_nothing_C \rangle$, $j_2 = \langle do\_nothing_H, buy_C \rangle$, $j_3 = \langle compensate_H, do\_nothing_C \rangle$.

<u>Propositional Variables:</u>
$\Phi = \{insulin_H, money_H, alive_H, insulin_C, money_C, alive_C\}$

Transitions/Pre-conditions/Interpretation are given in Table 1:

Table 1. Transitions/Pre-conditions/Interpretation

q\j	j_0	j_1	j_2	j_3	π (q)
q_0	q_5	q_4	–	–	$\{\text{alive}_H, \text{insulin}_C, \text{alive}_C\}$
q_1	q_7	q_6	–	–	$\{\text{alive}_H, \text{insulin}_C, \text{money}_C, \text{alive}_C\}$
q_2	q_9	q_8	–	–	$\{\text{money}_H, \text{alive}_H, \text{insulin}_C, \text{alive}_C\}$
q_3	q_{11}	q_{10}	–	–	$\{\text{money}_H, \text{alive}_H, \text{insulin}_C, \text{money}_C, \text{alive}_C\}$
q_4	q_{12}	–	–	–	$\{\text{insulin}_H, \text{alive}_H, \text{alive}_C\}$
q_5	–	–	–	–	$\{\text{insulin}_C, \text{alive}_C\}$
q_6	q_{13}	–	q_{14}	–	$\{\text{insulin}_H, \text{alive}_H, \text{money}_C, \text{alive}_C\}$
q_7	–	–	–	–	$\{\text{insulin}_C, \text{money}_C, \text{alive}_C\}$
q_8	q_{15}	–	–	q_{14}	$\{\text{insulin}_H, \text{money}_H, \text{alive}_H, \text{alive}_C\}$
q_9	–	–	–	–	$\{\text{money}_H, \text{insulin}_C, \text{alive}_C\}$
q_{10}	q_{18}	–	q_{17}	q_{16}	$\{\text{insulin}_H, \text{money}_H, \text{alive}_H, \text{money}_C, \text{alive}_C\}$
q_{11}	–	–	–	–	$\{\text{money}_H, \text{insulin}_C, \text{money}_C, \text{alive}_C\}$
q_{12}	q_{12}	–	–	–	$\{\text{insulin}_H, \text{alive}_H\}$
q_{13}	q_{13}	–	–	–	$\{\text{insulin}_H, \text{alive}_H, \text{money}_C\}$
q_{14}	q_{14}	–	–	–	$\{\text{insulin}_H, \text{alive}_H, \text{insulin}_C, \text{alive}_C\}$
q_{15}	q_{15}	–	–	–	$\{\text{insulin}_H, \text{money}_H, \text{alive}_H\}$
q_{16}	q_{16}	–	–	–	$\{\text{insulin}_H, \text{alive}_H, \text{insulin}_C, \text{money}_C, \text{alive}_C\}$
q_{17}	q_{17}	–	–	q_{16}	$\{\text{insulin}_H, \text{money}_H, \text{alive}_H, \text{insulin}_C, \text{alive}_C\}$
q_{18}	q_{18}	–	–	–	$\{\text{insulin}_H, \text{money}_H, \text{alive}_H, \text{money}_C\}$

3 Constructing the Arguments

In [2] we have proposed an argument scheme and associated critical questions to enable agents to propose, attack and defend justifications for action. Such an argument scheme follows Walton [9] in viewing reasoning about action (practical reasoning) as presumptive justification - *prima facie* justifications of actions can be presented as instantiations of an appropriate argument scheme, and then critical questions characteristic of the scheme used can be posed to challenge these justifications. The argument scheme we have developed is an extension of Walton's *sufficient condition scheme for practical reasoning* [9] and our argument scheme is stated as follows:

AS1 In the current circumstances R
We should perform action A
Which will result in new circumstances S
Which will realise goal G
Which will promote some value V.

In this scheme we have made Walton's notion of a goal more explicit by separating it into three elements: the state of affairs brought about by the action; the goal (the desired features in that state of affairs); and the value (the reason why those features are desirable). Our underlying idea in making this distinction is that the agent performs an action to move from one state of affairs to another. The new state of affairs may have many differences from the current state of affairs, and it may be that only some of them are significant to the agent. The significance of these differences is that they make the new state of affairs better with respect to some good valued by the agent. Note that typically the new state of affairs will be better through improving the lot of some *particular* agent: the sum of human happiness is increased only by increasing the happiness of some particular human. In this paper we take the common good of all agents as the aggregation of their individual goods. It may be that there are common goods which are not reflected in this aggregation: for example, if equality is such a common good, increasing the happiness of an already happy agent may diminish the overall common good. For simplicity, we ignore such possibilities here.

Now an agent who does not accept this presumptive argument may attack the contentious elements in the instantiation through the application of critical questions. We have elaborated Walton's original four critical questions associated with his scheme by extending them to address the different elements identified in the goal in our new argument scheme. Our extension results in sixteen different critical questions, as we have described in [2]. In posing such critical questions agents can attack the validity of the various elements of the argument scheme and the connections between them, and additionally there may be alternative possible actions, and side effects of the proposed action. Each critical question can be seen as an attack on the argument it is posed against and examples of such critical questions are: "Are the circumstances as described?", "Does the goal promote the value?", "Are there alternative actions that need to be considered?". The full list of critical questions can be found in [2].

To summarise, we therefore believe that in an argument about a matter of practical action, we should expect to see one or more *prima facie* justifications advanced stating, explicitly or implicitly, the current situation, an action, the situation envisaged to result from the action, the features of that situation for which the action was performed and the value promoted by the action, together with negative answers to critical questions directed at those claims. We now describe how this approach to practical reasoning can be represented in terms of an AATS.

In this particular scenario we recognise two values relative to each agent: life and freedom (the ability to act in a given situation). The value 'life' (L) is demoted when Hal or Carla cease to be alive. The value 'freedom' (F) is demoted when Hal or Carla cease to have money. The arcs in Figure 2 are labelled with the value demoted by a transition, subscripted to show the agent in respect of which it is demoted. We can now examine the individual arguments involved.

In all of $q_0 - q_3$, the joint action j_0 demotes the value 'life' in respect of Hal, whereas the action j_1 is neutral with regard to this value. We can instantiate argument scheme AS1 by saying where Hal has no insulin he should take Carla's to avoid those states where dying demotes the value 'life'.

A1: Where $\text{Insulin}_h = 0$, Take_h (i.e. j_1), To avoid $\text{Alive}_h = 0$, Which demotes L_h.

Argument A2 attacks A1 and it arises from q_0 where Hal taking the insulin leads to Carla's death and thus demotes the value 'life Carla'. By 'not take' we mean any of the other available actions.

A2 attacks A1: Where $Money_c = 0$, Not $Take_h$ (i.e. j_0 or j_2), To avoid $Alive_c = 0$, Which demotes L_c.

Argument A3 arises from q_2 where Carla's death is avoided by Hal taking the insulin and paying Carla compensation.

A3 attacks A2 and A5: Where $Insulin_h = 0$, $Take_h$ and $Compensate_h$ (i.e. j_1 followed by j_3), To achieve $Alive_c = 1$ and $Money_c = 1$, Which promotes L_c and F_c.

Argument A4 represents a critical question directed at A2 which challenges the factual premise of A2, that Carla has no money.

A4 attacks A2: $Money_c = 1$, (Known to Carla but not Hal)

Next argument A5 mutually attacks A3 and it also attacks A2. A5 states that where Hal has no insulin but he does have money, then he should take Carla's insulin and she should buy some more. The consequences of this are that Carla remains alive, promoting the value 'life Carla', and, Hal has money, promoting the value 'freedom Hal'.

A5 attacks A3 and A2: Where $Insulin_h = 0$ and $Money_h = 1$, $Take_h$ and Buy_c (i.e. j_1 followed by j_2), To achieve $Alive_c = 1$ and $Money_h = 1$, Which promotes L_c and F_h.

Argument A6 critically questions A5 by attacking the assumption in A5 that Carla has money.

A6 attacks A5: $Money_c = 0$ (Known to Carla but not Hal)

Another attack on A5 can be made by argument A7 stating that where Carla has money then she should not buy any insulin so as to avoid not having money, which would demote the value 'freedom Carla'.

A7 attacks A5: Where $Money_c = 1$, Not Buy_c (i.e. j_0 or j_1 or j_3), To avoid $Money_c = 0$, Which demotes F_c.

A8 is a critical question against A3 which states that where Hal does not have money, taking the insulin and compensating Carla is not a possible strategy.

A8 attacks A3: Where $Money_h = 0$, $Take_h$ and $Compensate_h$ (i.e. j_1 followed by j_3), Is not a possible strategy.

A8 is attacked by argument A9 which challenges the assumption in A8 that Hal has no money, and A9 is in turn attacked by A10 which challenges the opposite assumption, that Hal does have money.

A9 attacks A8 and A11: $Money_h = 1$ (Known to Hal but not Carla)

A10 attacks A9: $Money_h = 0$ (Known to Hal but not Carla)

Argument A11 attacks A1 in stating that where Hal does not have money but Carla does, then Hal should not take the insulin to avoid Carla being left with no money, which would demote the value of 'freedom Carla'.

A11 attacks A1: Where $Money_h = 0$ and $Money_c = 1$, Not $Take_h$ (i.e. j_0), To avoid $Money_c = 0$, Which demotes F_c.

Argument A12 can attack A5 by stating that in the situations where Hal does not have insulin, then he should take Carla's insulin but not compensate her. This would avoid him being left with no money, as when Hal has no money the value 'freedom Hal' is demoted.

A12 attacks A5: Where $Insulin_h = 0$, $Take_h$ and Not $Compensate_h$ (i.e. j_1 followed by j_0 or j_2), To avoid $Money_h = 0$, Which demotes F_h.

Finally, argument A13 attacks A2 by stating that where Hal has no insulin and no money he should take Carla's insulin and she should buy some. This would ensure that Carla stays alive, promoting the value 'life Carla'.

A13 attacks A2: Where $Insulin_h = 0$ and $Money_h = 0$, $Take_h$ and Buy_c (i.e. j_1 followed by j_2), To achieve $Alive_c = 1$, Which promotes L_c.

This concludes the description of the arguments and attacks that can be made by instantiating argument scheme AS1 and posing appropriate critical questions.

4 Evaluating the Arguments

In the previous section we identified the arguments that the agents in our problem situation need to consider. In order to evaluate the arguments and see which ones the agents will accept, we organise the arguments into a Value Based Argumentation Framework (VAF) [3]. VAFs extend the Argumentation Frameworks introduced by Dung in [6], so as to accommodate different audiences with different values and interests. The key notion in Dung's argumentation framework is that of a preferred extension (PE), a subset of the arguments in the framework which:

- is conflict free, in that no argument in the PE attacks any other argument in the PE;
- is able to defend every argument in the PE against attacks from outside the extension, in that every argument outside the PE which attacks an argument in the PE is attacked by some argument in the PE;
- is maximal, in that no other argument can be added to the PE without either introducing a conflict or an argument that cannot be defended against outside attacks.

In a VAF strengths of arguments for a particular *audience* are compared with reference to the *values* to which they relate. An audience has a *preference order* on the values of the arguments, and an argument is only *defeated for that audience* if its value is not preferred to that of its attacker. We then replace the notion of attack in Dung's PE by the notion of *defeat for an audience* to get the *PE for that audience*. We represent the VAF as a directed graph, the vertices representing arguments and labelled with an argument identifier and the value promoted by the argument, and the edges representing attacks between arguments. Attacks arise from the process of critical questioning, as described in the previous section, not from an analysis of the arguments themselves. The values promoted by the arguments are identified in the instantiations of the argument scheme presented in the previous section. The VAF for our problem scenario is shown in Figure 3. Note that two pairs of arguments, A4–A6 and A9–A10 relate to facts known only to Carla and Hal respectively. In order to bring these into a value based framework, we ascribe the value "truth" to statements of fact, and as in [3], truth is given the highest value preference for all audiences, since while we can choose what we consider desirable, we are constrained by the facts to accept what is true.

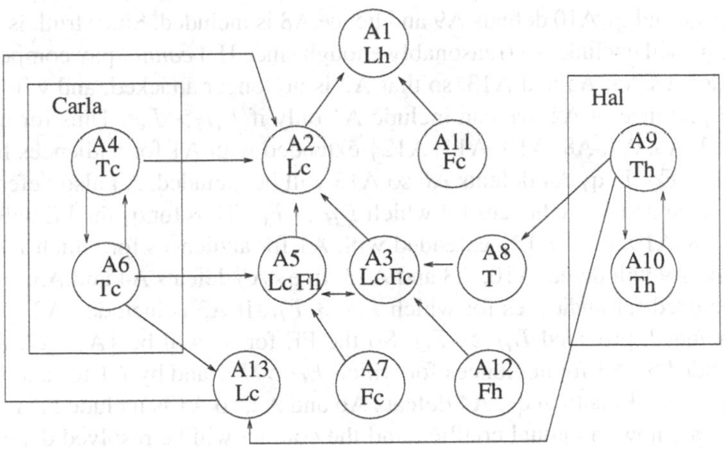

Fig. 3. VAF for the problem scenario

The questions posed by the problem scenario are whether Hal should take the insulin and whether Hal should compensate Carla. We answer these questions by finding the preferred extensions (PE) of the framework for various audiences. Note that the PE may contain both arguments providing reasons for performing an action and for not performing it. The actions which will be chosen are those supported by *effective* arguments, that is, those which do not feature in an unsuccessful attack. Thus in q_0, for example, A2, which provides a reason for Hal not to take the insulin, is not attacked and so will be in the PE. If, however, we prefer L_H to L_C, A1, which gives a reason for Hal to take the insulin will also be included. In such a case A2 is ineffective and so Hal should take the insulin, despite there being reasons against this action which cannot be countered through argument. If A1 is in the PE it is always effective since it attacks nothing, and so if A1 is present then Hal should take the insulin. If A3, which gives a reason for Hal to compensate Carla, is included it is also always effective since it always defeats A2, because its values are a superset of A2, and it must defeat A5 or be defeated by it. If both A1 and A3 are in the PE, Hal should take the insulin and compensate Carla. If A3, but not A1, is present Hal should take the insulin *only* if he then compensates Carla. What we must do therefore is to consider for which audiences A1 and A3 appear in the PE.

For this discussion we will assume that the agents are part of a common culture in which the value life is preferred to the value freedom. This seems reasonable in that life is a precondition for any exercise of freedom. There will therefore be no audience with a value order in which $F_A > L_A$, for any agent A, although of course it is possible for an agent to prefer its own freedom to the life of another.

First we note that {A7, A11, A12} are not attacked and so will appear in every PE. Immediately from this we see that A1 will not appear in any PE of an audience for which $F_C \geq L_H$, and that A3 will not appear in any PE of an audience for which $F_H \geq L_C$. A5 will never be defeated by A7, since $L_C > F_C$ for all audiences.

To proceed we must now resolve the factual issues which determine the conflicts A4–A6 and A9–A10. Thus we need to consider the initial states $q_0 - q_3$ separately.

In states q_0 and q_1 A10 defeats A9 and hence A8 is included. Since truth is the highest value this will exclude A3 (reasonably enough since Hal *cannot* pay compensation). In q_0 A6 defeats A4, A5 and A13, so that A2 is no longer attacked, and will be in the PE. In the presence of A2, we can include A1 only if $L_H > L_C$. Thus for q_0 the PE will be {A2, A6, A7, A8, A10, A11, A12} extended with A1 for audiences for which $L_H > L_C > F_C$. In q_1 A4 defeats A6 so A13 will be included. A4 also defeats A2 so A1 will be included for audiences for which $L_H > F_C$. Thus for q_1 the PE will be {A4, A13, A7, A8, A10, A11, A12} extended with A1 for audiences for which $L_H > F_C$. In q_2 and q_3 A9 will defeat A10, A8 and A13. In q_2 A6 defeats A4 and A5, so A3 will now be included for audiences for which $L_C > F_H$. If A3 is included A2 is defeated and A1 included, provided $L_H > F_C$. So the PE for q_2 will be {A6, A7, A9, A11, A12} extended by A3 for audiences for which $L_C > F_H$ and by A1 for audiences for which $L_H > F_C$. Finally in q_3, A4 defeats A6 and A2, so A1 is included if $L_H > F_C$. A5 and A3 are now in mutual conflict, and the conflict will be resolved depending on whether F_C or F_H is preferred. Thus the PE in q_3 will contain {A4, A7, A9, A11, A12}, extended by A1 if $L_H > F_C$, by A3 if $F_C > F_H$ and by A5 if $F_H > F_C$.

We can now summarise the status of A1 and A3 in Table 2.

Table 2. Status of A1 and A3

Initial State	A1 included if:	A3 included if:
q_0	$L_H > L_C > F_C$	never
q_1	$L_H > F_C$	never
q_2	$L_H > F_C$	$L_C > F_H$
q_3	$L_H > F_C$	$F_C > F_H$ *A5 included otherwise*

From this we can see that if the interests of Hal are ranked above those of Carla, Hal should take the insulin and not pay compensation, whereas if the interests of Carla are ranked above those of Hal, then Hal should take the insulin only if he pays for it. These two positions thus express what is prudentially right for Hal and Carla respectively.

From the standpoint of pure morality, however, people should be treated equally: that is $(L_H = L_C) > (F_H = F_C)$. Remember, that if the problem is considered in the abstract, one does not know if one will be the person who loses the insulin: one may find oneself playing the role of Hal or Carla, and so there is no reason to prefer one agent to the other. If this perspective is adopted, then Hal should take the insulin in all situations other than q_0, and is obliged to compensate only in q_2, since there are two PEs in q_3. We can see this as representing the morally correct judgement, the judgement that would be arrived at by a neutral observer *in full possession of the facts*.

However, the point about being in full possession of the facts is important. In practice we need to evaluate the conduct of the agents in the situations in which they find themselves. In our scenario Hal cannot know whether or not Carla is in a position to replace the insulin herself: for Hal, q_0 is epistemically indistinguishable from q_1, and q_2 is epistemically indistinguishable from q_3. Now consider Hal in q_2/q_3. He will of course take the insulin and justify this by saying that his life is more important than Carla's freedom of choice with regard to her money. In a society which rates L > F, this will be accepted. Thus Hal should take the insulin. If he then chooses to compensate, he can be sure of acting in a morally acceptable manner, since this is required in q_2 and appears in one of the alternative PEs in q_3. If, on the other hand, he does not compensate, while he may attempt justification in q_3 by saying that he saw no reason to prefer Carla's freedom of choice to his own, in q_2 he would have to argue that his freedom of choice is preferred to Carla's life. This justification will be rejected for the same reason that the justification for taking the insulin at all was accepted, namely that L > F. Morally, therefore, in q_2/q_3, Hal should take the insulin and compensate Carla.

Now consider q_0/q_1, where compensation is impossible. In q_1 taking the insulin is justifiable by L > F. In q_0, however, the justification is only $L_H > L_C$. Hal's problem, if this is not acceptable, is that he cannot be sure of acting in a morally correct manner, since he could take the insulin in q_1 and not take it in q_0. Our view is that taking the insulin should be seen as morally *excusable*, even in q_0 although not morally *correct*[1], since the possibility of the actual state being q_1 at least excuses the preference of Hal's own interests to Carla's. The alternative is to insist on Hal not taking the insulin in q_1, which could be explained only by $L_H \leq F_C$, and it seems impossibly demanding to expect Hal to prefer Carla's lesser interest to his own greater interest.

5 Moral, Prudential and Legal "Ought"

In our discussion in the previous section we saw that what an agent should do can be determined by the ordering the agent places on values. This ordering can take into

[1] This distinction is merely our attempt to capture some of the nuances that are found in everyday discussions of right and wrong. There is considerable scope to explore these nuances, which are often obscured in standard deontic logic. See, for example, the discussion in [8] which distinguishes: what is *required* (what morality demands); what is *optimal* (what morality recommends); the supererogatory (exceeding morality's demands); the morally indifferent; the permissible suboptimal; the morally significant; and the minimum that morality demands. Clearly a full consideration of these nuances is outside the scope of this paper, but we believe that our approach may offer some insight into this debate.

account, or ignore, which of the agents the values relate to. Prudential reasoning takes account of the different agents, with the reasoning agent preferring values relating to itself, whereas strict moral reasoning should ignore the individual agents and treat the values equally. In fact there are five possible value orders which respect L > F, and which order the agents consistently.

V01 *Morally correct*: values are ordered: within each value agents are treated equally, and no distinctions relating to agents are made. In our example, for Hal: $(L_H = L_C) > (F_H = F_C)$.
V02 *Self-Interested*: values are ordered as for moral correctness, but within a value an agent prefers its own interests. In our example, for Hal: $L_H > L_C > F_H > F_C$.
V03 *Selfish*: values are ordered, but an agent prefers its own interests to those of other agents: In our example, for Hal: $L_H > F_H > L_C > F_C$.
V04 *Noble*: values are ordered as for moral correctness, but within a value an agent prefers the other's interests. In our example, for Hal: $L_C > L_H > F_C > F_H$.
V05 *Sacrificial*: values are ordered, but an agent prefers the other's interests to its own. In our example, for Hal: $L_C > F_C > L_H > F_H$.

Note that the morally correct order is common to both agents, while the orders for self-interested Hal and noble Carla are the same, as are those for selfish Hal and sacrificial Carla.

Now in general an agent can determine what it should do by constructing the VAF comprising the arguments applicable in the situation and calculating the PE for that VAF using some value order. Using VO1 will give what it morally should do and VO3 what it prudentially should do.

It is, however, possible that there will not be a unique PE: this may be either because the value order cannot decide a conflict (as with A3 and A5 when using VO1 in q_3 above), or because the agent lacks the factual information to resolve a conflict (as with Hal with respect to A4 and A6 above). In this case we need to consider all candidate PEs. In order to justify commitment to an action the agent will need to use a value order which includes the argument justifying the action in all candidate PEs.

Consider q_3 and VO1: we have two PEs, {A1, A3, A4, A7, A9, A11, A12} and {A1, A4, A5, A7, A9, A11, A12}. A1 is in both and it is thus morally obligatory to take the insulin. A3 on the other hand is in one PE but not the other and so both compensate and not compensate are morally correct in q_3. It is possible to justify A3 by choosing a value order with $F_H > F_C$, or A5 by choosing a value order with $F_C > F_H$. Thus in q_3 a selfish or a self-interested agent will not compensate, whereas a noble or sacrificing one will. Either choice is, however, consistent with the morally correct behaviour. Next we must consider what is known by the reasoning agent. Consider Hal in q_2/q_3, where we have three PEs to take into account. The relevant PE for q_2 is {A1, A3, A6, A7, A9, A11, A12} and as A1 is in all three, taking the insulin is obligatory. To exclude A3 from the PE for q_2, the preference $F_H > L_C$ is required. Here legitimate self-interest cannot ground a choice: this preference is only in VO3, which means that only a selfish agent will not compensate. In q_2, however, failing to compensate is not consistent with morally correct behaviour, and an agent which made this choice would be subject to moral condemnation. VO2 cannot exclude A3 from the PE in q_2, and so cannot rule out compensation. Therefore, the agent must, to act morally, adopt VO4 or VO5, and compensate, even if the state turns out to be q_3.

In q_0/q_1, we have two PEs for Hal using VO1: from q_0 {A2, A6, A7, A8, A10, A11, A12} and from q_1 {A1, A4, A5, A7, A8, A10, A11, A12, A13}. Here A3 is always rejected, reflecting the fact that compensation is impossible. Hal must, however, still choose whether to take the insulin or not. This means that he must adopt a value order which either includes A1 in the PE for both q_0 and q_1, or which excludes it from both. A1 can be included in both given the preference $L_H > L_C$. A1 can, however, only be excluded from the PE for q_1 if $F_C > L_H$. VO4 does not decide the issue: thus Hal must choose between self-interest (VO2) and being sacrificial (VO5). Neither choice will be sure to be consistent with morally correct behaviour: VO2 will be wrong in q_0 and V5 will be wrong in q_1, where the sacrifice is an unnecessary waste. It is because it is unreasonable to require an agent to adopt VO5 (for Carla to expect Hal to do this would require her to adopt the selfish order VO1), that we say that it is morally excusable for Hal to take the insulin in q_0/q_1.

The above discussion suggests the following. An agent must consider the PEs relating to every state which it may be in. An action is justified only if it appears in every PE formed using a given value order.

- If VO1 justifies an action, that action is morally obligatory.
- If VO1 does not produce a justified action, then an action justified under VO2, VO4 or VO5 is morally permissible.
- If an action is justified only under VO3, then that action is prudentially correct, but not morally permissible.

Amongst the morally permissible actions we may discriminate according to the degree of preference given to the agent's own interests and we might say that: VO2 gives actions which are morally *excusable*, VO4 gives actions which are morally *praiseworthy*, and VO5 gives actions which are *supererogatory*, beyond the normal requirements of morality[2].

We may now briefly consider what might be appropriate legislation to govern the situation. We will assume that the following principle governs just laws: that citizens are treated equally under the law. This in turn means that the legislator can only use VO1, as any other ordering requires the ability to discriminate between the interests of the agents involved. We will also assume that the legislator is attempting to ensure that the best outcome (with regard to the interests of all agents) is reached from any given situation. Thus in our example, from q_0 the legislature will be indifferent between q_5 and q_{12}; from q_1 and q_2 they will wish to reach q_{14}; and from q_3 they will be indifferent between q_{16} and q_{17}. Now consider the following possible laws:

Law 1. Any agent in Hal's position should be obliged to take the insulin absolutely. This may lead to q_{14} if such an agent does not compensate in q_2, and so may not achieve the desired ends. Moreover, in q_0 this requires that q_{12} rather than q_5 be reached, which prefers the interests of agents in Hal's position to agents in Carla's position.

[2] Again, this is merely our suggestion for possible moral nuances.

Law 2. Any agent in Hal's position is forbidden to take the insulin unless he pays compensation. This fails to produce the desired outcome in q_1, where it leads to q_7.

Law 3. Any agent in Hal's position is permitted to take the insulin, but is obliged to compensate if he is able to. This will reach a desired outcome in all states, and is even-handed between agents in Hal and Carla's positions in q_0. In q_3, however, it favours the interests of agents in Carla's position over agents in Hal's position by determining which of the two agents ends up with money.

Law 4. Any agent in Hal's position is obliged to take the insulin and obliged to compensate if able to. This will reach a desired state in every case, but favours agents in Hal's position in q_0 and agents in Carla's position in q_3.

Thus if we wish to stick with the principle of not favouring the interests of either agent, we can only *permit* Hal to take the insulin and *permit* Hal to pay compensation: none of the proposed laws are at once even-handed and desirable in all of the possible situations. Under this regime we have no problem in q_0: the states reached are of equal value, and it is Hal, not the state, who chooses whose interest will be favoured. In q_1 we will reach a desired state provided Hal is not sacrificial. In q_2 we must rely on Hal not being selfish, and acting in a moral fashion. Finally in q_3 we reach a desired state and again Hal chooses whose interests will be favoured. Provided that we can expect agents to act in a morally acceptable, but not supererogatory, fashion, and so use VO2 or VO4, the desired outcomes will be reached. It may be, however, that the legislature will take the view that favouring Carla in q_3 is a price worth paying to prevent selfish behaviour on the part of Hal in q_2, and pass Law 3. This is a political decision, turning on whether the agents are trusted enough to be given freedom to choose and the moral responsibility that goes with such freedom. A very controlling legislature might even pass Law 4, which gives the agents no freedom of choice, but which reaches the desired state even when agents act purely in consideration of their own interests.

Finally we return to the initial observations of Hare: is it possible to crystallise our reasoning into "not too specific and not too detailed form"? What moral principle might Hal form? First moral principles which apply to particular states would be too specific. In practice Hal would never have sufficient knowledge of his situation to know which principle to apply. On the other hand, to frame a principle to cover all four states would be arguably too general, as it would ignore pertinent information. In states q_2/q_3, the appropriate moral principle is to take and compensate: this ensures that moral censure is avoided, and although it may be, if the state turns out to be q_3, that Carla's interests are favoured, Hal is free to make this choice, even if we believe that the state should not impose it. In q_0/q_1, the choice is not so clear: since moral correctness cannot be ensured, either taking or not taking the insulin is allowed. While taking it is morally excusable, and so an acceptable principle, Hal is free to favour Carla's interests over his own, provided that it is *his own choice* to do so. While Hal cannot be compelled, or even expected, to be sacrificial, he cannot be morally obliged to be self-interested either.

6 Concluding Remarks

In this paper we have described how agents can reason about what they ought to do in particular situations, and how moral principles can emerge from this reasoning. An important feature is how their choices are affected by the degree of consideration given to the interests of the other agents involved in the situation, which is captured by an ordering on the values used to ground the relevant arguments. Different value orders will attract varying degrees of moral praise and censure.

In future work we will wish to consider further the relation between the various kinds of "ought" we have identified here. In particular, it might be conjectured that reasoning with the morally reasonable value orders VO2 and VO4 will always lead to an outcome which is desirable when aggregating the interests of the agents involved. Another interesting line of inquiry would be to increase the number of agents involved, and to consider the effect of agents having different attitudes towards the others depending on their inter-relationships, modelling notions such as kinship, community and national groupings. A third interesting line of inquiry would be to see whether this approach gives insight into the emergence of norms of cooperation. Finally, we intend to fully formalise, in terms of an AATS, the instantiations of arguments in the form of our argument scheme, AS1, and the critical questions that accompany this scheme. This will enable our approach to be fully automatable and it forms the basis of our current work.

References

1. R. Alur, T. A. Henzinger, and O. Kupferman. Alternating-time temporal logic. *ACM Journal*, 49 (5):672–713, 2002.
2. K. Atkinson, T. Bench-Capon, and P. McBurney. A dialogue game protocol for multi-agent argument for proposals over action. *Journal of Autonomous Agents and Multi-Agent Systems*, 11(2):153–171, 2005.
3. T. J. M. Bench-Capon. Persuasion in practical argument using value based argumentation frameworks. *Journal of Logic and Computation*, 13 3:429–48, 2003.
4. C. G. Christie. *The Notion of an Ideal Audience in Legal Argument*. Kluwer Academic Press, 2000.
5. J. Coleman. *Risks and Wrongs*. Cambridge University Press, 1992.
6. P. M. Dung. On the acceptability of arguments and its fundamental role in nonmonotonic reasoning, logic programming and n-person games. *Artificial Intelligence*, 77:321–357, 1995.
7. R. M. Hare. *Freedom and Reason*. The Clarendon Press, Oxford, UK, 1963.
8. P. McNamara. Doing well enough: Towards a logic for common sense morality. *Studia Logica*, 57:167–192, 1996.
9. D. N. Walton. *Argument Schemes for Presumptive Reasoning*. Lawrence Erlbaum Associates, Mahwah, NJ, USA, 1996.
10. M. Wooldridge and W. van der Hoek. On obligations and normative ability: Towards a logical analysis of the social contract. *Journal of Applied Logic*, 3:396–420, 2005.

A Logical Architecture of a Normative System

Guido Boella[1] and Leendert van der Torre[2]

[1] Dipartimento di Informatica Università di Torino, Italy
guido@di.unito.it
[2] University of Luxembourg
leendert@vandertorre.com

Abstract. Logical architectures combine several logics into a more complex logical system. In this paper we study a logical architecture using input/output operations corresponding to the functionality of logical components. We illustrate how the architectural approach can be used to develop a logic of a normative system based on logics of counts-as conditionals, institutional constraints, obligations and permissions. In this example we adapt for counts-as conditionals and institutional constraints a proposal of Jones and Sergot, and for obligations and permissions we adapt the input/output logic framework of Makinson and van der Torre. We use our architecture to study logical relations among counts-as conditionals, institutional constraints, obligations and permissions. We show that in our logical architecture the combined system of counts-as conditionals and institutional constraints reduces to the logic of institutional constraints, which again reduces to an expression in the underlying base logic. Counts-as conditionals and institutional constraints are defined as a pre-processing step for the regulative norms. Permissions are defined as exceptions to obligations and their interaction is characterized.

1 Introduction

In this paper we are interested in logical architectures. The notion of an 'architecture' is used not only in the world of bricks and stones, but it is used metaphorically in many other areas too. For example, in computer science it is used to describe the hard- or software organization of systems, in management science it is used to describe the structure of business models in enterprise architectures [16], and in psychology and artificial intelligence it is used to describe cognitive architectures of agent systems like SOAR [15], ACT [2] or PRS [13]. Though architectures are typically visualized as a diagram and informal, there are also various formal languages to describe architectures, see, for example, [16]. The notion of architecture reflects in all these examples an abstract description of a system in terms of its components and the relations among these components. This is also how we use the metaphor in this paper. In logic and knowledge representation, architectures combine several logics into a more complex logical system.

Advantages of the architectural approach in logic are that logical subsystems can be analyzed in relation to their environment, and that a divide and conquer strategy can reduce a complex theorem prover to simpler proof systems. These advantages are related to the advantages of architectural approaches in other areas. For example, in

L. Goble and J.-J.C. Meyer (Eds.): DEON 2006, LNAI 4048, pp. 24–35, 2006.

computer science a devide and conquer strategy is used frequently to develop computer systems. In this area, the architectural approach is used also to bridge the gap between formal specification and architectural design, and to facilitate the communication among stakeholders discussing a system, using visual architectural description languages [16].

The logical architecture we introduce and discuss in this paper is based on an architecture we recently introduced for normative multiagent systems [7]. In Figure 1, components are visualized by boxes, and the communication channels relating the components are visualized by arrows. There are components for counts-as conditionals (CA), institutional constraints (IC), obligations (O) and permissions (P). Moreover, the norm base (NB) component contains sets of norms or rules, which are used in the other components to generate the component's output from its input. This component does not have any inputs, though input channels can be added to the architecture to represent ways to modify the norms. The institutional constraints act as a wrapper around the counts-as component to enable the connection with the other components, as explained in detail in this paper. The open circles are ports or interface nodes of the components, and the black circles are a special kind of merge nodes, as explained later too. Note that the architecture is only an abstract description of the normative system, focussing on the relations among various kinds of norms, but for example abstracting away from sanctions, control systems, or the roles of agents being played in the system.

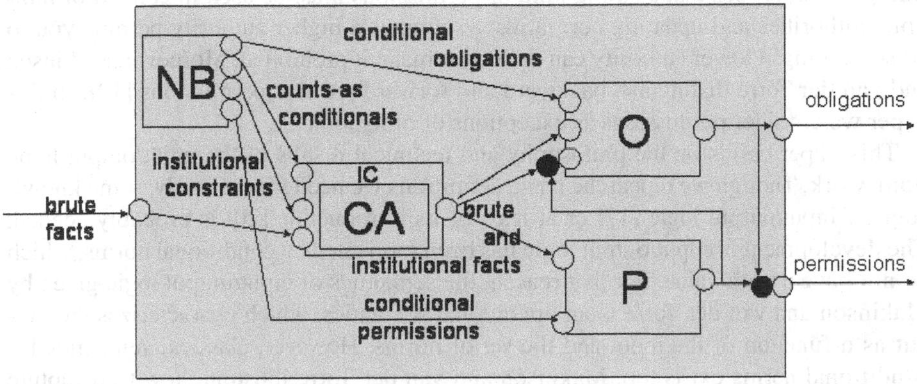

Fig. 1. A Logical Architecture of a Normative System

Figure 1 is a visualization of a logical architecture, where logical input/output operations correspond to the functionality of components. Considering a normative system as an input/output operation is not unusual. For example, inspired by Tarski's definition of deductive systems, Alchourrón and Bulygin [1] introduce normative systems with as inputs factual descriptions and as output obligatory and permitted situations. For counts-as conditionals and institutional constraints we adapt a proposal of Jones and Sergot [14], and for obligations and permissions we adapt the input/output logic framework of Makinson and van der Torre [17]. Moreover, we use Searle's distinction between regulative and constitutive norms [21], and brute and institutional facts.

The contribution of the paper thus lies not in the components, but in their mutual integration. Rather than following some highly general template for such an integration we stick closely to the contours of the specific domain: complex normative structures with multiple components. We focus on two kinds of interactions.

The main issue in the logical architecture is the relation among regulative and constitutive norms, that is, among on the one hand obligations and permissions, and on the other hand so-called counts-as conditionals. The latter are rules that create the possibility of or define an activity. For example, according to Searle, "the activity of playing chess is constituted by action in accordance with these rules. Chess has no existence apart from these rules. The institutions of marriage, money, and promising are like the institutions of baseball and chess in that they are systems of such constitutive rules or conventions." They have been identified as the key mechanism to normative reasoning in dynamic and uncertain environments, for example to realize agent communication, electronic contracting, dynamics of organizations, see, e.g., [6].

Extending the logical analysis of the relation between constitutive and regulative norms, we also reconsider the relation between obligations and permissions in our architecture. In the deontic logic literature, the interaction between obligations and permissions has been studied in some depth. Von Wright [22] started modern deontic logic literature by observing a similarity between the relation between on the one hand necessity and possibility, and on the other hand obligation and permission. He defined permissions as the absence of a prohibition, which was later called a weak permission. Bulygin [10] argues that a strong kind of permissions must be used in context of multiple authorities and updating normative systems: if a higher authority permits you to do something, a lower authority can no longer make it prohibited. Moreover, Makinson and van der Torre distinguish backward and forward positive permissions [19]. In this paper we consider permissions as exceptions of obligations.

This paper builds on the philosophy and technical results of the input/output logic framework. Though we repeat the basic definitions we need for our study, some knowledge of input/output logic [17] or at least of its introduction [20] is probably needed. The development of input/output logic has been motivated by conditional norms, which do not have a truth value. For that reason, the semantics of input/output logic given by Makinson and van der Torre is an operational semantics, which characterizes the output as a function of the input and the set of norms. However, classical semantics for conditional norms exists too. Makinson and van der Torre illustrate how to recapture input/output logic in modal logic, and thus give it a classical possible worlds semantics. More elegantly, as illustrated by Bochman [4], the operational semantics of input/output logic can be rephrased as a bimodel semantics, in which a model of a set of conditionals is a pair of partial models from the base logic (in this paper, propositional logic).

The layout of this paper is as follows. We first represent a fragment of Jones and Sergot's logic of counts-as conditionals as an input/output operation, then we represent their logic of institutional constraints as an input/output operation, and characterize their interaction. Thereafter we adapt Makinson and van der Torre's logic of input/output for multiple constraints, and we characterize the interaction among institutional constraints and obligations. Finally we introduce permissions as an input/output operation with multiple outputs, and we use them as exceptions to obligations.

2 Counts-as Conditionals (Constitutive Norms)

For Jones and Sergot [14], the counts-as relation expresses the fact that a state of affairs or an action of an agent "is a sufficient condition to guarantee that the institution creates some (usually normative) state of affairs". Jones and Sergot formalize this introducing a conditional connective $\Rightarrow_s$ to express the "counts-as" connection holding in the context of an institution s. They characterize the logic for $\Rightarrow_s$ as a conditional logic together with axioms for right conjunction,

$$((A \Rightarrow_s B) \wedge (A \Rightarrow_s C)) \supset (A \Rightarrow_s (B \wedge C))$$

left disjunction,

$$((A \Rightarrow_s C) \wedge (B \Rightarrow_s C)) \supset ((A \vee B) \Rightarrow_s C)$$

and transitivity.

$$((A \Rightarrow_s B) \wedge (B \Rightarrow_s C)) \supset (A \Rightarrow_s C)$$

Moreover, they consider other important properties not relevant here.

In this paper, we consider an input/output operation corresponding to the functionality of the counts-as component. This means that we restrict ourselves to the flat conditional fragment of Jones and Sergot's logic. Despite this restriction, we can still express the relevant properties Jones and Sergot discuss.[1] The input of the operation is a set of counts-as conditionals CA, a normative system or institution s, and the sufficient condition x, and an output is a created state of affairs y. Calling the operation out_{CA}, we thus write $y \in out_{CA}(CA, s, x)$. To relate the operation to the conditional logic, we equivalently write $(x, y) \in out_{CA}(CA, s)$, where (x, y) is read as "x counts as y".

Jones and Sergot propose a combined logic that incorporates besides the conditional logic also an action logic. We therefore assume a base action logic on which the input/output operation operates. This may be a logic of successful action as adopted by Jones and Sergot, according to which an agent brings it about that, or sees to it that, such-and-such is the case. But alternatively, using a simple propositional logic based on the distinction between controllable and uncontrollable propositions [9] most relevant properties can be expressed as well, and of course a more full-fledged logic of action can also be used, incorporating for example a model of causality.

Definition 1. *Let L be a propositional action logic with $\vdash$ the related notion of derivability and Cn the related consequence operation $Cn(x) = \{y \mid x \vdash y\}$. Let CA be a set of pairs of L, $\{(x_1, y_1), \ldots, (x_n, y_n)\}$, read as '$x_1$ counts as y_1', etc. Moreover, consider the following proof rules: conjunction for the output (AND), disjunction of the input (OR), and transitivity (T) defined as follows:*

$$\frac{(x, y_1), (x, y_2)}{(x, y_1 \wedge y_2)} AND \qquad \frac{(x_1, y), (x_2, y)}{(x_1 \vee x_2, y)} OR \qquad \frac{(x, y_1), (y_1, y_2)}{(x, y_2)} T$$

[1] When comparing their framework to a proposal of Goldman, Jones and Sergot mention (but do not study) irreflexive and asymmetric counts-as relations. With an extension of our logical language covering negated conditionals, these properties can be expressed too.

For an institution s, the counts-as output operator out_{CA} *is defined as closure operator on the set* CA *using the rules above, together with a silent rule that allows replacement of logical equivalents in input and output. We write*

$$(x, y) \in out_{CA}(CA, s)$$

Moreover, for $X \subseteq L$*, we write*

$$y \in out_{CA}(CA, s, X)$$

if there is a finite $X' \subseteq X$ *such that* $(\wedge X', y) \in out_{CA}(CA, s)$*, indicating that the output* y *is derived by the output operator for the input* X*, given the counts-as conditionals* CA *of institution s. We also write* $out_{CA}(CA, s, x)$ *for* $out_{CA}(CA, s, \{x\})$.

Example 1. If for some institution s we have $CA = \{(a, x), (x, y)\}$, then we have $out_{CA}(CA, s, a) = \{x, y\}$.

Jones and Sergot argue that the strengthening of the input (SI) and weakening of the output (WO) rules presented in Definition 2 are invalid, see their paper for a discussion. The adoption of the transitivity rule T for their logic is criticized by Artosi *et al.* [3]. Jones and Sergot say that "we have been unable to produce any counter-instances [of transitivity], and we are inclined to accept it". Neither of these authors consider to replace transitivity by cumulative transitivity (CT), see Definition 4.

Jones and Sergot give a semantics for their classical conditional logic based on minimal conditional models. For our reconstruction, the following questions can be asked (where the second could provide a kind of operational semantics analogous to the semantics of input/output logics [17], and the third to the semantics of input/output logic given in [4]):

- Given CA of s, and (x, y), do we have $(x, y) \in out_{CA}(CA, s)$?
- Given CA of s and x, what is $out_{CA}(CA, s, x)$?
- Given CA of s, what is $out_{CA}(CA, s)$?

The following theorem illustrates a simpler case.

Theorem 1. *With only the AND and T rule, assuming replacements by logical equivalents,* $out_{CA}(CA, s, X) = \{\wedge Y \mid Y \subseteq \cup_{i=0}^{\infty} out_{CA}^{i}(CA, s, X)\}$ *is calculated as follows.*
$out_{CA}^{0}(CA, s, X) = \emptyset$
$out_{CA}^{i+1}(CA, s, X) = out_{CA}^{i}(CA, s, X) \cup \{y \mid (\wedge X', y) \in CA, X' \subseteq out_{CA}^{i}(CA, s, X)\}$

With the OR rule the situation is more complicated due to reasoning by cases. However, as we show by Theorem 3 in Section 3, we do not need the semantics of this component to define a semantics of the whole normative system.

3 Institutional Constraints

Jones and Sergot's analysis of counts-as conditionals is integrated with their notion of so-called institutional constraints. Note that the term "constraints" is used here in another way than it is used in input/output logic for handling contrary-to-duty obligations and also permissions, as we discuss in the following section, because the input/output constraints, but not the institutional constraints, impose consistency requirements. We have chosen not to avoid the possible confusion, given the existence of the institutional constraints in the literature and the appropriateness of input/output constraints for what it is actually doing.

Jones and Sergot introduce the normal KD modality D_s such that $D_s A$ means that A is "recognized by the institution s". D_s is represented by a so-called wrapper around the counts-as component in our normative system architecture. In computer science, a wrapper is a mechanism to pre-process the input and to post-process the output of a component. there are various ways to formalize this idea in a logical architecture. In this section we formalize it by stating that the input/output relation of the counts-as component is a subset of the input/output relation of the institutional constraints component.

Jones and Sergot distinguish relations of logical consequence, causal consequence, and deontic consequence. We do not consider the latter, as they are the obligations and permissions which we represent by separate logical subsystems. An institutional constraint "if x then y" is represented by $D_s(x \rightarrow y)$.

Definition 2. *Let IC be a set of pairs of L, $\{(x_1, y_1), \ldots, (x_n, y_n)\}$, read as 'if x_1 then y_1', etc. Moreover, consider the following proof rules strengthening of the input (SI), weakening of the output (WO) and Identity (Id) defined as follows:*

$$\frac{(x_1, y)}{(x_1 \wedge x_2, y)} SI \qquad \frac{(x, y_1 \wedge y_2)}{(x, y_1)} WO \qquad \frac{}{(x, x)} Id$$

For an institution s, the institutional constraint output operator out_{IC} is defined as closure operator on the set IC using the three rules of Definition 1, together with the three rules above and a silent rule that allows replacement of logical equivalents in input and output.

The output of the institutional constraints can be obtained by a reduction to the base logic. Whereas the logic of counts-as conditionals is relatively complicated, the logic of institutional constraints is straightforward.

Theorem 2. $out_{IC}(IC, s, x) = Cn(\{x\} \cup \{x \supset y \mid (x, y) \in IC\})$

Proof. *(sketch). T follows from the other rules, and the property for the remaining rules follows from results on throughput in [17].*

Counts-as conditionals and institutional constraints are related by Jones and Sergot by the axiom schema:

$$(A \Rightarrow_s B) \supset D_s(A \supset B)$$

Using our input/output operations, the combined system of counts-as conditionals and institutional constraints are thus characterized by the following bridge rule.

Definition 3. *Let out_{IC+CA} be an operation on two sets of pairs of L, IC and CA, defined as out_{IC} (with all six rules discussed thus far) on the first parameter, together with the following rule:*

$$\frac{(x, y) \in out_{CA}(CA, s)}{(x, y) \in out_{IC+CA}(IC, CA, s)}$$

The following theorem shows that we can calculate the output of the joint system without taking the logic of counts-as conditionals into account.

Theorem 3. $out_{IC+CA}(IC, CA, s, X) = out_{IC}(IC \cup CA, s, X)$.

Proof. (sketch). Clearly we have $out_{IC+CA}(IC, CA, s, X) = out_{IC}(IC \cup out_{CA}(CA), s, X)$. *Since the proof rules of* out_{IC} *contain the proof rules of* out_{CA}*, the result follows.*

The limitation of Jones and Sergot approach, according to Gelati *et al.* [11], is that the consequences of counts-as connections follow non-defeasibly (via the closure of the logic for modality D_s under logical implication), whereas defeasibility seems a key feature of such connections. Their example is that in an auction if a person raises one hand, this may count as making a bid. However, this does not hold if he raises his hand and scratches his own head. There are many ways in which the logic of institutional constraints can be weakened, which we do not further consider here.

4 Obligations

There are many deontic logics, and there are few principles of deontic logic which have not been criticized. In this paper we do not adopt one particular deontic logic, but Makinson and van der Torre's framework [17] in which various kinds of deontic logics can be defined. Their approach is based on the concept of logic as a 'secretarial assistant', in the sense that the role of logic is not to formalize reasoning processes themselves, but to pre- and post-process such reasoning processes. Though a discussion of this philosophical point is beyond the scope of this paper, the idea of pre- and post-processing is well suited for the architectural approach.

A set of conditional obligations or rules is a set of ordered pairs $a \to x$, where a and x are sentences of a propositional language. For each such pair, the body a is thought of as an input, representing some condition or situation, and the head x is thought of as an output, representing what the rule tells us to be obligatory in that situation. Makinson and van der Torre write (a, x) to distinguish input/output rules from conditionals defined in other logics, to emphasize the property that input/output logic does not necessarily obey the identity rule. In this paper we also follow this convention. We extend the syntax of input/output logic with a parameter s for the institution to match Jones and Sergot's definitions.

Definition 4 (Input/output logic). *For an institution s, let O be a set of pairs of L,* $\{(a_1, x_1), \ldots, (a_n, x_n)\}$*, read as 'if* a_1 *then obligatory* x_1*', etc. Moreover, consider the following proof rules strengthening of the input (SI), conjunction for the output (AND), weakening of the output (WO), disjunction of the input (OR), and cumulative transitivity (CT) defined as follows:*

$$\frac{(a,x)}{(a \wedge b, x)}SI \quad \frac{(a, x \wedge y)}{(a,x)}WO \quad \frac{(a,x),(a,y)}{(a, x \wedge y)}AND$$

$$\frac{(a,x),(b,x)}{(a \vee b, x)}OR \quad \frac{(a,x),(a \wedge x, y)}{(a,y)}CT$$

The following output operators are defined as closure operators on the set O using the rules above, together with a silent rule that allows replacement of logical equivalents in input and output. Moreover, we write $(a, x) \in out_O(O, s)$ *to refer to any of these operations. We also write* $x \in out_O(O, s, A)$ *if there is a finite* $A' \subseteq A$ *with* $(\wedge A', x) \in out_O(O, s)$.
out_O^1: *SI+AND+WO (simple-minded output)*
out_O^2: *SI+AND+WO+OR (basic output)*
out_O^3: *SI+AND+WO+CT (reusable output)*
out_O^4: *SI+AND+WO+OR+CT (basic reusable output)*

Example 2. Given $O = \{(a, x), (x, z)\}$ the output of O contains $(x \wedge a, z)$ using the rule SI. Using also the CT rule, the output contains (a, z). (a, a) follows only if there is an identity rule in addition (when Makinson and van der Torre call it throughput [17]).

The institutional constraints (and thus the counts-as conditionals) can be combined with the obligations using iteration.

Definition 5. *For an institution s, let* $out_{IC+CA+O}$ *be an operation on three sets of pairs of L, IC, CA, and O, defined in terms of* out_{IC+CA} *and* out_O *using the following rule:*

$$\frac{(x, y) \in out_{IC+CA}(IC, CA, s), (y, z) \in out_O(O, s)}{(x, z) \in out_{IC+CA+O}(IC, CA, O, s)}$$

Theorem 4

$$out_{IC+CA+O}(IC, CA, O, s, x) = out_O(O, s, out_{IC+CA}(IC, CA, x))$$

In case of contrary-to-duty obligations, the input represents something which is inalterably true, and an agent has to ask himself which rules (output) this input gives rise to: even if the input should have not come true, an agent has to "make the best out of the sad circumstances" [12]. In input/output logics under constraints, a set of mental attitudes and an input does not have a set of propositions as output, but a set of sets of propositions. We can infer a set of propositions by for example taking the join (credulous) or meet (sceptical), or something more complicated. In this paper we use the meet to calculate the output of the obligation component. Moreover, we extend the definition to a set of constraints. Although we need only one constraint set for Definition 7, we will need arbitrary sets in the following section in order to integrate permissions.

Definition 6 (Constraints). *For an institution s, let O be a set of conditional obligations, and let* $\{C_1, \ldots, C_n\}$ *be a set of sets of arbitrary formulas, which we will call the "constraint set". For any input set A, we define maxfamily(*$O, s, A, \{C_1, \ldots, C_n\}$*) to be the family of all maximal* $O' \subseteq O$ *such that* $out_O(O', A)$ *is consistent with* C_i *for* $1 \leq i \leq n$. *Moreover, we define outfamily(*$O, s, A, \{C_1, \ldots, C_n\}$*) to be the family of outputs under input A, generated by elements of maxfamily(*$O, s, A, \{C_1, \ldots, C_n\}$*). The meet output under constraints is*

$$out_O^{\cap}(O, s, A, \{C_1, \ldots, C_n\}) = \cap outfamily(O, s, A, \{C_1, \ldots, C_n\})$$

We can adopt an output constraint (the output has to be consistent) or an input/output constraint (the output has to be consistent with the input). The following definition uses the input/output constraint, because the output of the obligation component is consistent with the output of the institutional constraints.

Definition 7

$$out^{\cap}_{\text{IC+CA+O}}(IC, CA, O, s, x) = out^{\cap}_{\text{O}}(O, s, \{out_{\text{IC+CA}}(IC, CA, s, x)\}, out_{\text{IC+CA}}(IC, CA, s, x))$$

See [17, 18] for the semantics of input/output logics, further details on its proof theory, its possible translation to modal or propositional logic, the extension with the identity rule, alternative constraints, and examples.

5 Permissions

Permissions are often defined in terms of obligations, called 'weak' permissions, in which case there are no conditional permissions in NB. When they are not defined as weak permissions, as in this paper, then the norm databse also contains a set of permissive norms [19]. Such 'strong' permissions are typically defined analogously to obligation, but without the AND rule. The reason AND is not accepted is that p as well as $\neg p$ can be permitted, but it does not make sense to permit a contradiction. Permissions are simpler than obligations, as the issue of contrary-to-duty reasoning is not relevant, and therefore we do not have to define constraints. Here we consider only the rules SI and WO. As the output of the permission component we do not take the union of the set indicated, as this would lose information that we need in the next integrative step.

Definition 8 (Conditional permission). *For an institution s, let conditional permissions P be a set of pairs of L, $\{(a_1, x_1), \ldots, (a_n, x_n)\}$, read as 'if a_1 then permitted x_1', etc. The output of the permission component is*

$$out_{\text{P}}(P, s, A) = \{Cn(x) \mid (\wedge A', x) \in P, A' \subseteq A\}$$

In the normative system, we merge the output of permission component X with the output of the obligation component Y to get the property that if something is obliged, then it is also permitted.

Definition 9. *Let $X \subseteq 2^L$ and $Y \subseteq L$. The merger of the two sets is defined as follows*

$$merge(X, Y) = \{Cn(x \cup Y) \mid x \in X\}$$

The combination of the counts-as conditionals and the permissions is analogous to the combination of the counts-as conditionals and obligations.

Definition 10

$$\frac{(x, y) \in out_{\text{IC+CA}}(IC, CA, s), (y, Z) \in out_{\text{P}}(P, s)}{(x, Z) \in out_{\text{IC+CA+P}}(IC, CA, P, s)}$$

Finally, we consider the output of the normative system. For obligations, the output of the institutional constraints is merged with the output of the permissions component as defined in Definition 9. It is an extension of the output of the obligation component given in Definition 7, where we did not consider permissions yet.

Definition 11. $out^{\cap}_{\text{IC+CA+P+O}}(IC, CA, O, P, s, X) =$
$out^{\cap}_{\text{O}}(O, merge(out_{\text{IC+CA+P}}(IC, CA, P, s, X), out_{\text{IC+CA}}(IC, CA, s, X)), s, out_{\text{IC+CA}}(IC, CA, s, X))$

We have now defined the output of each component in the normative system architecture visualized in Figure 1, and the final step is to define the output of the whole normative system. The first output is the set of obligations as given in Definition 11. The second output is the set of permissions, which combines the output of the permission component with the output of the obligation component. The permissions of the normative system are defined as follows.

Definition 12. $out_{\text{IC+CA+P+O+P}}(IC, CA, P, O, s, X) =$
$$merge(out_{\text{IC+CA+P}}(IC, CA, P, s, X), out^{\cap}_{\text{IC+CA+P+O}}(IC, CA, O, P, s, X))$$

6 Further Extensions

In a single component, feedback is represented by the cumulative transitivity rule, in the following sense [5]. If x is in the output of a, and y is in the output of $a \wedge x$, then we may reason as follows. Suppose we have as input a, and therefore as output x. Now suppose that there is a feedback loop, such that we have as input $a \wedge x$, then we can conclude as output y. Thus in this example, feedback of x corresponds with the inference that y is in the output of a.

Moreover, there may be feedback among components, leading to cycles in the network. As a generalization of the cumulative transitivity rule for the obligation component, we may add a feedback loop from the obligation component to the counts-as component, such that new institutional facts can be derived for the context in which the obligations are fulfilled. Likewise, we may add a feedback loop from obligation to permission. Since there is already a channel from permission to obligation, this will result in another cycle.

Another interesting feedback loop is from counts-as conditional to a new input of the norm database. In this way, the normative system can define how the normative system can be updated, see [6] for a logical model how constitutive norms can define the role and powers of agents in the normative system. This includes the creation of contracts, which may be seen as legal institutions, that is, as normative systems within the normative system [8].

A more technical issue is what happens when we create a feedback loop by connecting the permissions in the output of the normative system component to the constraints input of the obligation component.

7 Concluding Remarks

In the paper we have presented a logical architecture of normative systems, combining the logics of counts-as conditionals and institutional constraints of Jones and

Sergot with the input/output logics of Makinson and van der Torre. The contribution of the paper thus lies not in the components, but in their mutual integration. The results established are all fairly straightforward given familiarity with the background on input/output logics. Finally, we have discussed various ways in which the normative system can be further extended using timed streams, feedback loops and hierarchical normative systems.

The architecture presented is just one (rather natural) example, and there may be additional components, and other kinds of interactions which would also be worth studying using the same techniques. We believe that it may be worthwhile to study other logical components and other ways to connect the components, leading to other relations among counts-as conditionals, institutional constraints, obligations and permissions. An important contribution of our work is that it illustrates how such studies can be undertaken.

Besides the further logical analysis of the architecture of normative system, there are two other important issues of further research. First there is a need for a general theory of logical architectures, besides the existing work on combining logics and formal software engineering, along the line of logical input/output nets as envisioned in [20, 5]. Second, in agent based software engineering there is a need for a study whether the logical architecture developed here can be used for the design of architectures in normative multi-agent systems.

References

1. C. Alchourrón and Bulygin. *Normative systems.* Springer, 1972.
2. J.R. Anderson. *Rules of the mind.* Lawrence Ellbaum Associates, 1993.
3. A. Artosi, A. Rotolo, and S. Vida. On the logical nature of count-as conditionals. In *Procs. of LEA 2004 Workshop*, 2004.
4. A. Bochman. *Explanatory Nonmonotonic reasoning.* World Scientific, 2005.
5. G. Boella, J. Hulstijn, and L. van der Torre. Interaction in normative multi-agent systems. *Electronic Notes in Theoretical Computer Science*, 141(5):135–162, 2005.
6. G. Boella and L. van der Torre. Regulative and constitutive norms in normative multiagent systems. In *Proceedings of the Ninth International Conference on the Principles of Knowledge Representation and Reasoning (KR'04)*, pages 255–265, 2004.
7. G. Boella and L. van der Torre. An architecture of a normative system (short paper). In *Proceedings of AAMAS06*, 2006.
8. G. Boella and L. van der Torre. A game theoretic approach to contracts in multiagent systems. *IEEE Trans. SMC, Part C*, 36(1):68– 79, 2006.
9. C. Boutilier. Toward a logic for qualitative decision theory. In *Proceedings of KR'94*, pages 75–86, 1994.
10. E. Bulygin. Permissive norms and normative systems. In *Automated Analysis of Legal Texts*, pages 211–218. Publishing Company, Amsterdam, 1986.
11. J. Gelati, G. Governatori, N. Rotolo, and G. Sartor. Declarative power, representation, and mandate. A formal analysis. In *Procs. of JURIX 02.* IOS press, 2002.
12. B. Hansson. An analysis of some deontic logics. *Nôus*, 3:373–398, 1969.
13. F.F. Ingrand, M.P Georgeff, and A.S. Rao. An architecture for real-time reasoning and system control. *IEEE Expert*, 7(6), 1992.
14. A. Jones and M. Sergot. A formal characterisation of institutionalised power. *Journal of IGPL*, 3:427–443, 1996.

15. J.E. Laird, A. Newell, and Paul S. Rosenbloom. SOAR: an architecture for general intelligence. *Artificial Intelligence*, 33:1–64, 1987.
16. M. Lankhorst et al. *Enterprise Architecture At Work*. Springer, 2005.
17. D. Makinson and L. van der Torre. Input-output logics. *Journal of Philosophical Logic*, 29:383–408, 2000.
18. D. Makinson and L. van der Torre. Constraints for input-output logics. *Journal of Philosophical Logic*, 30:155–185, 2001.
19. D. Makinson and L. van der Torre. Permissions from an input/output perspective. *Journal of Philosophical Logic*, 32 (4):391–416, 2003.
20. D. Makinson and L. van der Torre. What is input/output logic? In *Foundations of the Formal Sciences II: Applications of Mathematical Logic in Philosophy and Linguistics*, volume 17 of *Trends in Logic*. Kluwer, 2003.
21. J. R. Searle. *Speech Acts: An Essay in the Philosophy of Language*. Cambridge University Press, Cambridge, United Kingdom, 1969.
22. G.H. von Wright. Deontic logic. *Mind*, 60:1–15, 1951.

Delegation of Power in Normative Multiagent Systems

Guido Boella[1] and Leendert van der Torre[2]

[1] Dipartimento di Informatica - Università di Torino - Italy
guido@di.unito.it
[2] University of Luxembourg
leendert@vandertorre.com

Abstract. In this paper we reconsider the definition of counts-as relations in normative multiagent systems: counts-as relations do not always provide directly an abstract interpretation of brute facts in terms of institutional facts. We argue that in many cases the inference of institutional facts from brute facts is the result of actions of agents acting on behalf of the normative systems and who are in charge of recognizing which institutional facts follow from brute facts. We call this relation delegation of power: it is composed of a counts-as relation specifying that the effect of an action of an agent is an institutional fact and by a goal of the normative system that the fact is considered as an institutional fact. This relation is more complex than institutional empowerment, where an action of an agent counts-as an action of the normative system but no goal is involved, and than delegation of goals, where a goal is delegated to an agent without giving it any power. With two case studies we show the importance of the delegation of power. Finally, we show how the new definition can be related with existing ones by using different levels of abstraction.

1 Introduction

It is well known that normative systems include not only regulative norms like obligations, prohibitions and permissions, but also constitutive norms stating what counts as institutional facts in a normative system.

In this paper we introduce a new notion, called *delegation of power* beside constitutive and regulative norms. Thus, the research questions of this paper are:

- What is delegation of power in a normative multiagent system?
- How does it relate to counts-as conditionals?
- How does it relates to regulative norms?

The notion of counts-as introduced by Searle [1] has been interpreted in deontic logic in different ways and it seems to refer to different albeit related phenomena [2]. For example, Jones and Sergot [3] consider counts-as from the constitutive point of view. According to Jones and Sergot , the fact that A counts-as B in context C is read as a statement to the effect that A represents conditions for guaranteeing the applicability of particular classificatory categories. The counts-as guarantees the soundness of that inference, and enables "new" classifications which would otherwise not hold.

An alternative view of the counts-as relation is proposed by Grossi *et al.* [4]: according to the classificatory perspective A counts-as B in context C is interpreted as: A is

L. Goble and J.-J.C. Meyer (Eds.): DEON 2006, LNAI 4048, pp. 36–52, 2006.

classified as B in context C. In other words, the occurrence of A is a sufficient condition, in context C, for the occurrence of B. Via counts-as statements, normative systems can establish the ontology they use in order to distribute obligations, rights, prohibitions, permissions, *etc.*

In [5, 6] we propose a different view of counts-as which focuses on the fact that counts-as often provides an abstraction mechanism in terms of institutional facts, allowing the regulative rules to refer to legal notions which abstract from details.

None of the above analyses, however, considers the motivational aspects behind constitutive norms required by agent theory. Constitutive norms are modelled as counts-as conditionals which allow to infer which institutional facts follow from brute facts and from existing institutional facts. E.g., a car counts as a vehicle for the traffic law. The inference from facts to institutional facts is considered as automatic, i.e., it is assumed not to need any agent or resource to perform it. Agent theory, instead, considers also the resources needed to perform inferences. Calculating the consequences following from some premises has a cost which must be traded off against the benefit of making the inferences. Thus in Georgeff and Ingrand [7] inferences are considered as actions which are planned and subject to decision processes as any other action: there must be an agent taking the decision and executing them.

According to resource bounded reasoning, the assumption made above on constitutive rules, even if useful in some circumstances, is not realistic. In many circumstances facts which in principle should be considered as institutional facts are not recognized as such. In such circumstances, the interpretation of a fact as an institutional fact may depend on the action of some agent who acts to achieve a goal of the normative system that a brute fact is interpreted as an institutional fact: we say that this agent has been *delegated the power* to interpret the fact as an institutional fact. In the next section two such examples are considered.

To answer the research questions of this paper we use our normative multiagent system framework [5, 8] which explicitly takes into account the activity of agents in the definition of sanction based obligations. The basic assumptions of our model are that beliefs, goals and desires of an agent are represented by conditional rules, and that, when an agent takes a decision, it recursively models the other agents interfering with it in order to predict their reaction to its decision as in a game. Most importantly, the normative system itself can be conceptualized as a socially constructed agent with whom it is possible to play games to understand what will be its reaction to the agent's decision: to consider its behavior as a violation and to sanction it. These actions are carried out by agents playing roles in the normative system, like judges and policemen. In the model presented in [5, 6], regulative norms are represented by the goals of the normative system and constitutive norms as its beliefs.

We relate the notion of delegation of power with our previous work on norms providing the definition of counts-as and of obligations at different levels of abstraction depending on whether agents are considered or not as acting for the normative system.

The paper is organized as follows. In Section 2 we show two case studies motivating the paper. In Section 3 we introduce delegation of power and in Section 4 we distinguish three different levels of abstraction. In Section 5 we introduce the formal model explained by examples. Comparison with related work and conclusion end the paper.

2 Motivating Examples

To illustrate the necessity of the notion of delegation of power we resort to two motivating case studies. The first one is a case which happened to one of the authors. The second case concerns how the law deals with the enforcement of obligations.

The first example is about traffic norms. Due to increased levels of pollution, on certain days, only ecological vehicles are allowed in major towns. In particular, cars with a catalytic converter count as ecological vehicles. One author of this paper bought many years ago one of the first catalytic cars. So he felt permitted to go around by car in the days when non-catalytic cars were prohibited. Instead, he was stopped by the police and fined. Why? The car was bought before the local law recognized catalytic cars as ecological vehicles. The police agreed that the car had a catalytic converter: they could see it, the car worked only with unleaded fuel, both the manual and the licence of the car said it has a catalytic converter. However, there was a missing rubber stamp by the office declaring that the car counts as an ecological vehicle. The problem is not simply that only catalytic cars bought after a certain date are considered as ecological. Rather, a catalytic car is not ecological unless an agent officialy recognizes it as such. The police has no power to consider the car as ecological, the evidence notwithstanding.

The moral is that even if a brute fact is present and could allow the recognition of an institutional fact, the institutional fact is the result of the action of some agent who is empowered to do that.

The second example concerns how obligations are dealt with by the law. Obligations represent the social order the law aims to achieve. However, specifying this order is not sufficient to achieve it. Thus, obligations are associated with other instrumental norms - to use Hart [9]'s terminology: the lack of the wanted state of affairs is considered as a violation and the violator is sanctioned. These tasks are distributed to distinct agents, like judges and policemen, who have to decide whether and how to fulfill them.

There is, however, an asymmetry between considering something as a violation and sanctioning. The latter can be a physical action like putting into jail, while the former has always an institutional character. So, while the sanction can be directly performed by a policeman, the recognition of a violation can only be performed indirectly by means of some action which counts as the recognition of a violation, e.g., a trial by a judge.

The two examples have some similarities and differences. Both in case of ecological cars and in case of violations an action of an agent is necessary to create the institutional fact. These cases can be modelled by a counts-as relation between the action of an agent (putting a stamp on the car licence or recognizing a violation) and the institutional fact (being an ecological vehicle or having violated an obligation), rather than by a direct counts-as relation between the brute facts and the institutional facts. But at first sight the two cases also have a difference: the recognition of a violation is wanted by the normative system to achieve its social order. In this case besides the counts-as rule between the action and the recognition as a violation there is also the goal of the normative system that this recognition contributes to the social order. In the next section we argue that indeed both cases should be modelled by means of a goal of the normative system and a counts-as relation between actions of agents acting for the normative system and institutional facts: they are both examples of delegation of power.

3 Goal Delegation and the Delegation of Power

In this section we show how both examples of Section 2 can be modelled in the same way, starting from the analysis of two apparently unrelated phenomena: goal delegation and institutional empowerment.

According to Castelfranchi [10], goal delegation is relying on another agent for the achievement of one own's goal: "in delegation an agent A needs or likes an action of another agent B and includes it in its own plan. In other words, A is trying to achieve some of its goals through B's behaviours or actions; thus A has the goal that B performs a given action/behaviour." This is not an institutional phenomenon but a basic capability of agents which enables them to interact with each other.

Institutional empowerment, instead, is by nature an institutional phenomenon which is based on the counts-as relation: an agent is empowered to perform an institutional action - a kind of institutional fact - if some of its actions counts as the institutional action. For example, a director can commit by means of his signature his institution to purchase some goods. Thus it is essentially related to counts-as rules, albeit restricted to actions of agents. Consider as a paradigmatic case the work by Jones and Sergot [3]

Bottazzi and Ferrario [11] argue that the two phenomena are related, as in cases like the two examples of Section 2: an agent which is institutionally empowered, is also delegated the goal of the institution of making true an institutional fact by exercising its power in the specified situations.

The connection between goal delegation and institutional empowerment is not a necessary one. For example, the agent in charge for sanctioning an obligation is delegated the goal of sanctioning, but there is no need of institutional powers in case of physical sanctions. Viceversa, the law institutionally empowers agents to stipulate private contracts which have the force of law, without being delegated by the law to do so, since contracting agents act for their own sake [5].

This connection, which we call *delegation of power*, can be used to explain the two examples above. In the case of cars, for the normative system, catalytic cars have to be considered as ecological vehicles. There are three possibilities: first, recognizing all catalytic cars as ecological vehicles by means of a constitutive norm. This solution, however, does not consider the actual performance of the inference and the possible costs related to it. Second, the normative system can rely on some agent to recognize catalytic cars as ecological vehicles. As said above, this can be done by means of a counts-as relation between an action of an agent and its effects. This solution, however, fails to account for the motivations that the agent should have to perform the action of recognizing ecological vehicles as such. Third, also a goal of the normative system is added to motivate its action: there is an agent who has the institutional power to recognize cars as ecological vehicles and the normative system has delegated it the goal that it does so in order to motivate it.

In the case of obligations, beside the counts-as relation between an action of judge and the institutional fact that a violation is recognized, we have the goal that the specified behavior is considered as a violation. The goal is an instrumental goal associated with the definition of obligation which aims at regulating how violations are prosecuted.

Thus to model delegation of power we need a model where:

- Both constitutive and regulative norms are modelled.
- Since we want to model goal delegation, mental attitudes are attributed to the normative system.
- Agents act inside the normative system.

All these features are present in our model of normative multiagent systems [5]. Our model is based on the so called agent metaphor: social entities like normative systems can be conceptualized as agents by attributing them mental attitudes like beliefs and goals. The cognitive motivations of the agent metaphor underlying our framework are discussed in [12].

Beliefs model the constitutive rules of the normative system, while goals model regulative rules. Thus, in the normative system the interaction between constitutive and regulative rules is the same as the interaction of beliefs and goals in an agent.

However, differently from a real agent, the normative system is a socially constructed agent. It exists only because of the collective acceptance by all the agents and, thus, it cannot act in the world. Its actions are carried out by agents playing roles in the normative system, like legislators, judges and policemen. It is a social construction used to coordinate the behavior of the agents.

Our model of roles, which allows to structure organizations in sub-organizations and roles to make a multiagent system modular and thus manage its complexity, is described in [13]. For space reason, we do not introduce roles explicitly in this paper.

In our model obligations are not only modelled as goals of the normative system, but they are also associated with the instrumental goals that the behavior of the addressee of the norms is considered as a violation and that the violation is sanctioned. Considering something as a violation and sanctioning are actions which can be executed by the normative system itself, or, at a more concrete level of detail, by agents playing roles in the normative system.

The counts-as relation in our model is modelled as a conditional belief of the normative system to provide an abstraction of reality in terms of institutional facts. Regulative norms can refer to this level, thus decoupling them from the details of reality. For example, it must be distinguished the institutional fact that traffic lights are red from the brute fact that red light bulbs in the traffic lights are on: in the extreme case the institutional fact can be true even if all the red bulbs are broken. As a consequence, as discussed in [6, 14], we do not accept the identity rule for counts-as.

In this paper, we consider how counts-as can be used to define delegation of power. Counts-as relations are not used in this case to directly connect brute facts to institutional facts, but only to express the (institutional) effects of actions of agents empowered by the normative system (in the same sense as the act of signing of the director counts as the commitment of the institution he belongs to).

In our model [5], constitutive rules have also other roles, in particular, they specify how the normative system can change. In this sense a normative multiagent system is a system in the sense that it specifies itself how it can change. Since it is a socially constructed agent, it cannot directly change itself. Rather it relies on the actions of agents playing roles in it, like legislators, which count as changes of the system.

4 Level of Abstractions in the Definition of Norms

Obligations and counts-as relations can be defined at different levels of abstraction. We identify three different levels of abstraction, which are suited for different applications. The abstraction dimension is the detail at which we consider agents acting for the normative system: at the higher abstraction level agents have no role, at the middle abstraction level only actions of the normative system are considered but agents are not considered; only at the more concrete level, where agents are in charge of the actual functioning of the normative system concerning regulative and constitutive rules, delegation of power enters the picture:

1. The higher level abstracts from the fact that violations, sanctions and institutional facts are the result of the action that an agent decides to perform. Thus obligations are defined as in Anderson's reduction [15]: the recognition of the violation and the sanction logically follow from the violation. This abstraction level for regulative rules is adopted also by [16, 17] and we use it in [18]. For constitutive rules, up to our knowledge, this is the only level considered.
2. The middle level abstracts from the role of agents in the normative system, but the normative system is in some way personified and is assumed to act in the world: thus the recognition of violations and sanctions are considered as the actions of the normative system itself. We adopt this level of representation for regulative norms in [5, 8]. Analogously, institutional facts follow from actions of the normative system: they are not anymore logical consequences of facts, but consequences of decisions of the normative systems which are traded-off against other decisions. They, thus, do not follow automatically, since the normative system can take a different decision due to conflicts with other goals or to lack of resources.
3. The lower level takes into account the actions of the agents in the normative system. Concerning regulative norms, some agents are delegated the goal to sanction violations and the goal and power of recognizing violations. I.e., they are delegated the power to do so. Concerning constitutive norms, the agents are delegated the goal to recognize some facts as institutional facts and the power to do so by means of their actions. I.e., they are delegated the power to do so. The problem of agents recognizing violations has been partially addressed in [19], but the recognition action was considered as a physical action like the sanction. In this paper we add the counts-as relation to the recognition of violations.

At the lower two levels it becomes possible to answer the question whether constitutive norms can be violated like it happens for regulative ones. A constitutive norm can be violated in the sense that the normative system or the agent who is delegated the goal to recognize the institutional fact and empowered to do so fails to achieve the delegated goal. In our first example the office could fail or refuse to recognize the car as an ecological vehicle. The reason can be the inability to perform the necessary actions, laziness, bribing, *etc.*, like it happens for regulative norms. Moreover, constitutive rules can be abused, in the sense that the delegated agent can exercise its power without being delegated to do so in the given circumstances. This possibility assumes that the institutional power can be exercised beyond the conditions under which it has been delegated the goal to exercise it.

5 The Formal Model

The definition of the agents is inspired by the rule based BOID architecture [20], though in our theory, and in contrast to BOID, obligations are not taken as primitive concepts. Beliefs, desires and goals are represented by conditional rules rather than in a modal framework. Intentions have been introduced as a form of bounded rationality: since an agent has not enough resources to make the optimal decision at each moment, he maintains its previous choices. In this paper we consider only one decision, so we do not need to introduce intentions to model decisions which persist over time.

5.1 Input/Output Logic

To represent conditional mental attitudes we take a simplified version of the input/output logics introduced in [21, 22]. A rule set is a set of ordered pairs $P \rightarrow q$. For each such pair, the body P is thought of as an input, representing some condition or situation, and the head q is thought of as an output, representing what the rule tells us to be believed, desirable, obligatory or whatever in that situation. In this paper, to keep the formal exposition simple, input and output are respectively a set of literals and a literal.

The development of input/output logic has been motivated by conditional norms, which do not have a truth value. For that reason, the semantics of input/output logic given by Makinson and van der Torre [21] is an operational semantics, which characterizes the output as a function of the input and the set of norms. However, it is easy to define a classical semantics for conditional norms too. Makinson and van der Torre illustrate how to recapture input/output logic in modal logic, and thus give it a classical possible worlds semantics. More elegantly, as illustrated by Bochman [23], the operational semantics of input/output logic can be rephrased as a bimodel semantics, in which a model of a set of conditionals is a pair of partial models from the base logic (in this paper, propositional logic).

Though the development of input/output logic has been motivated by the logic of norms, the same logic can be used for other conditionals like conditional beliefs and conditional goals – which explains the more general name of the formal system. Moreover, Bochman [23] also illustrates how the same logic is used for causal reasoning and various non-monotonic reasoning formalisms.

Definition 1 (Input/output logic). *Let X be a set of propositional variables, the set of literals built from X, written as $Lit(X)$, is $X \cup \{\neg x \mid x \in X\}$, and the set of rules built from X, written as $Rul(X) = 2^{Lit(X)} \times Lit(X)$, is the set of pairs of a set of literals built from X and a literal built from X, written as $\{l_1, \ldots, l_n\} \rightarrow l$. We also write $l_1 \wedge \ldots \wedge l_n \rightarrow l$ and when $n = 0$ we write $\top \rightarrow l$. For $x \in X$ we write $\sim x$ for $\neg x$ and $\sim(\neg x)$ for x. Moreover, let Q be a set of pointers to rules and $RD : Q \rightarrow Rul(X)$ is a total function from the pointers to the set of rules built from X.*

Let $S = RD(Q)$ be a set of rules $\{P_1 \rightarrow q_1, \ldots, P_n \rightarrow q_n\}$, and consider the following proof rules strengthening of the input (SI), disjunction of the input (OR), cumulative transitivity (CT) and Identity (Id) defined as follows:

$$\frac{p \rightarrow r}{p \wedge q \rightarrow r} SI \qquad \frac{p \wedge q \rightarrow r, p \wedge \neg q \rightarrow r}{p \rightarrow r} OR \qquad \frac{p \rightarrow q, p \wedge q \rightarrow r}{p \rightarrow r} CT \qquad \frac{}{p \rightarrow p} Id$$

The following output operators are defined as closure operators on the set S using the rules above.

out_1: SI (simple-minded output) out_3: SI+CT (reusable output)
out_2: SI+OR (basic output) out_4: SI+OR+CT (basic reusable output)

Moreover, the following four throughput operators are defined as closure operators on the set S. out_i^+: out_i+Id (throughput) We write $out(Q)$ for any of these output operations and $out^+(Q)$ for any of these throughput operations. We also write $l \in out(Q,L)$ iff $L \rightarrow l \in out(Q)$, and $l \in out^+(Q,L)$ iff $L \rightarrow l \in out^+(Q)$.

Example 1. Given $RD(Q) = \{a \rightarrow x, x \rightarrow z\}$ the output of Q contains $x \wedge a \rightarrow z$ using the rule SI. Using also the CT rule, the output contains $a \rightarrow z$. $a \rightarrow a$ follows only if there is the Id rule.

A technical reason to distinguish pointers from rules is to facilitate the description of the priority ordering we introduce in the following definition.

The notorious contrary-to-duty paradoxes such as Chisholm's and Forrester's paradox have led to the use of constraints in input/output logics [22]. The strategy is to adapt a technique that is well known in the logic of belief change - cut back the set of norms to just below the threshold of making the current situation inconsistent.

Definition 2 (Constraints). *Let $\geq$: $2^Q \times 2^Q$ be a transitive and reflexive partial relation on the powerset of the pointers to rules containing at least the subset relation and $RD : Q \rightarrow Rul(X)$ a function from the pointers to the set of rules. Moreover, let out be an input/output logic:*

- *$maxfamily(Q,P)$ is the set of $\subseteq$-maximal subsets Q' of Q such that $out(Q',P) \cup P$ is consistent.*
- *$preffamily(Q,P,\geq)$ is the set of $\geq$-maximal elements of $maxfamily(Q,P)$.*
- *$outfamily(Q,P,\geq)$ is the output under the elements of preffamily, i.e., $\{out(Q',P) \mid Q' \in preffamily(Q,P,\geq)\}$.*
- *$P \rightarrow x \in out_\cup(Q,\geq)$ iff $x \in \cup outfamily(Q,P,\geq)$*
  $P \rightarrow x \in out_\cap(Q,\geq)$ iff $x \in \cap outfamily(Q,P,\geq)$

Example 2. Let $RD(\{a,b,c\}) = \{a = (\top \rightarrow m), b = (p \rightarrow n), c = (o \rightarrow \neg m)\}$, $\{b,c\} > \{a,b\} > \{a,c\}$, where by $A > B$ we mean as usual $A \geq B$ and $B \not\geq A$.
$maxfamily(Q,\{o\}) = \{\{a,b\},\{b,c\}\}$,
$preffamily(Q,\{o\},\geq) = \{\{b,c\}\}$,
$outfamily(Q,\{o\},\geq) = \{\{\neg m\}\}$

The *maxfamily* includes the sets of applicable compatible pointers to rules together with all non applicable ones: e.g., the output of $\{a,c\}$ in the context $\{o\}$ is not consistent. Finally $\{a\}$ is not in *maxfamily* since it is not maximal, we can add the non applicable rule b. Then *preffamily* is the preferred set $\{b,c\}$ according to the ordering on set of rules above. The set *outfamily* is composed by the consequences of applying the rules $\{b,c\}$ which are applicable in o (c): $\neg m$.

Due to space limitations we have to be brief on details with respect to input/output logics, see [21, 22] for the semantics of input/output logics, further details on its proof theory, its possible translation to modal logic, alternative constraints, and examples.

5.2 Multiagent Systems

We assume that the base language contains boolean variables and logical connectives. The variables are either *decision variables* of an agent, which represent the agent's actions and whose truth value is directly determined by it, or *parameters*, which describe both the state of the world and *institutional facts*, and whose truth value can only be determined indirectly. Our terminology is borrowed from Lang *et al.* [24] and is used in discrete event systems, and many formalisms in operations research.

Given the same set of mental attitudes, agents reason and act differently: when facing a conflict among their motivations and beliefs, different agents prefer to fulfill different goals and desires. We express these agent characteristics by a priority relation on the mental attitudes which encode, as detailed in [20], how the agent resolves its conflicts. The priority relation is defined on the powerset of the mental attitudes such that a wide range of characteristics can be described, including social agents that take the desires or goals of other agents into account. The priority relation contains at least the subset-relation which expresses a kind of independence among the motivations.

Background knowledge is formalized by a set of effects E represented by rules.

Definition 3 (Agent set). *An agent set is a tuple* $\langle A, X, B, D, G, AD, E, \geq, \geq_E \rangle$, *where:*

- *the agents* A*, propositional variables* X*, agent beliefs* B*, desires* D*, goals* G*, and effects* E *are six finite disjoint sets.*
- B, D, G *are sets of mental attitudes. We write* $M = D \cup G$ *for the motivations defined as the union of the desires and goals.*
- *an agent description* $AD : A \rightarrow 2^{X \cup B \cup M}$ *is a total function that maps each agent to sets of variables (its decision variables), beliefs, desires and goals, but that does not necessarily assign each variable to at least one agent. For each agent* $b \in A$*, we write* X_b *for* $X \cap AD(b)$*, and* B_b *for* $B \cap AD(b)$*,* D_b *for* $D \cap AD(b)$*, etc. We write parameters* $P = X \setminus \cup_{b \in A} X_b$.
- *the set of effects* E *represents the background knowledge of all agents.*
- *a priority relation* $\geq: A \rightarrow 2^B \times 2^B \cup 2^M \times 2^M$ *is a function from agents to a transitive and reflexive partial relation on the powerset of the motivations containing at least the subset relation. We write* $\geq_b$ *for* $\geq (b)$.
- *a priority relation* $\geq_E: 2^E \times 2^E$ *is a transitive and reflexive partial relation on the powerset of effects containing at least the subset relation.*

Example 3. $A = \{\mathbf{a}\}$, $X_{\mathbf{a}} = \{drive\}$, $P = \{s, catalytic\}$, $D_{\mathbf{a}} = \{d_1, d_2\}$, $\geq_{\mathbf{a}} = \{d_2\} \geq \{d_1\}$. There is a single agent, agent **a**, who can drive a car. Moreover, it can be sanctioned and the car can be catalytic. It has two desires, one to drive (d_1), another one not to be sanctioned (d_2). The second desire is more important.

In a multiagent system, beliefs, desires, goals and effects are abstract concepts which are described by rules built from literals.

Definition 4 (Multiagent system). *A multiagent system, written as NMAS, is a tuple* $\langle A, X, B, D, G, AD, E, RD, \geq, \geq_E \rangle$, *where* $\langle A, X, B, D, G, AD, E, \geq, \geq_E \rangle$ *is an agent set, and the rule description* $RD : (B \cup M \cup E) \rightarrow Rul(X)$ *is a total function from the sets of beliefs, desires and goals, and effects to the set of rules built from* X. *For a set of pointers* $S \subseteq B \cup M \cup E$, *we write* $RD(S) = \{RD(q) \mid q \in S\}$.

Example 4 (Continued). $RD(d_1) = \top \rightarrow drive$, $RD(d_2) = \top \rightarrow \neg s$.

In the description of the normative system, we do not introduce norms explicitly, but we represent several concepts which are illustrated in the following sections. Institutional facts (I) represent legal abstract categories which depend on the beliefs of the normative system and have no direct counterpart in the world. $F = X \setminus I$ are what Searle calls "brute facts": physical facts like the actions of the agents and their effects. $V(x, \mathbf{a})$ represents the decision of agent $\mathbf{n}$ that recognizes x as a violation by agent $\mathbf{a}$. The goal distribution $GD(\mathbf{a}) \subseteq G_{\mathbf{n}}$ represents the goals of agent $\mathbf{n}$ the agent $\mathbf{a}$ is responsible for.

Definition 5 (Normative system). *A normative multiagent system, written as NMAS, is a tuple* $\langle A, X, B, D, G, AD, E, RD, \geq, \geq_E, \mathbf{n}, I, V, GD \rangle$ *where the tuple* $\langle A, X, B, D, G, AD, E, RD, \geq, \geq_E \rangle$ *is a multiagent system, and*

- *the normative system* $\mathbf{n} \in A$ *is an agent.*
- *the institutional facts* $I \subseteq P$ *are a subset of the parameters.*
- *the norm description* $V : Lit(X) \times A \rightarrow X_{\mathbf{n}} \cup P$ *is a function from the literals and the agents to the decision variables of the normative system and the parameters.*
- *the goal distribution* $GD : A \rightarrow 2^{G_{\mathbf{n}}}$ *is a function from the agents to the powerset of the goals of the normative system, such that if* $L \rightarrow l \in RD(GD(\mathbf{a}))$, *then* $l \in Lit(X_{\mathbf{a}} \cup P)$.

Agent $\mathbf{n}$ is a normative system with the goal that non catalytic cars are not driven.

Example 5 (Continued). There is agent $\mathbf{n}$, representing the normative system.
$P = \{s, V(drive, \mathbf{a}), catalytic\}$, $D_{\mathbf{n}} = G_{\mathbf{n}} = \{g_1\}$,
$RD(g_1) = \{\neg catalytic \rightarrow \neg drive\}$, $GD(\mathbf{a}) = \{g_1\}$.

The parameter $V(drive, \mathbf{a})$ represents the fact that the normative system considers a violation agent a's action of driving. It has the goal that non-ecological vehicles should not be driven by $\mathbf{a}$ and it has distributed this goal to agent $\mathbf{a}$.

In the following, we use an input/output logic *out* to define whether a desire or goal implies another one and to define the application of a set of belief rules to a set of literals; in both cases we use the out_3 operation since it has the desired logical property of not satisfying identity.

We now define obligations and counts-as at the three levels of abstraction.

Regulative norms are conditional obligations with an associated sanction. At the higher level of abstraction, the definition contains three clauses: the first two clauses state that recognitions of violations and sanctions are a consequence of the behavior of agent $\mathbf{a}$, as it is represented by the background knowledge rules E. For an obligation to be effective, the third clause states that the sanction must be disliked by its addressee.

Definition 6 (Obligation (level 1)). *Let NMAS be a normative multiagent system* $\langle A, X, B, D, G, AD, E, RD, \geq, \geq_E, \mathbf{n}, I, V, GD\rangle$.

Agent $\mathbf{a} \in A$ *is obliged to see to it that* $x \in Lit(X_\mathbf{a} \cup P)$ *with sanction* $s \in Lit(P)$ *in context* $Y \subseteq Lit(X)$ *in NMAS, written as* $NMAS \models O^1_{\mathbf{an}}(x, s|Y)$, *if and only if:*

1. $Y \cup \{\sim x\} \rightarrow V(\sim x, \mathbf{a}) \in out(E, \geq_E)$: *if* Y *and* x *is false, then it follows that* $\sim x$ *is a violation by agent* $\mathbf{a}$.
2. $Y \cup \{V(\sim x, \mathbf{a})\} \rightarrow s \in out(E, \geq_E)$: *if* Y *and there is a violation by agent* $\mathbf{a}$, *then it is sanctioned.*
3. $Y \rightarrow \sim s \in out(D_\mathbf{a}, \geq_\mathbf{a})$: *if* Y, *then agent* $\mathbf{a}$ *desires* $\sim s$, *which expresses that it does not like to be sanctioned.*

Example 6. Let: $E = \{e_1, e_2\}, D_\mathbf{a} = \{d_2\}$
$RD(e_1) = \{\neg catalytic, drive\} \rightarrow V(drive, \mathbf{a})$
$RD(e_2) = \{\neg catalytic, V(drive, \mathbf{a})\} \rightarrow s$
$RD(d_2) = \neg catalytic \rightarrow \sim s$
$NMAS \models O^1_{\mathbf{an}}(\neg drive, s \mid \neg catalytic)$, since:

1. $\{\neg catalytic, drive\} \rightarrow V(drive, \mathbf{a}) \in out(E, \geq_E)$
2. $\{\neg catalytic, V(drive, \mathbf{a})\} \rightarrow s \in out(E, \geq_E)$
3. $\neg catalytic \rightarrow \sim s \in out(D_\mathbf{a}, \geq_\mathbf{a})$

Constitutive norms introduce new abstract categories of existing facts and entities, called institutional facts. In [6] we formalize the counts-as conditional as a belief rule of the normative system $\mathbf{n}$. Since the condition x of the belief rule is a variable it can be an action of an agent, a brute fact or an institutional fact. So, the counts-as relation can be iteratively applied. In our model the counts-as relation does not satisfy the identity rule. See [6] for a discussion of the motivations.

Definition 7 (Counts-as relation (level 1)). *Let NMAS be a normative multiagent system* $\langle A, X, B, D, G, AD, E, RD, \geq, \geq_E, \mathbf{n}, I, V, GD\rangle$. *A literal* $x \in Lit(X)$ *counts-as* $y \in Lit(I)$ *in context* $C \subseteq Lit(X)$, $NMAS \models counts\text{-}as^1_\mathbf{n}(x, y|C)$, *iff* $C \cup \{x\} \rightarrow y \in out(B_\mathbf{n}, \geq_\mathbf{n})$: *if agent* $\mathbf{n}$ *believes* C *and* x *then it believes* y.

Example 7. $P \setminus I = \{catalytic\}, I = \{eco\}, X_\mathbf{a} = \{drive\}, B_\mathbf{n} = \{b_1\}, RD(b_1) = catalytic \rightarrow eco$

Consequently, $NMAS \models counts\text{-}as^1_\mathbf{n}(catalytic, eco|\top)$. This formalizes that for the normative system a catalytic car counts as an ecological vehicle. The presence of the catalytic converter is a physical "brute" fact, while being an ecological vehicle is an institutional fact. In situation $S = \{catalytic\}$, given $B_\mathbf{n}$ we have that the consequences of the constitutive norms are $out(B_\mathbf{n}, S, \geq_\mathbf{n}) = \{eco\}$ (since out_3 does not include Id).

At the middle level of abstraction, actions of the normative systems are added in the definition of the obligations: the recognition of a violation and sanctions. Since the actions undergo a decision process, desires and goals of the normative system are added. The first and central clause of our definition of obligation defines obligations of agents as goals of the normative system, following the "your wish is my command" metaphor.

It says that the obligation is implied by the desires of the normative system **n**, implied by the goals of agent **n**, and it has been distributed by agent **n** to the agent. The latter two steps are represented by $out(GD(\mathbf{a}), \geq_{\mathbf{n}})$.

The second and third clause can be read as the normative system has the goal that the absence of p is considered as a violation. The third clause says that the agent desires that there are no violations, which is stronger than that it does not desire violations, as would be expressed by $\top \rightarrow V(\sim x, a) \notin out(D_{\mathbf{n}}, \geq_{\mathbf{n}})$.

The fourth and fifth clause relate violations to sanctions. The fifth clause says that the normative system is motivated not to sanction as long as their is no violation, because otherwise the norm would have no effect. Finally, for the same reason the last clause says that the agent does not like the sanction. The second and fourth clause can be considered as instrumental norms [9] contributing to the achievement of the main goal of the norm.

Definition 8 (Obligation (level 2)). *Let NMAS be a normative multiagent system* $\langle A, X, B, D, G, AD, E, RD, \geq, \geq_E, \mathbf{n}, I, V, GD\rangle$. *Agent* $\mathbf{a} \in A$ *is obliged to see to it that* $x \in Lit(X_{\mathbf{a}} \cup P)$ *with sanction* $s \in Lit(X_{\mathbf{n}} \cup P)$ *in context* $Y \subseteq Lit(X)$ *in NMAS, written as* $NMAS \models O^2_{\mathbf{an}}(x, s|Y)$, *if and only if:*

1. $Y \rightarrow x \in out(D_{\mathbf{n}}, \geq_{\mathbf{n}}) \cap out(GD(\mathbf{a}), \geq_{\mathbf{n}})$: *if* Y *holds then agent* **n** *desires and has as a goal that* x, *and this goal has been distributed to agent* **a**.
2. $Y \cup \{\sim x\} \rightarrow V(\sim x, \mathbf{a}) \in out(D_{\mathbf{n}}, \geq_{\mathbf{n}}) \cap out(G_{\mathbf{n}}, \geq_{\mathbf{n}})$: *if* Y *holds and* $\sim x$, *then agent* **n** *has the goal and the desire* $V(\sim x, \mathbf{a})$: *to recognize it as a violation by agent* **a**.
3. $\top \rightarrow \neg V(\sim x, \mathbf{a}) \in out(D_{\mathbf{n}}, \geq_{\mathbf{n}})$: *agent* **n** *desires that there are no violations.*
4. $Y \cup \{V(\sim x, \mathbf{a})\} \rightarrow s \in out(D_{\mathbf{n}}, \geq_{\mathbf{n}}) \cap out(G_{\mathbf{n}}, \geq_{\mathbf{n}})$: *if* Y *holds and agent* **n** *decides* $V(\sim x, \mathbf{a})$, *then agent* **n** *desires and has as a goal that it sanctions agent* **a**.
5. $Y \rightarrow \sim s \in out(D_{\mathbf{n}}, \geq_{\mathbf{n}})$: *if* Y *holds, then agent* **n** *desires not to sanction. This desire of the normative system expresses that it only sanctions in case of violation.*
6. $Y \rightarrow \sim s \in out(D_{\mathbf{a}}, \geq_{\mathbf{a}})$: *if* Y *holds, then agent* **a** *desires* $\sim s$, *which expresses that it does not like to be sanctioned.*

The rules in the definition of obligation are only motivations, and not beliefs, because a normative system may not recognize that a violation counts as such, or that it does not sanction it: it is up to its decision. Both the recognition of the violation and the application of the sanction are the result of autonomous decisions of the normative system that is modelled as an agent.

The beliefs, desires and goals of the normative agent - defining the obligations - are not private mental states of an agent. Rather they are collectively attributed by the agents of the normative system to the normative agent: they have a public character, and, thus, which are the obligations of the normative system is a public information.

Example 8. Let: $\{g_1, g_2, g_4\} = G_{\mathbf{n}}, G_{\mathbf{n}} \cup \{g_3, d_2\} = D_{\mathbf{n}}, \{g_1\} = GD(\mathbf{a}), \{d_2\} = D_{\mathbf{a}}$

$RD(g_2) = \{\neg catalytic, drive\} \rightarrow V(drive, \mathbf{a})$ $RD(g_3) = \top \rightarrow \neg V(drive, \mathbf{a})$
$RD(g_4) = \{\neg catalytic, V(drive, \mathbf{a})\} \rightarrow s$

$NMAS \models O^2_{\mathbf{an}}(\neg drive, s \mid \neg catalytic)$, since:

1. $\neg catalytic \rightarrow \neg drive \in out(D_{\mathbf{n}}, \geq_{\mathbf{n}}) \cap out(GD(\mathbf{a}), \geq_{\mathbf{n}})$
2. $\{\neg catalytic, drive\} \rightarrow V(drive, \mathbf{a}) \in out(D_{\mathbf{n}}, \geq_{\mathbf{n}}) \cap out(G_{\mathbf{n}}, \geq_{\mathbf{n}})$
3. $\top \rightarrow \neg V(drive, \mathbf{a}) \in out(D_{\mathbf{n}}, \geq_{\mathbf{n}})$
4. $\{\neg catalytic, V(drive, \mathbf{a})\} \rightarrow s \in out(D_{\mathbf{n}}, \geq_{\mathbf{n}}) \cap out(G_{\mathbf{n}}, \geq_{\mathbf{n}})$
5. $\neg catalytic \rightarrow \sim s \in out(D_{\mathbf{n}}, \geq_{\mathbf{n}})$
6. $\neg catalytic \rightarrow \sim s \in out(D_{\mathbf{a}}, \geq_{\mathbf{a}})$

At the middle level of abstraction, the beliefs of the normative system represent only the connections between actions and the consequences of these actions for the normative system. The normative system has the desire and goal that the institutional fact y holds if the fact x holds in context C. The normative system believes that to make y true it has to perform an action z. Thus it is not sufficient the fact x holding in context C for the institutional fact y to be true: it is necessary also a decision to do z by the normative system.

Definition 9 (Counts-as relation (level 2)). *Let $NMAS$ be a normative multiagent system $\langle A, X, B, D, G, AD, E, RD, \geq, \geq_E, \mathbf{n}, I, V, GD\rangle$. A literal $x \in Lit(X)$ counts-as $y \in Lit(I)$ in context $C \subseteq Lit(X)$, $NMAS \models counts\text{-}as^2_{\mathbf{n}}(x, y|C)$, iff:*

1. *$C \wedge x \rightarrow y \in out(D_{\mathbf{n}}, \geq_{\mathbf{n}}) \cap out(G_{\mathbf{n}}, \geq_{\mathbf{n}})$: it is a desire and goal of the normative system that in context C the fact x is considered as the institutional fact y.*
2. *$\exists z \in X_{\mathbf{n}}$ such that $C \cup \{z\} \rightarrow y \in out(B_{\mathbf{n}}, \geq_{\mathbf{n}})$: there exists an action z of the normative system $\mathbf{n}$ such that if it decides z in context C then it believes that the institutional fact y follows (i.e., $counts\text{-}as^1_{\mathbf{n}}(z, y|C)$ at the first level of abstraction).*

Example 9. $P \setminus I = \{catalytic\}$, $I = \{eco\}$, $X_{\mathbf{a}} = \{drive\}$, $X_{\mathbf{n}} = \{stamp\}$
$D_{\mathbf{n}} = G_{\mathbf{n}} = \{d_3\}$, $RD(d_3) = catalytic \rightarrow eco$
$B_{\mathbf{n}} = \{b_1\}$, $RD(b_1) = stamp \rightarrow eco$

Consequently, $NMAS \models counts\text{-}as^2_{\mathbf{n}}(catalytic, eco|\top)$. This formalizes that the normative system wants that if a car is catalytic, then it is considered as an ecological vehicle and the normative believes that from system putting a stamp on a catalytic car licence follows the fact that the car is catalytic. In situation $S = \{catalytic\}$, given $B_{\mathbf{n}}$ we have that the consequences of the constitutive norms are $out(B_{\mathbf{n}}, S, \geq_{\mathbf{n}}) = \emptyset$ and thus the goal d_3 remains unsatisfied, while in situation $S' = \{catalytic, stamp\}$ they are $out(B_{\mathbf{n}}, S', \geq_{\mathbf{n}}) = \{eco\}$ and the goal d_3 is satisfied.

The institutional facts can appear in the conditions of regulative norms:

Example 10. A regulative norm which forbids driving non catalytic cars can refer to the abstract concept of ecological vehicle rather than to catalytic converters: $O^2_{\mathbf{an}}(\neg drive, s \mid \neg eco)$.

As the system evolves, new cases can be added to the notion of ecological vehicle by means of new constitutive norms, without changing the regulative norms about it. E.g., if a car has fuel cells, then it is an ecological vehicle: $fuelcell \rightarrow eco \in RD(B_{\mathbf{n}})$.

This level of abstraction supposes that the normative system is an agent acting in the world. This abstraction can be detailed by introducing agents acting on behalf of the normative system: the normative system wants that an agent a makes the institutional fact y true if x holds in context C and believes that the effect of action z of agent a is the institutional fact y.

Before introducing the more concrete level of abstraction in obligations we discuss the third level of constitutive norms which is based on the notion of delegation of power.

Definition 10 (Counts-as relation (level 3) and delegation of power). *Let NMAS be a normative multiagent system* $\langle A, X, B, D, G, AD, E, RD, \geq, \geq_E, \mathbf{n}, I, V, GD\rangle$.

$a \in A$ is an agent, $z \in X_a$ an action of agent a, $x \in Lit(X)$ is a literal built out of a variable, $y \in Lit(I)$ a literal built out of an institutional fact, $C \subseteq Lit(X)$ the context. Agent a has been delegated the power to consider x in context C as the institutional fact y, $NMAS \models delegated_{\mathbf{n}}(a, z, x, y|C)$, iff:

1. $C \wedge x \rightarrow y \in out(D_{\mathbf{n}}, \geq_{\mathbf{n}}) \cap out(GD(a), \geq_{\mathbf{n}})$: *it is a desire of the normative system and a goal distributed to agent a that in context C the fact x is considered as the institutional fact y.*
2. $\exists z \in X_a$ *such that* $C \cup \{z\} \rightarrow y \in out(B_{\mathbf{n}}, \geq_{\mathbf{n}})$: *there exists an action z of agent a such that if it decides z then the normative system believes that the institutional fact y follows (i.e., counts-as$^1_{\mathbf{n}}(z, y|C)$ at the first level of abstraction).*

If $NMAS \models delegated_{\mathbf{n}}(a, z, x, y|C)$, then $NMAS \models counts\text{-}as^3_{\mathbf{n}}(x, y|C)$.

Example 11. $b \in A, P \setminus I = \{catalytic\}, I = \{eco\}, X_{\mathbf{a}} = \{drive\}, X_b = \{stamp\}$
$D_{\mathbf{n}} = GD(b) = \{d_3\}, RD(d_3) = catalytic \rightarrow eco$
$B_{\mathbf{n}} = \{b_1\}, RD(b_1) = stamp \rightarrow eco$

Thus, $NMAS \models delegated_{\mathbf{n}}(b, stamp, catalytic, eco|\top)$. Note that with respect to Example 9, the goal d_3 is distributed to agent b and *stamp* is an action of agent b.

We can now define obligations where agents have been delegated the power of recognizing violations by means of actions which count as such. Differently from the obligation of level 2, clause 2 distributes a goal to agent b who is in charge of recognizing violations and whose action z is believed by the normative system $\mathbf{n}$ to be the recognition of a violation (clause 7).

Definition 11 (Obligation (level 3)). *Let NMAS be a normative multiagent system* $\langle A, X, B, D, G, AD, E, RD, \geq, \geq_E, \mathbf{n}, I, V, GD\rangle$. *Agent $\mathbf{a} \in A$ is obliged to see to it that $x \in Lit(X_{\mathbf{a}} \cup P)$ with sanction $s \in Lit(X_b \cup P)$ in context $Y \subseteq Lit(X)$ in NMAS, written as $NMAS \models O^3_{\mathbf{an}}(x, s|Y)$, if and only if $\exists b \in A$ and a decision variable $z \in X_b$ such that Definition 8 holds except that:*

2. $Y \cup \{\sim x\} \rightarrow V(\sim x, \mathbf{a}) \in out(D_{\mathbf{n}}, \geq_{\mathbf{n}}) \cap out(GD(b), \geq_{\mathbf{n}})$: *if Y holds and $\sim x$ is true , then agent $\mathbf{n}$ has distributed the goal $V(\sim x, \mathbf{a})$: that it is recognized as a violation in context Y.*
7. $Y \cup \{z\} \rightarrow V(\sim x, \mathbf{a}) \in out(B_{\mathbf{n}}, \geq_{\mathbf{n}})$: *from action z of agent b is believed to follow the recognition of the violation.*

From clause 2 and 7 it follows that agent b has been delegated the power to recognize violations by means of its action z.

$$NMAS \models \exists b \in A, z \in X_b \; delegated_{\mathbf{n}}(b, z, \sim x, V(\sim x, \mathbf{a}) \mid Y)$$

Note that clause 2 of the definition above is like the first clause of an obligation $O_{bn}(V(\sim x, \mathbf{a}), s' \mid Y \cup \{\sim x\})$. The model can thus be extended with obligations towards agents which have to take care of the procedural aspects of law, like prosecuting violations and sanctioning violators. These additional obligations are discussed in [19] and provide a motivation for the prosecuting and sanctioning agents. In the Italian law, for example, it is obligatory for an attorney to start a prosecution process when he comes to know about a crime (art. 326 of *Codice di procedura penale*).

6 Conclusions and Related Work

In this paper we introduce the notion of delegation of power which elaborates the counts-as relation extending it to cope with some real situations. We show that counts-as relations in some cases depend on the action of agents which are in charge of recognizing facts as institutional facts. Moreover, we show that these agents are motivated to do so by a goal delegated to them by the normative system. If these two conditions are true we say that the agents have been delegated a power. Once we define the delegation of power relation, we can use it to extend our sanction based definition of obligations in order to model agents which prosecute violations.

Our model allows to distinguish three levels of abstractions: at the higher level of abstraction violations, sanctions and institutional facts follow without the intervention of any agent. At the middle level the normative system acts to satisfy its goal to recognize violations, to sanction and to establish institutional facts. At the most concrete level, agents are in charge of achieving the goals of the normative system and are empowered to do so.

The notion of empowerment in normative multiagent systems is widely discussed, but it has not been related yet with the notion of goal delegation.

Pacheco and Santos [25], for example, discuss the delegation of obligations among roles. In particular, they argue that when an obligation is delegated, a corresponding permissions must be delegated too. This rationality constraint inside an institution parallels our notion of delegation of power: when the goal to make true an institutional fact is delegated, the agent must be empowered to do so too. Moreover, in our model we can add to the notion of delegation of power also the permission for the delegated agent to perform the action which counts as the delegated institutional fact. This can be done using the definition of permission given in [8].

Pacheco and Santos consider the delegation process among roles rather than among agents. This feature can be added to our model too, using the framework for roles we discuss in [13]. Note that our model of roles describes roles by means of beliefs and goals; it is, thus, compatible with the distribution of goals to agents described by clause 2 of Definition 11.

Gelati *et al.* [26] combine obligations and power to define the notion of mandate in contracts: "a mandate is a proclamation intended to create the obligation of exercising a declarative power". However, they do not apply their analysis to the definition of constitutive rules but to the normative positions among agents.

Comparison with other models of counts-as is discussed in [6] and [14].

Future work is studying the relation between regulative rules and delegation of power: defining how it is possible to create global policies [8] obliging or permitting other agents to delegate their power. Finally, abstraction in the input/output logic framework has been left for lions or input/output networks. In such networks each black box corresponding to an input/output logic is associated with a component in an architecture. A discussion can be found in [14].

References

1. Searle, J.: The Construction of Social Reality. The Free Press, New York (1995)
2. Grossi, D., Meyer, J.J., Dignum, F.: Counts-as: Classification or constitution? an answer using modal logic. In: Procs. of DEON'06, Berlin, Springer Verlag (2006)
3. Jones, A., Sergot, M.: A formal characterisation of institutionalised power. Journal of IGPL **3** (1996) 427–443
4. Grossi, D., Dignum, F., Meyer, J.: Contextual terminologies. In: Procs. of CLIMA'05, Berlin, Springer Verlag (2005)
5. Boella, G., van der Torre, L.: A game theoretic approach to contracts in multiagent systems. IEEE Transactions on Systems, Man and Cybernetics - Part C (2006)
6. Boella, G., van der Torre, L.: Constitutive norms in the design of normative multiagent systems. In: Procs. of CLIMA'05, Berlin, Springer Verlag (2006)
7. Georgeff, M., Ingrand, F.F.: Decision-making in an embedded reasoning system. In: Procs. of 11th IJCAI. (1989) 972–978
8. Boella, G., van der Torre, L.: Security policies for sharing knowledge in virtual communities. IEEE Transactions on Systems, Man and Cybernetics - Part A (2006)
9. Hart, H.: The Concept of Law. Clarendon Press, Oxford (1961)
10. Castelfranchi, C.: Modeling social action for AI agents. Artificial Intelligence **103(1-2)** (1998) 157–182
11. Bottazzi, E., Ferrario, R.: A path to an ontology of organizations. In: Procs. of EDOC Int. Workshop on Vocabularies, Ontologies and Rules for The Enterprise (VORTE 2005). (2005)
12. Boella, G., van der Torre, L.: From the theory of mind to the construction of social reality. In: Procs. of CogSci'05, Mahwah (NJ), Lawrence Erlbaum (2005) 298–303
13. Boella, G., van der Torre, L.: Organizations as socially constructed agents in the agent oriented paradigm. In: LNAI n. 3451: Procs. of ESAW'04, Berlin, Springer Verlag (2004) 1–13
14. Boella, G., van der Torre, L.: An architecture of a normative system. In: Procs. of DEON'06, Berlin, Springer Verlag (2006)
15. Anderson, A.: A reduction of deontic logic to alethic modal logic. Mind **67** (1958) 100–103
16. Meyer, J.J.C.: A different approach to deontic logic: Deontic logic viewed as a variant of dynamic logic. Notre Dame Journal of Formal Logic **29(1)** (1988) 109–136
17. Artikis, A., Sergot, M., Pitt, J.: An executable specification of an argumentation protocol. In: Procs. of 9th International Conference on Artificial Intelligence and Law, ICAIL 2003, New York (NJ), ACM Press (2003) 1–11
18. Boella, G., van der Torre, L.: Δ: The social delegation cycle. In: LNAI n.3065: Procs. of ΔEON'04, Berlin, Springer Verlag (2004) 29–42
19. Boella, G., van der Torre, L.: Norm governed multiagent systems: The delegation of control to autonomous agents. In: Procs. of IEEE/WIC IAT'03, IEEE Press (2003) 329– 335
20. Broersen, J., Dastani, M., Hulstijn, J., van der Torre, L.: Goal generation in the BOID architecture. Cognitive Science Quarterly **2(3-4)** (2002) 428–447
21. Makinson, D., van der Torre, L.: Input-output logics. Journal of Philosophical Logic **29** (2000) 383–408
22. Makinson, D., van der Torre, L.: Constraints for input-output logics. Journal of Philosophical Logic **30(2)** (2001) 155–185

23. Bochman, A.: Explanatory Nonmonotonic Reasoning. World Scientific Publishing, London (UK) (2005)
24. Lang, J., van der Torre, L., Weydert, E.: Utilitarian desires. Autonomous Agents and Multi-agent Systems **5(3)** (2002) 329–363
25. Pacheco, O., Santos, F.: Delegation in a role-based organization. In: Procs. of ΔEON'04, Berlin, Springer Verlag (2004) 209–227
26. Gelati, J., Rotolo, A., Sartor, G., Governatori, G.: Normative autonomy and normative co-ordination: Declarative power, representation, and mandate. Artificial Intellingence and Law **12(1-2)** (2004) 53–81

Strategic Deontic Temporal Logic as a Reduction to ATL, with an Application to Chisholm's Scenario

Jan Broersen

Department of Information and Computing Sciences
Utrecht University
Utrecht, The Netherlands
broersen@cs.uu.nl

Abstract. In this paper we extend earlier work on deontic deadlines in CTL to the framework of alternating time temporal logic (ATL). The resulting setting enables us to model several concepts discussed in the deontic logic literature. Among the issues discussed are: conditionality, ought implies can, deliberateness, settledness, achievement obligations versus maintenance obligations and deontic detachment. We motivate our framework by arguing for the importance of temporal order obligations, from the standpoint of agent theory as studied in computer science. In particular we will argue that in general *achievement* obligations cannot do without a deadline condition saying the achievement has to take place before it. Then we define our logic as a reduction to ATL. We demonstrate the applicability of the logic by discussing a possible solution to Chisholm's paradox. The solution differs considerably from other known temporal approaches to the paradox.

1 Introduction

In agent theory, as studied in computer science, we are interested in designing logical models that describe how agents can reason about and decide what to do, given their obligations, permissions, abilities, desires, intentions, beliefs, etc. Decisions have a temporal aspect, namely, they are about what to do in the *future*, and they deal with conditional information, namely, they have to result from considering and reasoning about *hypothetical* circumstances. The deontic ATL operators we consider in this paper are both *conditional* and *temporal*. Their syntactical form is $O_A(\rho \leq \delta : \xi_A)$. The intuitive interpretation of the operator is that *if* the agents in the set A achieve δ, they are obliged to achieve ρ at the same time or before that, under penalty of suffering the negative condition ξ_A. A good example of such an obligation is the following: according to Dutch traffic regulations one has to indicate direction before one turns off. In this example, δ is 'turning off', ρ is 'indicating direction' and ξ can be the circumstance of being vulnerable for being fined by a police officer. Obligations $O_A(\rho \leq \delta : \xi_A)$ are thus *conditional* on the actual occurrence of δ and are *temporal* in the sense that the achievement ρ has to precede the condition δ. Readers familiar with the deontic logic literature will recognize that another example is the second sentence of Chisholm's original paradoxical scenario: 'if one helps, first one has to tell'. In section 9 we discuss formalizations of Chisholm's scenario in our formalism.

One might wonder why we think obligations expressed as $O_A(\rho \leq \delta : \xi_A)$ are so important. Let us explain. Obligations that guide agents in the actions they select for

L. Goble and J.-J.C. Meyer (Eds.): DEON 2006, LNAI 4048, pp. 53–68, 2006.

performing are always *about* the future. Obligations about the past are not interesting for agents having to decide what to do, because we may safely assume that agents do not *control* the past. We adopt terminology from BDI-theory, and say that the *reference* time of the obligations we are interested in is the future, while the *validity* time is the present[1]. The latter emphasizes that we want our logic to model the reasoning of an agent that has to decide what to do 'now', considering the obligations he has about the future. A rough classification of obligations whose reference time is the future is the division in *achievement* obligations and *maintenance* obligations. Similar terminology is used by Cohen and Levesque [8] who distinguish between *achievement goals* and *maintenance goals*. In an achievement obligation, the objective is to achieve something in the future that is not already (necessarily) true now. For a 'maintenance obligation' the objective is to preserve the truth of a condition that *is* already true now. Our main interest in this paper will be with achievement obligations, since as we will see in section 6, maintenance obligations can be rewritten into equivalent achievement obligations. So, for agent theory as studied in computer science achievement obligations are the most interesting type of norms. Now we will argue in section 2 that achievement obligations are close to meaningless without a condition δ before whose occurrence the achievement ρ has to be realized, which explains why obligations of the form $O_A(\rho \leq \delta : \xi_A)$ are central to our investigations.

In some of the previous work on this subject [4], we referred to the condition δ as a 'deadline' of an obligation $O_A(\rho \leq \delta : \xi_A)$. That was partly because there we studied this type of modality in the purely temporal framework of CTL. Here we use ATL for the temporal component. ATL has elements of logics of Agency. In [5] we showed how to embed Coalition Logic (CL), which is a subset of ATL, in the STIT framework of Horty [11][2]. Since ATL can be seen as a logic of (strategic) ability it enables us to define different notions of control over conditions. And adding information about control over the condition δ (or, to be more precise, absence of control over $\neg\delta$, which is something else) is actually what can turn a conditional temporal order obligation into a real deadline obligation, as we explain in section 8. So obligations $O_A(\rho \leq \delta : \xi_A)$ as such should not be referred to as 'deadline' obligations. They are conditional temporal order obligations, which can be made into deadline obligations, by adding that agents A do not control avoidance of the deadline *condition* δ.

2 Why Achievement Obligations Need a 'Deadline' Condition

Dignum et al [9, 17] stress the importance of providing deadlines (which they do *not* view as particular kinds of *conditionals*, like we do) for obligations from practical considerations. And indeed, in the environments where software agents are envisioned

[1] The distinction between validity time and reference time for logics that contain a temporal modality as one of the logic components, was, for instance, formulated by Lindström and Rabinowicz [13] in the context of temporal epistemic logics. But it equally applies to temporal motivational logics. And we belief that a failure to distinguish the two concepts is the source of a lot of confusion.

[2] And a paper on embedding ATL as a whole into the *strategic* version of Horty's STIT formalism is under review.

to operate, they will have to deal with deadlines. Think about agents negotiating certain deadlines, to coordinate their behavior. For instance, agents may engage in mutual agreements by means of contracts, which typically contain deadlines.

However, there are more fundamental reasons to equip achievement obligations with deadline conditions. A fundamental assumption in deontic logic is that norms (obligations) can be violated, and that we should be able to reason about this circumstance. Now, if we think about the situation where an agent has an achievement obligation φ for, say, 'stop smoking', and we want to make the temporal component of φ explicit, then we cannot model this by writing for φ a formula like $OF stop\_smoking$ ($O\psi$ for 'it is obliged that ψ' and $F\chi$ for 'some time in the future χ'), because we cannot *violate* this obligation. At any future point, the agent can claim that although he has not stopped smoking yet, he will eventually, at some point even further in the future. Most deontic logicians would agree that obligations that cannot be violated are no obligations at all. So what it takes for an achievement obligation to really be an obligation is reference to a *condition* under which it is violated. And this is exactly what a deadline condition is: a condition giving rise to a violation if the achievement has not been accomplished before. If this condition corresponds to a given point is some time metric, we have a genuine deadline obligation. But if this condition is an abstract proposition δ, we have a special kind of conditional obligation, namely, a temporal order obligation.

A possible objection against the above line of reasoning is that sometimes there do seem to be ways in which to violate an achievement obligation without a deadline. Before explaining the objection, we need to point out that obligations play multiple roles in rational agent modelling. There are two abstraction levels on which they play a role. First there are achievement obligations incurred and represented by agents. These obligations should have a deadline, since otherwise they cannot be violated which means that they do not influence the agents decision making. But there are also achievement obligations that function as specifications for the agents behavior as seen by an agent designer. A good example is the formal property of *fairness*. For instance, a designer might specify that his agent is obliged to distribute its deliberation efforts fairly over its set of goals. The design of the agent may violate this fairness obligation. But note that in general this is not something the agent itself is concerned with. The agent is designed as it is, and it cannot choose to have another design. And thus it cannot violate an achievement obligation without a deadline. Or can he? Of course, we can imagine that an agent indeed is able to alter its own design, thereby violating a certain achievement obligation. For instance, we might claim that an agent violates the achievement obligation to shake hands with president Bush someday by cutting off his own hands. A similar objection is that an agent might perform something irrevocable in its environment. For instance, an agent can be said to violate the achievement obligation to bring back the book to the library some day by burning it[3].

We acknowledge these as valid objections against our claim that achievement obligations need deadlines. However, there are still some good reasons to claim that deadline conditions are crucial for achievement obligations. For instance, an agent cannot destroy

[3] What if we are able to reproduce exactly the same book? Then, apparently burning did not *count as* irrevocably destroying the possibility to comply.

the possibility to be able to stop smoking, so at least for some achievement obligations a deadline is necessary. Furthermore, one seems to have a contrived interpretation of an obligation like 'having to return a book to the library some day' if one claims that it does not imply that one *actually* has to bring it back but *only* that one should not irrevocably destroy it.

Our suggestion, is that we can study the difference between, for instance the smoking example and the library example as a difference in the interaction with abilities. In section 8 we will mention several ways in which obligations act with ability. One interaction with ability is that we can choose to define that an obligation for achieving φ implies that an agent is obliged to keep open the possibility of reaching φ (we will however not discuss this interaction in section 8). If we choose to include this property, indeed, in the library example we violate the achievement obligation to bring back the book eventually, by burning the book. However, for the smoking example there is no difference: it is hard to imagine how an agent can violate an obligation to keep open the possibility to stop smoking.

3 Reduction Using Negative Conditions

As explained, the syntax of the central modality we study in this paper is $O_A(\rho \leq \delta : \xi_A)$ (although we permit ourselves a brief generalization to a conditional variant $O_A(\rho \leq \delta : \xi_A \mid \eta)$ in section 9). The O stands for obligation, A is a group of agents having the obligation, ρ is the condition to be achieved, δ the condition functioning as the deadline for the achievement, and ξ_A a condition necessarily true in case the deadline obligation is violated. We think of ξ_A as a condition that is in some sense negative for the group of agents A. Negative conditions typically play a role in the semantics of notions like obligation, commitment and intention. Goals, desires, wants, wishes, objectives, etc. are typically associated with positive conditions. Our approach differs from most others working with negative condition [3, 15] in that we make the negative condition explicit in the syntax of the obligation modalities. In standard deontic logic (SDL) [18], the negative conditions are implicit in the modal semantics: modally accessible worlds are optimal worlds where no violations occur. In Anderson's reduction for SDL [3], the negative conditions are explicit in the object language through a propositional constant *Viol*. In this paper we go one step further by explicitly giving the negative conditions ξ_A as parameters for the obligation operator $O_A(\rho \leq \delta : \xi_A)$. This has many advantages. For instance, it gives us the machinery to specify that certain obligations are incurred as the result of violating other obligations. That is, we can nest obligations by using formulas of the form $O_A(\rho \leq \delta : \xi_A)$ for ξ_A. However, we do not consider such nestings in this paper. The most important reason for making the negative conditions explicit, is the advantage this has in the study of Chisholm's scenario in section 9. It enables us to view the choice of which *temporal* obligation to comply to as a decision about which *non-temporal* negative condition to prefer. Obviously, such decisions can only be made relative to a preference order over negative conditions. In this paper such an order is not made explicit, since here we are only concerned with a possible definition for temporal obligations *in terms* of negative conditions. Possible logical structures for negative conditions themselves are not explored.

Also, allowing general negative conditions ξ_A enables us to discuss some principles of deontic logic. For instance, in case $\xi_A = \perp$, the norm is 'regimented'. Actually this means that the norm is no longer a norm, since it is no longer possible to violate it; the norm simply becomes a hard constraint on the behavior of agents. Thus $O_A(\rho \leq \delta : \perp)$ means that A can only do ρ before δ, and nothing else. In terms of strategies: the agent has no strategy where eventually he meets a δ without having met a ρ first. The case $\xi_A = \top$ gives a dual effect: now the penalty is always true, which means that it looses its meaning as a divider between 'the good' and 'the bad'. As a result, $O_A(\rho \leq \delta : \xi_A)$ looses its normative meaning; any point of achieving ρ or δ becomes as good as any other. This means that $O_A(\rho \leq \delta : \top)$ has to be generally valid, as is the case for our definitions for the operator.

Our aim is to define the semantics of the modality $O_A(\rho \leq \delta : \xi_A)$ entirely by constructing reductions to formulas of alternating time temporal logic (ATL) [1, 2] talking about negative conditions. ATL is a temporal logic of agency with a game theoretic semantics in terms of *strategies*. The deadline obligations we define will then be strategic obligations in the sense that they limit the possible strategies of agents by associating negative conditions with courses of agent choices that do not comply to the obligation.

Reducing to ATL has many advantages. We can use the logical machinery of ATL (axiomatization, model checking algorithms), to do reasoning. We can check properties of deontic logics by translating and proving them in ATL. Finally, we can do planning with obligations and deadlines using a satisfiability checker for ATL. We do not have to be too afraid that ATL, as a formal system, might be to weak to encode interesting properties, since it has been shown to have an exponential time complete complexity.

4 ATL

We present ATL ([1, 2]) here using a non-standard, but concise and intuitive syntax and semantics.

4.1 Core Syntax, Abbreviations and Intuitions

Definition 1. *Well-formed formulas of the temporal language $\mathcal{L}_{ATL}$ are defined by:*

$$\begin{aligned} \varphi, \psi, \ldots &:= p \mid \neg\varphi \mid \varphi \wedge \psi \mid \langle[A]\rangle\eta \mid [\langle A\rangle]\eta \\ \eta, \theta, \ldots &:= \eta U^{ee} \theta \end{aligned}$$

where $\varphi, \psi, \ldots$ represent arbitrary well-formed formulas, $\eta, \theta, \ldots$ represent temporal path formulas, the p are elements from an infinite set of propositional symbols $\mathcal{P}$, and A is a subset of a finite set of agent names E (we define $\overline{A} \equiv_{def} E \setminus A$). We use the superscript 'ee' for the until operator to denote that this is the version of 'the until' where φ is not required to hold for the present, nor for the point where ψ, i.e., the present and the point where ϕ are *both* excluded. Roughly, $\langle[A]\rangle\eta$ is read as 'A can ensure η', and the dual $[\langle A\rangle]\eta$ is read as 'A cannot avoid η'. A more precise explanation, revealing the existential and universal quantification over strategies in both these operators (which explains our choice of syntax using a combination of sharp and square brackets for both operators) is as follows:

$\langle [A] \rangle (\varphi U^{ee} \psi)$: agents A have a strategy that, whatever strategy is taken by agents $\overline{A}$, ensures that eventually, at some point m, the condition ψ will hold, while φ holds from the next moment until the moment before m

$[\langle A \rangle](\varphi U^{ee} \psi)$: for all strategies of agents A the agents $\overline{A}$ have a strategy such that eventually, at some point m, the condition ψ will hold, while φ holds from the next moment until the moment before m

We use standard propositional abbreviations, and also define the following operators as abbreviations.

Definition 2

$\langle [A] \rangle X\varphi \equiv_{def} \langle [A] \rangle (\bot U^{ee} \varphi)$	$[\langle A \rangle] X\varphi \equiv_{def} [\langle A \rangle](\bot U^{ee} \varphi)$
$\langle [A] \rangle F\varphi \equiv_{def} \varphi \vee \langle [A] \rangle (\top U^{ee} \varphi)$	$[\langle A \rangle] F\varphi \equiv_{def} \varphi \vee [\langle A \rangle](\top U^{ee} \varphi)$
$\langle [A] \rangle G\varphi \equiv_{def} \neg [\langle A \rangle] F \neg \varphi$	$[\langle A \rangle] G\varphi \equiv_{def} \neg \langle [A] \rangle F \neg \varphi$
$\langle [A] \rangle (\varphi U^{e} \psi) \equiv_{def} \varphi \wedge \langle [A] \rangle (\varphi U^{ee} \psi)$	$[\langle A \rangle](\varphi U^{e} \psi) \equiv_{def} \varphi \wedge [\langle A \rangle](\varphi U^{ee} \psi)$
$\langle [A] \rangle (\varphi U \psi) \equiv_{def} \langle [A] \rangle (\varphi U^{e} (\varphi \wedge \psi))$	$[\langle A \rangle](\varphi U \psi) \equiv_{def} [\langle A \rangle](\varphi U^{e} (\varphi \wedge \psi))$
$\langle [A] \rangle (\varphi U_w \psi) \equiv_{def} \neg [\langle A \rangle](\neg \psi U \neg \varphi)$	$[\langle A \rangle](\varphi U_w \psi) \equiv_{def} \neg \langle [A] \rangle (\neg \psi U \neg \varphi)$

The informal meanings of the formulas are as follows (the informal meanings in combination with the $[\langle A \rangle]$ operator follow trivially):

$\langle [A] \rangle X\varphi$: agents A have a strategy to ensure that at any next moment φ will hold

$\langle [A] \rangle F\varphi$: agents A have a strategy to ensure that eventually φ will hold

$\langle [A] \rangle G\varphi$: agents A have a strategy to ensure that holds globally

$\langle [A] \rangle (\varphi U^{e} \psi)$: agents A have a strategy to ensure that, eventually, at some point m, the condition ψ will hold, while φ holds from now until the moment before m

$\langle [A] \rangle (\varphi U \psi)$: agents A have a strategy to ensure that, eventually, at some point the condition ψ will hold, while φ holds from now until then

$\langle [A] \rangle (\varphi U_w \psi)$: agents A have a strategy to ensure that, if eventually ψ will hold, then φ holds from now until then, or forever otherwise

4.2 Model Theoretic Semantics

The intuition behind ATL models is that agents have choices, such that the non-determinism of each choice is *only* due to the choices other agents have at the same moment. Thus, the simultaneous choice of al agents together, always brings the system to a unique follow-up state. In other words, if an agent would know what the choices of other agents would be, given his own choice, he would know exactly in which state he arrives.

Definition 3. *An ATL model $\mathcal{M} = (S, C, \pi)$, consists of a non-empty set S of states, a total function $C : A \times S \mapsto 2^{2^S}$ yielding for each agent and each state a set of choices (informally: 'actions') under the condition that the intersection of each combination of choices for separate agents gives a unique next system state (i.e., for each s, the* *function* $RX(s) = \{\bigcap_{a \in A} Ch_a \mid Ch_a \in C(a, s)\}$ *yields a non-empty set of singleton sets representing the possible follow-up states of s), and, finally, an interpretation function* π *for propositional atoms.*

Note that from the condition on the function C it follows that the choices for each individual agent at a certain moment in time are a partitioning of the set of all choices possible for the total system of agents, as embodied by the relation $\mathcal{R}^{sys} = \{(s, s') \mid s \in S \text{ and } \{s'\} \in RX(s)\}$. And, also note that this latter condition does not entail the former. That is, there can be partitions of the choices for the total system that do not correspond to the choices of some agent in the system.

Definition 4. *A strategy* α_a *for an agent a, is a function* $\alpha_a : S \mapsto 2^S$ *with* $\forall s \in S : \alpha_a(s) \in C(a, s)$*, assigning choices of the agent a to states of the ATL model.*

Often, strategies are defined as mappings $\alpha_a : S^+ \mapsto 2^S$, from finite *sequences* of states to choices in the final state of a sequence. However, to interpret ATL, this is not necessary, because ATL is not expressive enough to recognize by which sequence of previous states a certain state is reached. More in particular, without affecting truth of any ATL formula, we can always transform an ATL model into one where $\mathcal{R}^{sys}$ is tree-like. On tree structures it is clear right away that a mapping from states to choices in that state suffices, since any state can only be reached by the actions leading to it.

The strategy function is straightforwardly extended to sets of agents.

Definition 5. *A full path* σ *in M is an infinite sequence*[4] $\sigma = s_0, s_1, s_2, \ldots$ *such that for every* $i \geq 0$, $s_i \in S$ *and* $(s_i, s_{i+1}) \in \mathcal{R}^{sys}$*. We say that the full path* σ *starts at s if and only if* $s_0 = s$*. We denote the state* s_i *of a full path* $\sigma = s_0, s_1, s_2, \ldots$ *in* $\mathcal{M}$ *by* $\sigma[i]$*.*

A full path σ complies *to a strategy* α_A *of a set of agents A if and only if for every* $n \geq 0$, $\sigma[n+1] \in \alpha_A(\sigma[n])$*. We denote the set of full paths complying to a strategy* α_A *by* $\Sigma(\alpha_A)$*.*

Definition 6. *Validity* $\mathcal{M}, s \models \varphi$*, of an ATL-formula* φ *in a world s of a model* $\mathcal{M} = (S, C, \pi)$ *is defined as:*

$$
\begin{array}{ll}
\mathcal{M}, s \models p & \Leftrightarrow s \in \pi(p) \\
\mathcal{M}, s \models \neg\varphi & \Leftrightarrow \textit{not } \mathcal{M}, s \models \varphi \\
\mathcal{M}, s \models \varphi \wedge \psi & \Leftrightarrow \mathcal{M}, s \models \varphi \textit{ and } \mathcal{M}, s \models \psi \\
\mathcal{M}, s \models \langle[A]\rangle\eta & \Leftrightarrow \exists \alpha_A \textit{ s. t. } \forall \sigma \in \Sigma(\alpha_A) \textit{ with } \sigma[0] = s : \mathcal{M}, \sigma[0], \sigma \models \eta \\
\mathcal{M}, s \models [\langle A\rangle]\eta & \Leftrightarrow \forall \alpha_A : \exists \sigma \in \Sigma(\alpha_A) \textit{ with } \sigma[0] = s \textit{ s. t. } \mathcal{M}, \sigma[0], \sigma \models \eta \\
\mathcal{M}, \sigma[0], \sigma \models \varphi U^{ee} \psi & \Leftrightarrow \exists n > 0 \textit{ s. t.} \\
& \quad (1)\ \mathcal{M}, \sigma[n] \models \psi \textit{ and} \\
& \quad (2)\ \forall i \textit{ with } 0 < i < n : \mathcal{M}, \sigma[i] \models \varphi
\end{array}
$$

Validity on a ATL model $\mathcal{M}$ is defined as validity in all states of the model. If φ is valid on an ATL model $\mathcal{M}$, we say that $\mathcal{M}$ is a model for φ. General validity of a formula

4 Alternatively, we may drop the requirement that $\mathcal{R}^{sys}$ is serial, and add a maximality condition to the notion of 'full path'.

φ is defined as validity on all ATL models. The logic ATL is the subset of all general validities of $\mathcal{L}_{ATL}$ over the class of ATL models.

5 Conditional Temporal Order Obligations

In this section we define operators for temporal order obligations as reductions to ATL-formulas talking about negative conditions. We use ATL-formulas indexed with a set of agents, i.e, ξ_A, to denote negative conditions. The central observation linking obligations $O_A(\rho \leq \delta : \xi_A)$ to ATL, is the following:

> *$O_A(\rho \leq \delta : \xi_A)$ holds if and only if it is not the case that the group of agents A has a strategy to achieve δ, to avoid ρ at all moments until δ occurs for the first time, and avoid the negative condition ξ_A at the point where δ.*

In other words, if A want to achieve δ at some future point, they have to make sure that before that they achieve ρ, because otherwise the negative condition ξ_A will be valid at the point where δ. We can rewrite this formally as a truth condition on ATL models:

Definition 7 (temporal order obligations)

$$\begin{aligned} \mathcal{M}, s \models O_A(\rho \leq \delta : \xi_A) \Leftrightarrow\ & \nexists \alpha_A, \forall \sigma \in \Sigma(\alpha_A) \textit{ with } \sigma[0] = s, \exists j : \\ & \textit{such that} \\ & \forall 0 \leq i < j : \mathcal{M}, \sigma[i] \models \neg\rho \wedge \neg\delta \textit{ and } \mathcal{M}, \sigma[j] \models \neg\rho \wedge \delta \\ & \textit{and} \\ & \mathcal{M}, \sigma[j] \models \neg\xi_A \end{aligned}$$

This says: *if* at some future point δ occurs, than A has no way of ensuring that, if ρ has not occurred before the point where δ occurs for the first time, there is not a negative condition ξ_A at the point where δ. This means that if A *do* have a strategy to avoid the negative condition while not doing ρ before δ, they do not have the obligation.

Under the above definition, in case only *some* strategies of agents may lead to negative conditions if they do not ensure that ρ is achieved before δ, the agents are not *obliged* to achieve ρ before δ. This situation actually constitutes a kind of conditionality other than the conditionality with respect to deadline conditions δ. Modelling it would require an operator $O_A(\rho \leq \delta : \xi_A \mid \eta)$, where η is a temporal formula denoting the subset of paths the obligation holds on. Note that the original obligation reappears as the case where η equals $\top$. This kind of conditionality (which is not further explored in this paper) can be modelled using the more expressive variant ATL^*. We leave this extension and a discussion on the different kinds of conditionality that can be defined for temporal deontic operators for a future paper.

A second aspect of definition 7 that has to be explained is that it by no means implies that an obligation requires that δ becomes true eventually (which is why it is conditional on δ). However, we do have that if A cannot avoid that δ might never become true, they cannot have a strategy that ensures that at some point δ will hold (and where if ρ has not been done before, there is not a negative condition), which means that they are obliged every ρ before δ (validity 10 in proposition 2 reflects this). This seems rather counter intuitive. However, in section 8 we define a deliberate version of the operator for which this property is eliminated.

The third thing to discuss is that intuitively, a strategic notion of obligation should distinguish between the strategies that are good and the strategies that are bad. However, our definition suggests that we can define the obligation in terms of what strategies agents *have*. The link between these two views is the use of the negative conditions and the conditionality with respect to occurrence of the condition δ. Actually we can view the definition as distinguishing between good and bad strategies in the following sense: the strategies in which an agent eventually meets the condition δ without having achieved the condition ρ before, are the bad strategies, all the others are the good ones.

We can circumscribe the truth condition of definition 7 as an ATL formula. We have the following proposition:

Proposition 1. *A formula $O_A(\rho \leq \delta : \xi_A)$ is true at some point of an ATL model if and only if the point satisfies the ATL formula $(\delta \wedge (\neg\rho \rightarrow \xi_A)) \vee \neg\langle[A]\rangle((\neg\rho \wedge \neg\delta)U^e(\delta \wedge \neg\rho \wedge \neg\xi_A))$.*

Proof. We only give an impression of the proof. The present is not controlled by any strategy. If δ holds presently, and ρ does not hold presently, there is a violation presently. In the truth condition this corresponds to the case $j = 0$, and in the formula to $\delta \wedge (\neg\rho \rightarrow \xi_A)$. Equivalence is easy to see. For moments other than the present, the equivalence follows almost directly from the semantics of the ATL operators involved.

6 Maintenance Obligations with a Relief Condition

In the introduction we explained what maintenance obligations are. Where achievement obligations for a property ρ naturally come with a property δ functioning as a *deadline condition*, maintenance properties φ come with a property ψ functioning as a *relief condition*: if the relief condition occurs, the obligation to maintain φ no longer holds. We can define maintenance obligations $O_A(\varphi \leftarrow\!\!\Diamond\ \psi : \xi_A)$ in terms of achievement obligations as follows:

Definition 8

$$O_A(\varphi \leftarrow\!\!\Diamond\ \psi : \xi_A) \equiv_{def} O_A(\psi \leq \neg\varphi : \xi_A)$$

The rationale for the definition is as follows. An agent can comply to obligations $O_A(\rho \leq \delta : \xi_A)$ in two different ways: (1) he can look at it as having to do ρ before he does δ, but he can also (2) look at it as having to preserve $\neg\delta$ as long as he has not achieved ρ. Note that for a maintenance obligation $O_A(\varphi \leftarrow\!\!\Diamond\ \psi : \xi_A)$, the negative condition occurs at the first point where φ is no longer maintained, provided this point is before ψ. In section 9 we will use a maintenance obligation to model one of the sentences of Chisholm's scenario.

7 More Logical Properties

In this section we mention some logical properties of the defined obligation operator.

Proposition 2. *The following schemas are valid:*

$$\models O_A((\rho \wedge \chi) \leq \delta : \xi_A) \rightarrow O_A(\rho \leq \delta : \xi_A) \quad (1)$$
$$\models O_A(\rho \leq \top : \xi_A) \wedge O_A(\chi \leq \top : \zeta_A) \rightarrow O_A((\rho \wedge \chi) \leq \top : \xi_A \vee \zeta_A) \quad (2)$$
$$\models O_A(\top \leq \delta : \xi_A) \quad (3)$$
$$\models O_A(\gamma \leq \gamma : \xi_A) \quad (4)$$
$$\models O_A(\rho \leq \bot : \xi_A) \quad (5)$$
$$\models \neg O_A(\bot \leq \top : \bot_A) \quad (6)$$
$$\models \neg(O_A(\rho \leq \top : \xi_A) \wedge O_A(\neg\rho \leq \top : \xi_A)) \quad (7)$$
$$\models O_A(\rho \leq \delta : \top) \quad (8)$$
$$\models O_A(\rho \leq \delta : \bot) \rightarrow O_A(\rho \leq \delta : \xi_A) \quad (9)$$
$$\models [\langle A \rangle]G\neg\delta \rightarrow O_A(\rho \leq \delta : \xi_A) \quad (10)$$
$$\models O_A(\rho \leq \delta : \xi_A) \rightarrow [\langle A \rangle](O_A(\rho \leq \delta : \xi_A) U_w (\rho \vee \delta)) \quad (11)$$
$$\models O_A(\rho \leq \delta : \xi_A) \rightarrow O_A(O_A(\rho \leq \delta : \xi_A) \leftrightarrow\!\diamond (\rho \vee \delta) : \xi_A) \quad (12)$$
$$\models \xi_A \rightarrow O_A(\rho \leq \top : \xi_A) \quad (13)$$
$$\models O_A(\rho \leq \xi_A : \xi_A) \quad (14)$$

Proposition 3. *The following schemas are not valid:*

$$\not\models O_A(\rho \leq \delta : \xi_A) \wedge O_A(\delta \leq \gamma : \zeta_A) \rightarrow O_A(\rho \leq \gamma : \xi_A \vee \zeta_A) \quad (15)$$
$$\not\models O_A(\rho \leq \delta : \xi_A) \rightarrow O_A(\rho \leq (\delta \wedge \gamma) : \xi_A) \quad (16)$$
$$\not\models O_A(\rho \leq \delta : \xi_A) \wedge O_A(\rho \leq \gamma : \xi_A) \rightarrow O_A(\rho \leq (\delta \vee \gamma) : \xi_A) \quad (17)$$
$$\not\models O_A(\rho \leq \delta : \xi_A) \wedge O_A(\rho \leq \gamma : \xi_A) \rightarrow O_A(\rho \leq (\delta \wedge \gamma) : \xi_A) \quad (18)$$
$$\not\models O_A(\rho \leq \delta : \xi_A) \wedge O_A(\chi \leq \delta : \xi_A) \rightarrow O_A((\rho \wedge \chi) \leq \delta : \xi_A) \quad (19)$$
$$\not\models O_A(\rho \leq \delta : \xi_A) \rightarrow O_A(\delta \leq \rho : \xi_A) \quad (20)$$
$$\not\models O_A(\bot \leq \delta : \xi_A) \qquad \not\models \neg O_A(\bot \leq \delta : \xi_A) \quad (21)$$
$$\not\models \neg O_A(\bot \leq \top : \xi_A) \qquad \not\models O_A(\bot \leq \top : \xi_A) \quad (22)$$
$$\not\models \neg(O_A(\rho \leq \delta : \xi_A) \wedge O_A(\neg\rho \leq \delta : \xi_A)) \quad (23)$$
$$\not\models O_A(\rho \leq \top : \xi_A) \qquad \not\models \neg O_A(\rho \leq \top : \xi_A) \quad (24)$$

We have no opportunity here to discuss these properties. In stead we briefly discuss some more logical issues.

The logical properties for maintenance obligations with a relief condition follow easily from the properties for achievement obligations with a deadline condition.

Many of the above properties concern properties of single paths within arbitrary strategies. Therefore we were able to give most of the proofs using an LTL theorem prover [12].

An interesting question is whether we can see classical non-temporal obligations (such as the ones of SDL) as limit cases of temporal order obligations. Intuitively it should be the case that if we 'substitute' the most common temporal connotations of general obligations in the temporal deontic operators, we get standard deontic operators back. In our opinion, the most likely substitution for this purpose is $\delta = \top$. We have the following theorem:

Theorem 1. *The logic of $O_a(\rho \leq \top : \bot)$ is standard deontic logic (the modal logic KD)*[5].

Proof. Substitution in the definition for O gives $\neg\langle[A]\rangle((\neg\rho \wedge \neg\top)U^e(\top \wedge \neg\rho \wedge \neg\bot))$. This reduces to $[\langle a\rangle]X\rho$. Since there is only one agent, system actions and actions of a are identical. The seriality condition on system actions ensures modal property D. K follows from the fact that for one agent, the ATL structure is based on a classical Kripke frame. From this it also follows that the logic is exactly KD, since this frame satisfies no additional properties.

8 Interactions with Ability: Deadlines and Deliberate Versions

In the previous sections, we did not consider the issue whether or not the conditions ρ, δ and ξ were actually 'under control' of groups of agents A. However, as is well known from the deontic literature, issues like 'ought implies can', 'settledness' and 'power' take a central place in it. In this section we study some interactions of obligations and 'control'.

First we discuss the issue of control over the condition δ. We called the obligations 'temporal order obligations' exactly because we did not exclude that δ was indeed under control of the group of agents A. In contrast, a deadline obligation can be viewed as a temporal order obligation where the agents A do *not* control δ. However we have to be very careful with what we mean. Actually, not controlling δ should not be understood as agents A not having a strategy for $F\delta$ (consequently they also would not have a strategy to violate without negative consequences, and thus would be obliged anything before ρ). Not controlling δ should be understood as not having a strategy for $G\neg\delta$. The difference with conditional temporal order obligations is thus that agents A cannot avoid their duty by pushing a deadline forward indefinitely, that is, they do not control $\neg\delta$. We can imagine that a temporal deadline $D(\delta, n)$ for n time units is defined as (X^n represents n nestings of the next operator.):

Definition 9

$$D(\delta, n) \equiv_{def} \langle[\emptyset]\rangle X^n(\delta \wedge \langle[\emptyset]\rangle G\neg\delta) \wedge \bigwedge_{0\leq i<n} \langle[\emptyset]\rangle X^i\neg\delta$$

The ATL formula $D(\delta, n)$ says that on all paths, after n steps δ is true, while δ is never true before or after that. Clearly, in case of a temporal deadline of this kind, no set of agents A can have a strategy for $G\neg\delta$. In this circumstance the temporal order obligation becomes a real deadline obligation: δ is sure to happen in n time units, and agents

[5] Like in conditional deontic logics, the logic of $O(\varphi \mid \top)$ is often also SDL.

do not have a strategy to avoid ρ at all points before δ and not experience a negative condition at δ. We may thus introduce deadline obligations by conjuncting temporal order obligations with formulas $D(\delta, n)$. Note that for deadline obligations, there is no longer any conditionality with respect to δ, since δ is sure to happen at a given point in the future.

Now we return to variants that are due to different possibilities for the control of achievements ρ. To give content to his concept of 'categorical imperative', Kant suggested the principle of 'ought implies can'. Kant's principle also makes sense in the present, more profane context. Rational agents are assumed to be realistic, which means that they will not let their decisions be influenced by obligations for conditions ρ they cannot achieve before δ anyway. Obligation variants that incorporate this property can be defined as:

Definition 11

$$O_A^{oc}(\rho \leq \delta : \xi_A) \equiv_{def} O_A(\rho \leq \delta : \xi_A) \wedge \langle [A] \rangle (\neg \delta U \rho)$$

For agent theory, Kant's dictum can be supplemented with a second principle concerning the interaction of obligation and ability. We might call this second principle 'ought implies can avoid'. This relates to a problem with the definition of O that has been signaled many times before in deontic logic. It is sometimes called the problem of 'settledness' [10, 14]. The issue is that any obligation O for which compliance is settled, or, in other words, temporally inevitable, is true. In particular we have the property $\rho \rightarrow O_A(\rho \leq \delta : \xi_A)$, which is an instance of the more general property 9 of section 7. We avoid the property (and some others that are non-intuitive, such as property 10 of section 7) by defining *deliberate* versions of the obligation operators:

Definition 12

$$O_A^{dl}(\rho \leq \delta : \xi_A) \equiv_{def} O_A(\rho \leq \delta : \xi_A) \wedge \neg O_A(\rho \leq \delta : \bot)$$

The formula $\neg O_A(\rho \leq \delta : \bot)$ says that it is not the case that O is an obligation for which a violation is impossible (i.e., an obligation for which the negative condition cannot become true). In other words, agents do have a strategy *not* to comply to the obligation. However, if they do so, there will be a negative condition. So, now the obligation is conditional on the possibility not to comply. Thus, agents can only have an obligation to achieve something if they have the *choice* not to do so. i.e., when it is not already settled.

The two principles of 'ought implies can' and 'ought implies can avoid' come down to the requirement that choices are not empty and have alternatives. Incorporating these principles in the definitions avoids counter intuitive properties like always having the obligation to achieve tautologies ($O_A(\top \leq \delta : \xi_A)$). But in deontic logic, properties like $O\top$ have actually been defended (it is, for instance, a property of SDL). We think that for the applications of deontic logic in agent theory, they should be excluded. An artificial agent having to deal with obligations is only interested reasoning about obligations that influence his decisions. If there is nothing to choose, either because the set of choices is empty or there is only one alternative, the obligations mean nothing to the agent.

To conclude this section, we want to point out that not all properties of section 7 hold for the variants in this section. To save space, we did not elaborate on the effects of the interaction with abilities on the logical properties. However, we do need to mention the following property that holds for the deliberate variant:

Proposition 4

$$\models O_A^{dl}(\rho \leq \delta : \xi_A) \wedge O_A^{dl}(\delta \leq \gamma : \zeta_A) \rightarrow O_A^{dl}(\rho \leq \gamma : \xi_A \vee \zeta_A)$$

This property is crucial in our discussion on the modelling of a Chisholm's scenario in section 9.

9 Modelling Chisholm's Scenario

The original formulation of Chisholm's problematic scenario is [7]:

1. it ought to be that a certain man go to the assistance of his neighbors
2. it ought to be that if he does go he tell them he is coming
3. if he does not go then he ought not to tell them he is coming
4. he does not go

The modelling task we pursue in this section is to find a logical formalization that:

- faithfully reflects the natural language meaning, including the temporal aspects (the temporal order in sentence 2, the future directedness of all obligations, the present as the validity time of all obligations, etc.),
- is consistent,
- has no logically redundant sentences,
- derives that A ought not to tell.

As explained in the introduction, we are interested in obligations whose validity time is the present and whose reference time is the future, since these are the obligations an agent has to account for when making a decision about what to do. In particular we will interpret all sentences of Chisholm's scenario as sentences being valid *presently* while the 'regulating force' of the obligations involved refers to the *future*. Note that this differs from many other temporal interpretations of Chisholm's sentences. For instance, [6] discusses also a backwards looking interpretation that considers a setting where we know for a fact that the man did not help, that the obligation to help has been violated, and whether or not the agent told that he would come. However, most temporal interpretations of the scenario have been particularly aimed at using time to avoid the looming inconsistencies of a-temporal interpretations. For instance, Prakken and Sergot [16] suggest that temporalization can avoid 'pragmatic oddities', such as the one consisting of the obligation to help in combination with the obligation not to tell, by stipulating that the validity times of these obligations are disjoint. Following their line of reasoning, the oddity should be solved by interpreting the scenario in such a way that the obligation to help is valid until the moment it is violated, while from that point on the obligation not to tell is valid. We do not regard that as a solution, since we want a solution where the validity time of all obligations is the present.

Let us give our formalization first, before explaining the formulas.

$$O_A^{dl}(\text{help} \leq \text{too\_late} : \xi_A) \wedge D(\text{too\_late}, n) \tag{25}$$

$$O_A^{dl}(\text{tell} \leq \text{help} : \zeta_A) \tag{26}$$

$$\langle[\emptyset]\rangle(\neg\text{help } U \text{ too\_late}) \rightarrow O_A^{dl}(\neg\text{tell} \leftarrow\!\!\diamond \text{ too\_late} : \eta_A) \tag{27}$$

$$\langle[\emptyset]\rangle(\neg\text{help } U \text{ too\_late}) \tag{28}$$

In the first sentence we have to model that the man is obliged to go to the assistance of his neighbors. As said, we want to interpret this as an obligation about the future: the man is obliged to help at some future point. However, as explained in section 2, we cannot simply model this as an obligation of the form $OFhelp$. Such obligations are vulnerable for indefinite deferral, and there is no reason for the man to start helping soon. So, if we want to interpret the obligation as an achievement obligation, we have to bring in a condition δ before which the helping needs to take place. Since the sentence does not explicitly refer to such a condition, we simply model it as the condition *too_late* and define that *too_late* is true in exactly n time units. One might argue that we are introducing a concept that is not in the natural language description of the obligation. However, we claim that this is the only way we can make sense of the sentence if we interpret it as an achievement obligation. Although from the natural language description we cannot know the exact value of n, in our opinion it is safe to assume it is a parameter playing a role in the intuitive interpretation of the sentence as an achievement obligation.

As mentioned in the introduction, the second sentence is an outstanding example of the kind of obligations we can model in our formalism. The obligation to tell, with the present as its validity time, is conditional on the condition of helping, while the telling has to precede the helping. We know of no other temporal deontic formalism that can model this sentence as faithful as the present one.

It has been argued that the third sentence should have the same form as the second sentence, since both are conditionals. However, we argue that for our future directed interpretation this is not a sensible requirement. In particular, the second sentence is an achievement obligation, while the third is a maintenance obligation (see section 6): from the present until the moment where it is too late to help, the man has the obligation to preserve the condition of not telling, that is, if he will not help. This conditionality with respect to not helping is simply modelled using a material implication[6] expressing dependency on the condition whether presently it is known for a fact that the man is not going to help.

To interpret the fourth sentence as a fact about the future, we model it as an ATL expression saying that no strategies are possible that possibly result in the man actually helping before it is too late. We acknowledge that this is not necessarily the most intuitive choice. First of all it would contradict (and not violate) the formula modelling the first sentence if this would be a variant that incorporates 'ought implies can' (see definition 11). Second, modelling the sentence as a fact about the future is problematic

[6] Actually a conditional obligation of the form $O_A^{dl}(\neg\text{tell} \leftarrow\!\!\diamond \text{ too\_late} : \eta_A) \mid (\neg\text{help } U \text{ too\_late}))$ as briefly mentioned in section 5 would be a better choice here. But this would not affect the main idea behind the solution to Chisholm's scenario.

as such. Intuitively, one should always keep open the possibility that the man will help. Therefore it would be much better to model the fourth sentence as an *intention*. Actually intentions can be suitably modelled as self-directed obligations, which means we can express them in the present formalism. We leave this for future research.

The above formalization is consistent, does not contain logical dependencies, and stays close to the natural language sentences. We now investigate whether it gives rise to the right conclusions. With the formulas modelling the first two sentences, together with the logical principle of proposition 4 for deliberate obligations, we derive $O_A(\text{tell} \leq \text{too\_late} : \xi_A \vee \zeta_A)$. Deriving a 'new' obligation from the first two sentences has been called 'deontic detachment' in the literature. But note that it is a rather special kind of deontic detachment specific for temporal order obligations.

With the formulas modelling the last two sentences, we derive $O_A(\neg\text{tell} \leftarrow\!\!\diamond \text{too\_late} : \eta_A)$. Obviously, this conflicts with the obligation derived through deontic detachment. But there is no inconsistency, not even when we use one and the same negative condition for all obligations involved (or when $\xi_A \leftrightarrow \zeta_A \leftrightarrow \eta_A$). What the conflicting information tells us is that we cannot avoid one of the negative conditions $\xi_A \vee \zeta_A$ or η_A becoming true at some point before *too_late*: we cannot at the same time achieve 'telling' and preserve 'not telling': a choice has to be made. Of course one of the requirements for a solution to the scenario is that this choice should be 'not telling': we should be able to conclude that given the above modelling of the scenario, the obligation not to tell is 'relevant', while the obligation to tell is not. This is seen as follows. The agent will want to avoid the negative conditions. And in this case there is a best way to do that. Given the information in sentence 4 that there will be no helping before it is too late, we can derive that negative condition ξ_A is sure to occur at the point *too_late*. This means that the derived obligation $O_A(\text{tell} \leq \text{too\_late} : \xi_A \vee \zeta_A)$ is not interesting for the agent to base its decision on: trying to obey it is pointless, because its negative condition is valid anyway. This leaves the obligation $O_A(\neg\text{tell} \leftarrow\!\!\diamond \text{too\_late} : \eta_A)$ as the relevant one: the agent will want to avoid the negative condition η_A, and thus should not tell.

10 Conclusion

In this paper we argued that achievement obligations need a deadline condition that functions as a point where a possible violation of the obligation is payed for. We named the resulting conditional obligations 'temporal order obligations'. We showed how to define several semantics for temporal order obligations by giving characterizations of these modalities in plain ATL. This has as an advantage that all logic machinery already developed for ATL is applicable. The resulting framework is quite rich: we showed that it enables us to investigate issues like 'ought implies can', 'ought implies can avoid', deliberateness and deontic detachment. We mentioned logical properties of the defined operators, discussed their conditionality aspect, and demonstrated its applicability by modelling Chisholm's famous scenario.

Many issues had to be left for future research. In particular the generalization to obligations $O_A(\rho \leq \delta : \xi_A \mid \eta)$ could not be explained in detail, despite its possible relevance for Chisholm's scenario. Also intentions, which are also relevant for the scenario, had to be left aside. Among the other issues we are planning to investigate in the present framework are concepts like 'power','responsibility' and 'counts as'. For the longer term, we would also like to investigate the relation between deontic semantics and game equilibria.

References

1. Rajeev Alur, Thomas A. Henzinger, and Orna Kupferman. Alternating-time temporal logic. In *FOCS '97: Proceedings of the 38th Annual Symposium on Foundations of Computer Science (FOCS '97)*, pages 100–109. IEEE Computer Society, 1997.
2. Rajeev Alur, Thomas A. Henzinger, and Orna Kupferman. Alternating-time temporal logic. *Journal of the ACM*, 49(5):672–713, 2002.
3. A.R. Anderson. A reduction of deontic logic to alethic modal logic. *Mind*, 67:100–103, 1958.
4. J.M. Broersen, F. Dignum, V. Dignum, and J.-J Meyer. Designing a deontic logic of deadlines. In A. Lomuscio and D. Nute, editors, *Proceedings 7th International Workshop on Deontic Logic in Computer Science (DEON'04)*, volume 3065 of *Lecture Notes in Computer Science*, pages 43–56. Springer, 2004. DOI: 10.1007/b98159.
5. J.M. Broersen, A. Herzig, and N. Troquard. From coalition logic to stit. In *Proceedings LCMAS 2005*, Electronic Notes in Theoretical Computer Science. Elsevier, 2005.
6. J.M. Broersen and L.W.N. van der Torre. Semantic analysis of chisholm's paradox. In K. Verbeeck, K. Tuyls, A. Nowe, B. Manderick, and B. Kuijpers, editors, *Proceedings of the 17th Belgium-Netherlands Artificial Intelligence Conference*, pages 28–34, 2005.
7. R.M. Chisholm. Contrary-to-duty imperatives and deontic logic. *Analysis*, 24:33–36, 1963.
8. P.R. Cohen and H.J. Levesque. Intention is choice with commitment. *Artificial Intelligence*, 42(3):213–261, 1990.
9. F. Dignum and R. Kuiper. Obligations and dense time for specifying deadlines. In *Proceedings of thirty-First HICSS, Hawaii*, 1998.
10. J.A. van Eck. A system of temporally relative modal and deontic predicate logic and its philosophical applications. *Logique et Analyse*, 100:339–381, 1982.
11. J.F. Horty. *Agency and Deontic Logic*. Oxford University Press, 2001.
12. U. Hustadt and B. Konev. Trp++ 2.0: A temporal resolution prover. In F. Baader, editor, *Proceedings of the 19th International Conference on Automated Deduction (CADE-19)*, volume 2741 of *Lecture Notes in Artificial Intelligence*, pages 274–278. Springer, 2003.
13. S. Lindström and W. Rabinowicz. Unlimited doxastic logic for introspective agents. *Erkenntnis*, 50:353–385, 1999.
14. B. Loewer and M. Belzer. Dyadic deontic detachment. *Synthese*, 54:295–318, 1983.
15. J.-J.Ch. Meyer. A different approach to deontic logic: Deontic logic viewed as a variant of dynamic logic. *Notre Dame Journal of Formal Logic*, 29:109–136, 1988.
16. H. Prakken and M.J. Sergot. Contrary-to-duty obligations and defeasible reasoning. *Studia Logica*, 57:91–115, 1996.
17. F. Dignum H. Weigand V. Dignum, J.J. Meyer. Formal specification of interaction in agent societies. In *2nd Goddard Workshop on Formal Approaches to Agent-Based Systems (FAABS), Maryland*, 2002.
18. G.H. von Wright. Deontic logic. *Mind*, 60:1–15, 1951.

Acting with an End in Sight

Mark A. Brown*

Philosophy Department
Syracuse University
Syracuse, NY, USA
mabrown@syr.edu

Abstract. Supplementing an account of actions offered by Horty and Belnap [8] makes it more suitable for use in deontic logic. I introduce a new tense operator, *for a while in the immediate future*, provide for action terms as well as action formulas, and introduce an intention function into our models. With these changes, we are able to (a) explore means/ends relations involving actions, (b) make room for one agent to enable another to act, and (c) provide a means for distinguishing intended from unintended consequences. In combination, these improvements make it possible to consider collaborative action aimed at a goal, within a setting open to detailed normative scrutiny of ends, means, actions and intentions.

1 Introduction

Deontic logic is inevitably concerned with actions, since actions can create obligations, permissions, or prohibitions, fulfill or violate obligations, exploit permissions, and in other ways transform normative situations. It is therefore natural to partner a deontic logic with, or indeed build it upon, a suitable logic of action. Recently Governatori and Rotolo [6] have noted three deficiencies in the logics of action most often used for such purposes, defects which prevent deontic logics based on such systems from respecting some important normative subtletics:

1. they cannot distinguish intended from unintended consequences of actions, and thus cannot distinguish between deliberately bringing something about, and merely bringing it about[1];

* A portion of this paper is based on my presention prepared for the conference on Norms, Knowledge and Reasoning in Technology, 3–4 June, 2005, Boxmeer, Netherlands. I am grateful to the organizers of that conference for the opportunity to participate, and particularly to Jesse Hughes, whose helpful comments and criticisms have motivated some of the further work reflected in this paper. I am grateful, as well, to the anonymous referees, whose helpful suggestions and comments have led to substantial improvements.

[1] Governatori and Rotolo use the phrase 'seeing to it' to mean *intentionally bringing it about*. This introduces a source of potential confusion into the discussion, since the phrase 'sees to it that' is now established in the literature as synonymous with 'bring it about that', and thus as devoid of any intimation of intention. Accordingly, I have rephrased their objection.

L. Goble and J.-J.C. Meyer (Eds.): DEON 2006, LNAI 4048, pp. 69–84, 2006.

2. they provide no way to discuss the means of achieving a result, and thus cannot be sensitive to the distinction between normatively acceptable and normatively unacceptable ways of achieving an end[2];
3. they cannot distinguish between direct and indirect actions, and thus cannot distinguish between doing something oneself, and influencing others to do it.

These criticisms seem entirely apt. Moreover, to these we may add another rather shocking criticism (not mentioned in [6]) which, if it does not apply to all logics of action, at least applies to my favorite versions, namely existing accounts based on branching time models:

4. existing logics of action based on branching time models do not, on close examination, offer a sense of action in which it can be said that an agent performs simple actions such as opening a door.

I will take these points up in reverse order. In section 2, after a brief review of the most useful of the existing logics of action based on branching time, I will discuss my reasons for holding that point 4 above applies. In section 3, I propose a step in the direction of remedying that defect and note that the resulting improved logic of action also provides new possibilities of addressing criticism 3. In section 4, I will argue that with this improved system we are in a better position to address criticism 2 above as well. In section 5, I will show how names for actions can be introduced into our modal language, placing us in a still better position to address criticism 2. In section 6, I provide a way to incorporate agents' intentions into the system so as to address criticism 1. Section 7 will sum up.

2 Logics of Action Based on Branching Time

Existing logics of action that are based on a theory of branching time (as explored in, for example, Thomason [10]) stem primarily from the work of Nuel Belnap and his various students and collaborators ([1–3], [7], [8], [11]). The version introduced by Horty [7] and Horty and Belnap [8] is particularly important for our discussion.

Early work on the logic of action (e.g. Pörn [9]) observed that what we do is not merely something that happens. This shows up in normative contexts. I am not responsible for the fact that it is raining—that just happens, and I have no choice in the matter. But I *am* responsible for the fact that I left my umbrella at home—that didn't just happen, it was a matter with respect to which I had, and made, some effective choice. Belnap holds that genuine action, as contrasted with mere occurrence of some event, requires some exercise of freedom of choice

---

[2] Governatori and Rotolo suggest that dynamic logic has an advantage over the usual modal systems, because it provides terms for actions. As I show in Section 5, modal systems can provide such terms too, so this is not a sound basis for preferring dynamic logic treatments.

on the part of the agent involved. But the availability of genuine choices implies indeterminism about the future: if I genuinely have a choice about whether to take my umbrella with me or not, then there are details of the future that depend on which choice I make; and to the degree that such choices are freely available, the course of the future is correspondingly undetermined.

This leads naturally to a view of time as branching into the future, with branching occurring at those moments when agents have genuine choices[3], and with different branches corresponding to the different courses which history will take, depending on the agent's choice. On some branches, having chosen to leave my umbrella behind, I get wet. On others, having chosen to bring it with me, I remain dry.

The past, we suppose, is fixed, so although time branches into the future there is no branching backwards into the past. Thus the basic picture is as in Figure 1, with time represented as flowing from bottom to top:

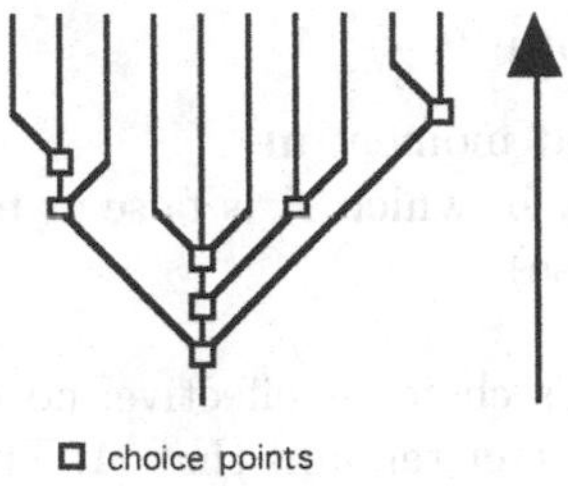

Fig. 1. Branching

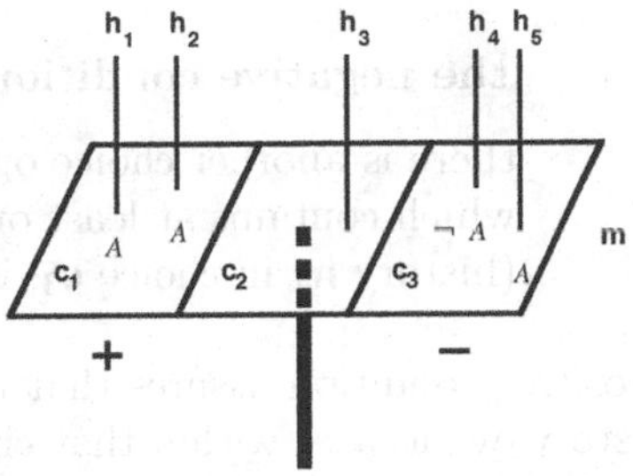

Fig. 2. Choices

A history is any complete, non-backtracking path from past to future (bottom to top, in the figures) through this tree. A total of 10 histories are represented in Figure 1.

In making choices, we are presumably not choosing a single future branch—to do so would mean that we were rendering the world deterministic from that moment on. Rather, then, we are choosing a class of possible futures, with other factors (including the choices of others, our own as yet unmade choices, and perhaps random events) narrowing that class gradually as time goes on. Thus, if we look in more detail at a single choice point, an "exploded view" might be represented as in Figure 2.

Here, at moment $\mathbf{m}$, the unnamed agent is depicted as having three choices, $\mathbf{c}_1$, $\mathbf{c}_2$, and $\mathbf{c}_3$, with choice $\mathbf{c}_1$ including histories $\mathbf{h}_1$ and $\mathbf{h}_2$, choice $\mathbf{c}_2$ including $\mathbf{h}_3$, and choice $\mathbf{c}_3$ including $\mathbf{h}_4$ and $\mathbf{h}_5$. To accommodate all these details, then, frames for such a logic of action will consist of a class $\mathbf{M}$ of moments, a partial ordering relation $<$ among moments, a class $\mathbf{A}$ of agents, and a choice function $\mathbf{C}$

[3] And perhaps at other moments as well, if there are other kinds of genuinely undetermined events, e.g. atomic decay.

which, for each moment/agent pair, partitions the class of histories through that moment into the choices available to that agent at that time.[4]

I will focus on the simplest plausible logic of action within the class of logics based on branching time, a system devised by Jeff Horty, using what he calls the deliberative[5] stit, or *dstit*, operator, with 'stit' standing as an acronym for "sees to it that".

For an agent **a** with name α, I'll represent this as the operator $\triangle_{\alpha}$. Then in the situation represented in Figure 2 we can say that at moment **m** in history $\mathbf{h_1}$, the agent **a** *deliberatively sees to it that A* is true[6], and can record this by writing that $\mathbf{m}, \mathbf{h_1} \models \triangle_{\alpha} A$, because two conditions are met:

\+ **the positive condition: reliability**

all the histories in the same choice as $\mathbf{h_1}$
(i.e. all the histories within $\mathbf{c_1}$, in this case)
are ones in which A is true at moment **m**; and

– **the negative condition: freedom**

there is another choice open to **a** at moment **m**
which contains at least one history in which A is false at **m**
(history $\mathbf{h_4}$ in choice $\mathbf{c_3}$, in this case).

The positive condition assures that the agent's choice is effective: no matter which history eventuates within that choice, the fact remains that A. The negative condition assures that the agent's choice is significant: a different choice might have had a significantly different outcome.

We can give satisfaction conditions for a number of useful temporal operators. Two of these (with their duals, defined as usual) are important here—a future tense operator and a historical possibility operator:

$\mathbf{m}, \mathbf{h} \models \diamondplus A$ *iff* for some $\mathbf{m}^*$ later than **m** in history **h**, we have $\mathbf{m}^*, \mathbf{h} \models A$
$\mathbf{m}, \mathbf{h} \models \blacklozenge A$ *iff* for some $\mathbf{h}^*$ through **m**, we have $\mathbf{m}, \mathbf{h}^* \models A$.

Combining the historical possibility operator $\blacklozenge$ with the *dstit* operator gives us an ability operator $\blacklozenge\triangle_{\alpha}$.

---

4 Suitable constraints must be imposed on $<$ and **C** to assure that time branches only forwards and that agents do not yet have choices which discriminate between histories which will for a time thereafter remain undifferentiated.

5 The use of the term 'deliberative' here was presumably intended to indicate that the action in question is one taken as a *part* of deliberation, not that it is an action taken as a *result* of deliberation. It is thus not intended to indicate that the action is deliberate, and in particular does not mean that the indicated outcome was intended. In general, operators introduced by Belnap and his co-workers are intended to be "austere", i.e. devoid of any psychological content such as belief, desire, and intention. In section 6 I will depart from this tradition by introducing intentions.

6 Hereafter, I will often abbreviate 'that A is true' to simply 'that A'.

Among results which fall out fairly easily from the semantics, we have:

$$\models \Delta_\alpha A \rightarrow A; \qquad \text{(however: } \not\models A \rightarrow \Delta_\alpha A);$$
$$\models \Delta_\alpha \Delta_\alpha A \leftrightarrow \Delta_\alpha A;$$
$$\models \Delta_\alpha A \rightarrow \blacklozenge \Delta_\alpha A.$$
$$\models \Delta_\alpha \neg \Delta_\alpha \neg \Delta_\alpha A \leftrightarrow \Delta_\alpha A;$$

To see the interest of this last result, note that in order to express the claim that a refrains from seeing to it that A, it seems appropriate to say that a sees to it that a doesn't see to it that A, i.e. $\Delta_\alpha \neg \Delta_\alpha A$. Thus we record that a refrains from refraining from doing it, by writing that $\Delta_\alpha \neg \Delta_\alpha \neg \Delta_\alpha A$. The equivalence in this third result, then, shows that, in the sense of action expressed by the Δ operator, refraining from refraining from doing something amounts to doing it. These are pretty results—charming to the logician's eye.

Horty's *dstit* operator is extremely useful, but it has an odd feature. In the sense of action (and the correlative sense of ability) which it directly captures, it is at best extremely unlikely that I am able to open the door. To see this, consider again the situation depicted in Figure 2. At the moment m, agent a has three choices (two would be sufficient to illustrate my point, more would not interfere with the outcome). If a adopts choice c_1, then (depending on circumstances not under a's control) time either continues as in history h_1, or else as in history h_2, but in either case A is true. On the other hand, choice c_3 is also available, and if it is adopted, history may continue as in h_4, in which case A is false. Any other choices are, for our purposes, irrelevant, and alternative histories issuing from c_3 do not affect matters. As long as all the histories issuing from c_1 are ones in which A is true, the positive condition is met for $\Delta_\alpha A$ to be satisfied at the moment m along h_1, and as long as at least one history issuing from c_3 (or from c_2) is one in which A is false, the negative condition is met. So this illustrates a situation in which we have $\mathsf{m}, \mathsf{h}_1 \models \Delta_\alpha A$.

But what sort of a claim can A express? If A expresses the claim that the door is open, then the door is not both open and shut at the very same moment and so we cannot satisfy both the positive and the negative condition for the truth of $\Delta_\alpha A$. If instead A expresses the claim that the door *will* be open, and is thus expressible in the form $\Diamond_F B$, then it is almost certainly the case that A is true at m along *every* history because surely, no matter what a does or doesn't do, that door will get opened eventually along history h_4 (or any other), if only by the wind or by someone else coming through some years from now. As a result, it is very unlikely that the negative condition for the truth of $\Delta_\alpha A$ will be met. If that is so, it would be equally unlikely that the formula $\blacklozenge \Delta_\alpha A$ can adequately express the claim that a *can* open the door. Yet an agent's being able to open the door doesn't seem at all unlikely. Supposing A to be of the form $\boxed{F} B$ fares no better, since then A is almost certainly false in each history.

The problem here seems to be that with Δ we have only the possibility of considering what is true either as a present aspect of the moment m or as a feature of the choice's outcome placed sometime in the indefinite future. What we would like instead of either the immediate present or the indefinite future, it seems, is some way to focus on the immediate future, the near-term result of an agent's choice. Of course if time were discrete, we could look at what would become true at the very next moment after **m**, along each of the various histories through **m**. However, if we can't be sure that time is discrete we can't be sure there is a next moment after **m**; and if, as in fact seems appropriate for discussions of human agency, we actually wish to assume that time is densely ordered along all histories, then there will be no next moment after **m** along any history.

3 An Improved Logic of Action

Reflecting on such matters (and influenced also by reflection on what I take to be some unsatisfactory features of Belnap's *astit* operator[7], which I won't take the time to discuss here) has led me to consider developing a new stit operator—I'll designate it Δ^{p} for *pstit* (with *p* for *progressive tense*, or perhaps for *process*)—focusing on choices among sets of transitions[8], and more specifically among sets of immediate transitions, from pasts to future branch points, rather than choices among sets of histories. This would emphasize the notion that actions are, in the sense presented by Xu, events, and more specifically processes.

To make a long story short, however, pursuing that line of investigation eventually reveals that there is an equivalent but much simpler approach, namely to introduce a new tense operator, with the sense *for a while in the immediate future*, and use it in combination with Horty's *dstit* operator.[9] Keeping in mind that the moments along a given history may well be densely ordered, so that there will likely be no immediately next moment after a given moment **m**, we can give truth conditions for such a tense operator as follows:

> $\mathbf{m},\mathbf{h} \models \boxed{W} A$ *iff* in history **h** there is some moment $\mathbf{m}^*$, later than **m**, such that at each moment $\mathbf{m}'$ after **m**, up to and including $\mathbf{m}^*$: $\mathbf{m}',\mathbf{h} \models A$.

Now the combined modality $\Delta_{\alpha}\boxed{W}$ can do the work I had expected a new operator Δ^{p}_{α} to do, and we can use the formula $\blacklozenge\Delta_{\alpha}\boxed{W}A$ to express the correlative claim that α *is able to see to it that (for a while in the immediate future)* A *will be true.*

However, it will make the next discussion easier to follow and to assess, if we simplify our notation as much as possible. Accordingly, let Δ^{p}_{α} serve to abbreviate $\Delta_{\alpha}\boxed{W}$, and let us even omit the subscript when only a single agent is

---

[7] While the *astit* operator does not fail in the way the *dstit* does, still it does not fill our needs here, since the *astit* expresses a claim in the past perfect tense: The agent *has seen to it that* A. We need a present tense, or present progressive tense, operator.

[8] Transitions were introduced by Belnap and exploited by Xu [11]. In the final analysis, the details of the definition turn out to be unimportant for our work here, though they were a significant influence in drawing me to my current views.

[9] So we come back to the *dstit* after all, which is certainly a testimonial to the fundamental importance of Horty's *dstit*. Still, if we, like sheep, have gone astray, we return to the fold older and wiser and—more to the point—better equipped by virtue of our new tense operator.

in question. So $\triangle^{\mathrm{p}} A$ will express the claim that the (tacit) agent sees to it that (for a while) A will be true.

The behavior of the *pstit* operator $\triangle^{\mathrm{p}}$ is quite different from that of the *dstit* operator $\triangle^{\mathrm{d}}$. Some of the similarities and differences are indicated in Table 1 (where the agent is taken to be the same throughout).

Table 1. Comparing the two operators

	schema	for the *pstit* operator $\triangle^{\mathrm{p}}$	for the *dstit* operator $\triangle^{\mathrm{d}}$
1	$\triangle A \rightarrow A$	invalid	valid
2	$A \rightarrow \triangle A$	invalid	invalid
3	$\triangle A \rightarrow \Diamond^{F} A$	valid	invalid
4	$\triangle A \rightarrow \neg\triangle\neg A$	valid	valid
5	$\triangle\triangle A \leftrightarrow \triangle A$	invalid	valid
6	$\triangle A \rightarrow \Diamond^{\bullet}\triangle A$	valid	valid
7	$\triangle\Diamond^{\bullet}\triangle A \rightarrow \Diamond^{\bullet}\triangle A$	invalid	trivially valid
8	$\triangle\neg\triangle\neg\triangle A \leftrightarrow \triangle A$	invalid	valid
9	$(\triangle A \wedge \triangle B) \rightarrow \triangle(A \wedge B)$	valid	valid

Reflecting on line 1 of the table, we recall that it sounded reasonable to say that if I see to it that A, then A is true, but that on closer examination we saw that this was not so reasonable when the moment at which A is to be true is the very same moment as the moment of choice. A careful reading of the *pstit* formula $\triangle^{\mathrm{p}}_{\alpha} A \rightarrow A$ would interpret it as saying that if I am (by my choice this very moment) seeing to it that A will become true then A is already true, and that sounds quite unreasonable. What will be correct, however, is to say, as in line 3 of the table, that if I am (by my choice this very moment) seeing to it that A, then A *will* be true. This schema for the *dstit* is invalid, because any future implications of the use of the *dstit* operator must arise from within the formula A to which it is applied, not from the use of the operator itself, since the operator only looks at the one moment of evaluation.

Of course the two operators agree, as line 2 indicates, that something's being so does not by itself imply that any agent has done, is doing, or will do anything to make it so. In line 4 they agree, as well, that it is not possible in a single act to both make something true and make it false at the very same time.

Line 5 shows that although there is no difference, for the *dstit* operator, between seeing to it that one sees to it that A, on the one hand, and simply seeing to it that A, on the other, no such reduction works for the *pstit.* From left to right, we have an instance of the generally invalid schema 1. But the right to left half of the equivalence also fails for the *pstit,* because my current choice may ensure that A is true for a short interval in the immediate future, but during that interval there may be no choice points, and so no chance to act.

Note that in the composite modality $\triangle^{\mathrm{d}}_{\alpha}\triangle^{\mathrm{p}}_{\alpha}$ the inner operator must apply to its complement continuously through an interval of time, and thus must express the activity of sustaining the truth of its complement, for a while, from a certain point on. This would be applicable in the case of holding the door open, and also in the case of (being engaged in the process of) opening the door, as contrasted with (the achievement of) the door's being open.[10] I can see to it that I keep the door open, by seeing to it that I see to it that the door is open.

In line 6, we see that the two operators agree on the thought that what an agent *does* do, that agent *can* do. It should be noted, however, lest this usage be misinterpreted, that in both cases the sense of *can* involved is exceedingly fleeting—it implies nothing about lasting talents, capacities, or opportunities, and speaks only of what is open to the agent at the very moment of evaluation.[11]

Turning to line 7 of the table, we should note that the consequent of the schema could be replaced using almost any modality composable by combining $\neg$, $\triangle$ and $\blacklozenge$, and the results would be the same. The exceptions: modalities equivalent to the antecedent simply by virtue of the fact that $\blacklozenge\blacklozenge$ is equivalent to $\blacklozenge$, and ones entailed by such equivalents under schema 6. Apart from such trivial consequences of the **S5** character of $\blacklozenge$ and the validity in line 6, no modality of this sort entails any other, for the *pstit*. For the *dstit*, however, it makes no difference what the consequent might be, since the antecedent is logically false. If, at moment $\mathbf{m}$, there is any history at all along which $\blacktriangle_{\alpha} A$ is true, then $\blacklozenge\blacktriangle_{\alpha} A$ is true along every history through $\mathbf{m}$, and as a result the negative condition for $\blacktriangle_{\alpha}\blacklozenge\blacktriangle_{\alpha} A$ cannot possibly be met.

This result and the result in line 5 illustrate the value of the new *pstit* operator. We now have a whole host of distinct modalities formed from combinations of action and ability, and available to perform distinct roles in our reasoning. We shall put these to use in Section 4.

In line 8 of the table we find that, in contrast to the sense of action given by the *dstit* operator, the *pstit* operator does not support the conclusion that refraining from refraining from an action is identical with simply performing that action.[12] Again the equivalence fails in both directions, and again it is the fact that a succession of times—not just a single moment—are involved in the *pstit* version, that creates the critical contrast with the *dstit*.

In line 9 of the table, we see that on either account, action is agglomerative: if the choice made at a given moment assures the truth of each of two claims, it assures the truth of their conjunction. The converse, however, is not valid in either account, because one of the conjuncts could be a logical truth, for which the negative condition could never be satisfied.

There is another important contrast—one which could not be expressed in the compressed format of the table—, namely that although any formula of the form

---

[10] This prompts the thought that previous accounts of refraining are probably inadequate, and that in at least some cases perhaps refraining involves actively *sustained* non-action, not just active momentary non-action.

[11] This meets a criticism, recently raised by Elgesem [5], of certain logics of ability, including the system in Brown [4].

[12] At least not in the $\triangle\neg\triangle$ account of refraining normally used in similar logics of action. If a more subtle account of refraining grows out of our present work, the matter will need to be reconsidered; but the prospects for the so-called *refref* thesis seem dim.

$\triangle^{d}_{\alpha}\blacklozenge\triangle^{d}_{\beta}\boldsymbol{A}$ will be logically false, it is perfectly possible for formulas of the form $\triangle^{p}_{\alpha}\blacklozenge\triangle^{p}_{\beta}\boldsymbol{A}$ (or, for that matter, ones of the form $\triangle^{d}_{\alpha}\blacklozenge\triangle^{p}_{\beta}\boldsymbol{A}$) to be true. Thus I cannot enable you (or enable myself, either) to do anything, according to the *dstit*, while according to the *pstit* it is conceivable that I can. This opens the door (so to speak) to addressing the third of the criticisms listed in the Introduction. Although this does not quite amount to talk of influencing the actions of other agents, it does at least make room for enabling such actions.

4 Actions as Means to Ends

Let us consider now what all this can do for us in expressing claims, and reasoning, about means and ends. Let us first consider the case of a single agent **a**, acting alone to secure some end state expressed by a formula $\boldsymbol{A}$. And since we will no longer be concerned with the *dstit* except as a component of the *pstit*, let us again prune our notation to bring out the essentials as sharply as possible, by using the simple expression $\triangle\boldsymbol{A}$ as an abbreviation for $\triangle^{p}_{\alpha}\boldsymbol{A}$, i.e. for $\triangle^{d}_{\alpha}\boxed{\mathrm{W}}\boldsymbol{A}$. Finally, in the same spirit of parsimony, let us abbreviate the associated ability operator by using $\nabla\boldsymbol{A}$ as an abbreviation for $\blacklozenge\triangle_{\alpha}\boldsymbol{A}$, i.e. for $\blacklozenge\triangle^{d}_{\alpha}\boxed{\mathrm{W}}\boldsymbol{A}$.

Using this abbreviated notation, we can say that the action reported by $\triangle\boldsymbol{A}$, which assures the truth of $\boldsymbol{A}$ for the immediate future, surely counts as a means to this end—the goal of having $\boldsymbol{A}$ true. But since this is so, it seems right to say, as well, that the action reported by $\triangle\triangle\boldsymbol{A}$ is also a means to that end, since it assures the truth of $\triangle\boldsymbol{A}$ for the immediate future, which will in turn secure the truth of $\boldsymbol{A}$ for a while thereafter.[13] And it might well happen that, at a given moment, it was not within **a**'s ability to secure the truth of $\boldsymbol{A}$ immediately and directly, yet it might well be possible for **a** to see to it that she will in the immediate future see to it that $\boldsymbol{A}$. That is to say, it might well be true that $\nabla\triangle\boldsymbol{A}$, yet false that $\nabla\boldsymbol{A}$. More likely still, it may well be true that $\nabla\nabla\boldsymbol{A}$, yet false that $\nabla\boldsymbol{A}$, i.e. that **a** is in a position to make it possible to see to it that $\boldsymbol{A}$, though not yet in a position to see to it that $\boldsymbol{A}$. I cannot just now hit the nail with a hammer, because I have no hammer in my hand, but the hammer is within reach, so I am in a position to grasp the hammer, thereby putting myself in a position to hit the nail.

With thoughts like these in mind, we see that we can initiate an iterative account of actions as means to an end $\boldsymbol{A}$, via the associated formulas asserting the performance of those actions.

(1) $\triangle\boldsymbol{A}$ is a means to $\boldsymbol{A}$;
(2) if $\triangle\boldsymbol{B}$ is a means to $\boldsymbol{A}$ then $\triangle\nabla\boldsymbol{B}$ is a means to $\boldsymbol{A}$;
(3) if $\triangle\boldsymbol{B}$ is a means to $\boldsymbol{A}$ then $\triangle\triangle\boldsymbol{B}$ is a means to $\boldsymbol{A}$.

---

[13] In such cases the "inner" action will have to be a sustaining action (e.g. holding the door open) or a progressive action (e.g. increasing the amount by which the door is open) as contrasted with an instantaneous action (e.g. making first contact with the door-opening button) in order for it to be possible to perform the action continuously over an interval of time.

Thus we have $\triangle \boldsymbol{A}, \triangle\triangle \boldsymbol{A}, \triangle\nabla \boldsymbol{A}, \triangle\triangle\triangle \boldsymbol{A}, \triangle\triangle\nabla \boldsymbol{A}, \triangle\nabla\triangle \boldsymbol{A}, \cdots$, all as various means to $\boldsymbol{A}$.[14]

In clause 1, we recognize that when achieving the end is something that can be accomplished by a single simple act, then performing that action is a means to the end. In clause 2, we recognize that achieving the ability to achieve the means to an end will also count. This will be a common situation, and indeed it might seem that it encompasses clause 3 as a special case. But in some cases, no matter what we do just now, we will still be able to achieve the end. So we can sometimes procrastinate without compromising our capacity to achieve the end. In such cases we cannot, strictly speaking, see to it that we will be able to achieve the end, because the negative condition is not satisfied: there's no alternative choice available just now which would risk *not* being able to achieve the end. In such situations, it would be inappropriate to describe any temporizing action as a means to the end. Nonetheless, in some such situations, there may be actions we can take which will not count as temporizing. For example I desire to complete a certain task before noon. There is still plenty of time, so if I so desire, I can go out for a coffee break now, without compromising my ability to finish my work before noon. However, I can also proceed directly to finish the task, and postpone or forego the coffee break. Taking the coffee break will not count as a means to the end, but proceeding to finish the task now rather than later certainly will, and it is actions such as this that are acknowledged as means to an end in clause 3.

Actions that fall under clause 2 may be of two sorts, which may be worth distinguishing in our thought, though they are not yet directly distinguished in the notation. There are cases in which it is not currently true that $\nabla \boldsymbol{B}$, and I perform an action that makes this become true for a while in the immediate future. There are also cases in which it is already true that $\nabla \boldsymbol{B}$, but where my action now will be relevant to whether it remains true for a while in the immediate future or not. In this latter kind of case, it seems more natural to describe the action reported in the formula $\triangle\nabla \boldsymbol{B}$ by saying that I non-trivially sustain the possibility of achieving the end, while in the former kind of case it seems more appropriate to say I achieve and then sustain that possibility. Both are equally covered by the formula $\triangle\nabla \boldsymbol{B}$; the only difference between the cases is that in one case $\nabla \boldsymbol{B}$ is not yet true, and in the other it is already true, though at risk for the future.

It will be helpful, as we shall soon see, to introduce a special notation $\rhd$ for this concept of non-trivially sustaining the truth of a formula, and we can define it as follows:

$$\rhd_\alpha \boldsymbol{A} \qquad \textit{iff} \qquad \boldsymbol{A} \wedge \triangle\!\!\!\!\!{\scriptscriptstyle P}\;\,_\alpha \boldsymbol{A}.$$

There are cases—most cases, I have no doubt—in which the end to be achieved is compound. Suppose, then, that the end to be achieved is conjunctive, of the

[14] One might consider saying, as well, that abilities such as are expressed by $\nabla \boldsymbol{A}, \nabla\triangle \boldsymbol{A}, \nabla\nabla \boldsymbol{A}, \cdots$ are also means to $\boldsymbol{A}$. That would perhaps not be unreasonable, but I prefer for present purposes to consider only actions, not potentials for action, as themselves being means to an end.

form $\boldsymbol{A} \wedge \boldsymbol{B}$. (We must assume $\boldsymbol{A}$ and $\boldsymbol{B}$ are consistent, else this cannot reasonably be an end goal.) Actions directed toward achieving the truth of one of the conjuncts would then normally be considered contributions to achieving the truth of the conjunction, provided they don't interfere with achieving the truth of the other conjunct. In such a case, of course, after making progress towards achieving the truth of one conjunct, we must sustain that gain as we take steps towards achieving the truth of the other conjunct.

Suppose, for example, that we wish to bring about the truth of the conjunction $\boldsymbol{A} \wedge \boldsymbol{B}$, and that there is a two step process by which we can make $\boldsymbol{A}$ true, and a three-step process by which we can make $\boldsymbol{B}$ true. One way in which we might proceed, in securing the truth of the conjunction, would be to pass through the following stages (where we again drop the subscript for the agent, which is assumed to be fixed for the duration of our current discussion):

stage 1	$\triangledown\triangledown \boldsymbol{A} \wedge \triangledown\triangledown\triangledown \boldsymbol{B}$	our starting position,
stage 2	$\triangle\triangledown \boldsymbol{A} \wedge \rhd\triangledown\triangledown\triangledown \boldsymbol{B}$	doing the first step towards $\boldsymbol{A}$,
stage 3	$\rhd\triangledown \boldsymbol{A} \wedge \triangle\triangledown\triangledown \boldsymbol{B}$	doing the first step towards $\boldsymbol{B}$,
stage 4	$\rhd\triangle \boldsymbol{A} \wedge \triangle\triangledown \boldsymbol{B}$	doing the second step towards $\boldsymbol{B}$,
stage 5	$\rhd\triangledown \boldsymbol{A} \wedge \triangle \boldsymbol{B}$	doing the final step towards $\boldsymbol{B}$,
stage 6	$\triangle \boldsymbol{A} \wedge \rhd \boldsymbol{B}$	doing the final step towards $\boldsymbol{A}$,
stage 7	$\boldsymbol{A} \wedge \boldsymbol{B}$	our final position.

The final action, in line 6, could also be described by the formula $\triangle(\boldsymbol{A} \wedge \boldsymbol{B})$, which is entailed by $\triangle \boldsymbol{A} \wedge \rhd \boldsymbol{B}$. The formula $\triangle(\boldsymbol{A} \wedge \boldsymbol{B})$ is perhaps more informative for purposes of assessing the overall project, but less informative for purposes of indicating the details of the process chosen. Similarly, each of the steps 2–5 could be described as the performance of a single action. Doing this for the whole process, we would get this description:

stage 1	$\triangledown\triangledown \boldsymbol{A} \wedge \triangledown\triangledown\triangledown \boldsymbol{B}$	$\triangledown\triangledown \boldsymbol{A} \wedge \triangledown\triangledown\triangledown \boldsymbol{B}$
stage 2	$\triangle\triangledown \boldsymbol{A} \wedge \rhd\triangledown\triangledown\triangledown \boldsymbol{B}$	$\triangle(\triangledown \boldsymbol{A} \wedge \triangledown\triangledown\triangledown \boldsymbol{B})$
stage 3	$\rhd\triangledown \boldsymbol{A} \wedge \triangle\triangledown\triangledown \boldsymbol{B}$	$\triangle(\triangledown \boldsymbol{A} \wedge \triangledown\triangledown \boldsymbol{B})$
stage 4	$\rhd\triangledown \boldsymbol{A} \wedge \triangle\triangledown \boldsymbol{B}$	$\triangle(\triangledown \boldsymbol{A} \wedge \triangledown \boldsymbol{B})$
stage 5	$\rhd\triangledown \boldsymbol{A} \wedge \triangle \boldsymbol{B}$	$\triangle(\triangledown \boldsymbol{A} \wedge \boldsymbol{B})$
stage 6	$\triangle \boldsymbol{A} \wedge \rhd \boldsymbol{B}$	$\triangle(\boldsymbol{A} \wedge \boldsymbol{B})$
stage 7	$\boldsymbol{A} \wedge \boldsymbol{B}$	$\boldsymbol{A} \wedge \boldsymbol{B}$

Obviously other routes to the same end were possible in the circumstances, but they could be described using the same logical tools.

Each of the stages in this process except the first and last should be considered a means to achieving the end state, described by $\boldsymbol{A} \wedge \boldsymbol{B}$. But now we see that we must add some clauses to our iterative definition in order to cover such cases. Clause 1 of our iterative definition establishes that stage 6 (as described in the right hand column) is a means to achieving stage 7. But no combination of clauses 1–3 will explain why we should say that stage 5 is a means to achieving either stage 6 or stage 7. To cover that, we must acknowledge that:

(4) if $\triangle \boldsymbol{C}$ is a means to $\boldsymbol{A}$ then $\triangle(\triangledown \boldsymbol{C} \wedge \boldsymbol{B})$ is a means to $\triangle(\boldsymbol{A} \wedge \boldsymbol{B})$.

And surely we should add:

(5) if $\triangle \boldsymbol{C}$ is a means to $\boldsymbol{A}$ then $\triangle(\triangle \boldsymbol{C} \wedge \boldsymbol{B})$ is a means to $\triangle(\boldsymbol{A} \wedge \boldsymbol{B})$.

Similarly, to cover transitions such as that from stage 4 to stage 5, we need:

(6) if $\triangle \boldsymbol{C}$ is a means to $\boldsymbol{B}$ then $\triangle(\boldsymbol{A} \wedge \nabla \boldsymbol{C})$ is a means to $\triangle(\boldsymbol{A} \wedge \boldsymbol{B})$;
(7) if $\triangle \boldsymbol{C}$ is a means to $\boldsymbol{B}$ then $\triangle(\boldsymbol{A} \wedge \triangle \boldsymbol{C})$ is a means to $\triangle(\boldsymbol{A} \wedge \boldsymbol{B})$.

Between them, these principles cover all the transitions in our sample case, provided we add two very general principles which we should acknowledge anyway:

(8) if $\boldsymbol{A}$ is a means to $\boldsymbol{B}$ and $\boldsymbol{B}$ is a means to $\boldsymbol{C}$, then $\boldsymbol{A}$ is a means to $\boldsymbol{C}$;
(9) if $\blacklozenge\neg\boxed{\scriptstyle W}\boldsymbol{A}$ and $\boldsymbol{B} \models \boldsymbol{A}$ then $\triangle \boldsymbol{B}$ is a means to $\boldsymbol{A}$.

Clause 8 obviates the necessity for clause 3 above, of course, but it is helpful for expository purposes to have clause 3 explicit.

Clause 9 says that (unless $\boldsymbol{A}$ expresses something which is absolutely certain to become true, in the circumstances) if $\boldsymbol{B}$ entails $\boldsymbol{A}$, then seeing to it that $\boldsymbol{B}$ becomes true is one means to $\boldsymbol{A}$. The need for guarding against guaranteed values of $\boldsymbol{A}$ is particularly evident in the case of logical truths: nothing we can do can properly be considered a means of achieving the truth of a tautology, though any $\boldsymbol{B}$ which we can achieve will certainly entail that logical truth.

When we move to the common situation in which more than one agent is involved, the expression of the principles indicated above—and their extensions to multiple agents—becomes less compact, of course, but not fundamentally more complex. For example, clause 2 above can be generalized to:

(2*) if $\triangle\!\!\!\!\!{\scriptscriptstyle P}\;_{\alpha}\boldsymbol{B}$ is a means to $\boldsymbol{A}$ then $\triangle\!\!\!\!\!{\scriptscriptstyle P}\;_{\beta}\blacklozenge\triangle\!\!\!\!\!{\scriptscriptstyle P}\;_{\alpha}\boldsymbol{B}$ is a means to $\boldsymbol{A}$.

This is appropriate irrespective of whether it is true that $\boldsymbol{\alpha} = \boldsymbol{\beta}$. Your getting the hammer for me to use in hitting the nail counts as a means to the end of the nail's being in its intended resting place, and counts just as much as my getting it for my own use to that end.

We could go on to cases of group action involving simultaneous action of the members of a set $\boldsymbol{\Gamma}$ of agents. This would require us to modify the syntax and truth conditions for the $\triangle\!\!\!\!\!{\scriptscriptstyle P}\;$ operator to allow it to take a group of agents as index, but the modification is easy, and proceeds along lines already thoroughly explored for the *dstit*, so I will not go into that here.

5 Naming Actions

At this point, it would be desirable to provide ourselves with a way of talking about the actions themselves. So far we have been dealing with formulas which depend for their truth on the performance of actions which the formulas do not name. The formulas describe, and thus correspond to, (types of) states, so our treatment via formulas treats means and ends both as states. Having the hammer in hand is a means to having the nail properly in the wood. But it must be agreed that it is more natural to think of means and ends as both actions, rather than states: getting the hammer is a means to sinking the nail properly in place. And it is even possible to think of the action as a means and the resulting state as its end, though when we express this we seem to revert naturally to speaking of the action as a means to *achieving* the state, thus treating the end as itself an action, or at least a completed process of some sort.

How, within the metaphysical framework provided by our models, are we to identify actions? One simple answer is that an action is, at its most basic level, a choice made by (or at least available to) an agent at a moment. But that is perhaps too simple, at least for the purposes at hand. When we say my picking up the hammer is a means to my pounding the nail, we don't intend to refer to anything quite so specific as a particular choice available to me at one particular moment. Rather we mean that *kind* of choice, whenever it might be available to me.[15] What characterizes the kind of choice in question? It's outcome state, apparently: my having the hammer in hand, in our example. So where up until now I have spoken as if a formula $\triangle^{p}_{\alpha}\boldsymbol{A}$ might express a means to $\boldsymbol{B}$, or to achieving (the truth of) $\boldsymbol{B}$, now I want to suggest that it is the *kind* of choice whose selection could be correctly reported by the formula $\triangle^{p}_{\alpha}\boldsymbol{A}$, i.e. by $\triangle^{d}_{\alpha}\boxed{W}\boldsymbol{A}$, that would constitute the action which was a means to $\boldsymbol{B}$. Let me introduce the notation $\boldsymbol{\delta_{\alpha} A}$ to name the action reportable by the formula $\triangle^{d}\boldsymbol{A}$, i.e. to correspond to the gerundial noun phrase: **a**'s (*deliberatively*) *seeing to it that* $\boldsymbol{A}$. Its extension would then be the set of choices available to **a** at various moments (various choice points) which, if selected, would make the formula $\triangle^{d}_{\alpha}\boldsymbol{A}$ true at those choice points. In the same fashion, we can also provide a progressive tense operator $\boldsymbol{\overline{\delta}_{\alpha}}$ based in the corresponding way on $\triangle^{p}_{\alpha}$.

To provide a home for these new locutions in our language, we need to provide a way to fit them into formulas, and that means, in effect, providing a verb with which such nouns as $\boldsymbol{\delta_{\alpha} A}$ and $\boldsymbol{\overline{\delta}_{\alpha} A}$ could be combined to form sentences. So let me introduce the performance verb $\boldsymbol{\pi}$ for this purpose, via the following equivalences:

$$\mathbf{m},\mathbf{h} \models \pi\delta_{\alpha}A \quad \textit{iff} \quad \mathbf{m},\mathbf{h} \models \triangle^{d}_{\alpha}A$$
$$\mathbf{m},\mathbf{h} \models \pi\overline{\delta}_{\alpha}A \quad \textit{iff} \quad \mathbf{m},\mathbf{h} \models \triangle^{p}_{\alpha}A$$

together with the condition that, applied to any other term except a $\boldsymbol{\delta}$-term, $\boldsymbol{\pi}$ produces a falsehood (only actions—not people or hammers, for example—can be performed in the sense intended here).

6 Intentions

A single action may fit many descriptions, and may have many outcomes. As I open the door, I also create a draft. The choice I made in doing so may have

[15] Perhaps we should limit this to some relevant time period. I will leave considerations of the relevant time period aside for the moment, however—our subject is difficult enough as it is, so I'll be content to tackle one matter at a time.

assured both outcomes, and both outcomes might have been avoidable had I chosen differently, and so the same choice might be the basis for the truth of both the formula $\triangle_\alpha \boldsymbol{A}$ and the formula $\triangle_\alpha \boldsymbol{B}$, where $\boldsymbol{A}$ expresses the claim *the door is opening*, and $\boldsymbol{B}$ expresses the claim *there is a draft.* As a result it may seem that (at that time and place) $\overline{\boldsymbol{\delta}}_\alpha \boldsymbol{A} = \overline{\boldsymbol{\delta}}_\alpha \boldsymbol{B}$. Yet the opening of the door might have been intended, and the creation of a draft an unintended, and even perhaps an unwanted, side effect. Surely this difference will often be important. How, then, are we to take account of such differences in intentions?

In a way, we already have, and yet in a way we have not. If, as I suggested in section 5, we take the action to be a *set* of choices characterized by a certain outcome, then the set of choices which (at various choice points) would result in the door's opening will be different from the set of those which would result in there being a draft. Opening the door will not in all cases involve creating a draft, and there are other ways of creating a draft without opening the door. So already, we can expect that in fact $\overline{\boldsymbol{\delta}}_\alpha \boldsymbol{A} \neq \overline{\boldsymbol{\delta}}_\alpha \boldsymbol{B}$ except when there is some strong connection (probably mutual entailment) between $\boldsymbol{A}$ and $\boldsymbol{B}$. So when we say that **a** performs the action $\overline{\boldsymbol{\delta}}_\alpha \boldsymbol{A}$, we should be permitted to deny that **a** performed the action $\overline{\boldsymbol{\delta}}_\alpha \boldsymbol{B}$, without fear of inconsistency, even if both $\triangle_\alpha \boldsymbol{A}$ and $\triangle_\alpha \boldsymbol{B}$ are true. However, the upshot of this is that the performance verb $\boldsymbol{\pi}$ we introduced in Section 5 isn't fully satisfactory, since if both $\triangle_\alpha \boldsymbol{A}$ and $\triangle_\alpha \boldsymbol{B}$ are true then both $\boldsymbol{\pi}\overline{\boldsymbol{\delta}}_\alpha \boldsymbol{A}$ and $\boldsymbol{\pi}\overline{\boldsymbol{\delta}}_\alpha \boldsymbol{B}$ will be, as well, according to the semantic account we offered a moment ago.

Properly understood, the performance verb $\boldsymbol{\pi}$ still makes a genuine contribution to our language: $\boldsymbol{\pi}\overline{\boldsymbol{\delta}}_\alpha \boldsymbol{A}$ expresses the claim that **a** performs the action $\overline{\boldsymbol{\delta}}_\alpha \boldsymbol{A}$ only in the weak sense that agent **a** selects one of the choices constituting the action $\overline{\boldsymbol{\delta}}_\alpha \boldsymbol{A}$. But we also want to be able to express the stronger claim that **a** chose *that* action (and no other) to perform. For this stronger sense of performance, let us introduce the verb $\boldsymbol{\Pi}$ into our formal language. Barring something comparable to a successful naturalistic theory of mental states, it is difficult to see how we can provide a semantics for the strong performance verb $\boldsymbol{\Pi}$ in which the truth conditions for $\boldsymbol{\Pi}\boldsymbol{\delta}_\alpha \boldsymbol{A}$ and $\boldsymbol{\Pi}\overline{\boldsymbol{\delta}}_\alpha \boldsymbol{A}$ would depend only on the sort of simple more-or-less atomic and more-or-less purely physical facts which we normally tacitly presuppose are the contents of the simplest formulas of the language—facts such as those about what is where at a given time, and related more complex, but purely naturalistic, facts such as whether the door is open. In contexts such as this, we don't normally think of the atomic formulas of the language as expressing claims about mental states, for example.[16]

The consequence of this reflection is that we cannot expect to base the truth conditions for formulas involving $\boldsymbol{\Pi}$ solely on the information which the valuation provides about the truth values of atomic formulas at the various points of evaluation, together with the semantic account of the other logical constants of our language. To deal with $\boldsymbol{\Pi}$, our models will need to be made more complex

[16] There are reasons for this presupposition, of course, associated with the expectation that logic should treat all expressions of propositional attitudes as involving operators of some sort.

in one of two ways. One way would be to augment the frame with a function from agents and points of evaluation to (intended) actions, or something of the sort. Another would be to modify the notion of a valuation so that in addition to assigning truth values to atomic formulas, it also assigned values directly to Π-formulas. The latter strategy, apart from being slightly distasteful, might be complex to pursue, since it seems likely to require some delicate constraints on the valuation to secure the proper logical relations between Π-formulas and related Π-free formulas.

So to the components already mentioned for our models, we add one more in the frame: an intention function I which, given any agent, any moment, and any history through that moment, specifies which action, if any[17], the agent intends at that moment to perform, by pursuing the choice within which that history falls. We may wish to impose the constraint that the action be one which has as a member the choice of which this history is an element, although perhaps a case could be made for leaving open the possibility that (speaking loosely, at least) the action the agent performs is not always in fact the one the agent intends, in the sense that sometimes $\neg\pi\delta_\alpha A$ even though the agent intends that it be true that $\pi\delta_\alpha A$. If we do impose this constraint on I, then we will have:

$$\mathsf{m},\mathsf{h} \models \Pi\delta_\alpha A \textit{ iff } \mathsf{I}(\alpha,\mathsf{m},\mathsf{h}) = \|\delta_\alpha A\|$$

(where $\|\delta_\alpha A\|$ is the extension of $\delta_\alpha A$).

If, instead, we choose not to impose the constraint on I, then we will have:

$$\mathsf{m},\mathsf{h} \models \Pi\delta_\alpha A \textit{ iff } \mathsf{I}(\alpha,\mathsf{m},\mathsf{h}) = \|\delta_\alpha A\| \text{ and } \mathsf{m},\mathsf{h} \models \pi\delta_\alpha A.$$

Analogous truth conditions would apply for $\Pi\overline{\delta}_\alpha A$.

7 Summing Up

With these resources all in place, we are now free to combine this logic of action with additional normative resources in ways that will allow us to place means in relation to ends, to differentiate normatively acceptable means from normatively unacceptable means, to speak of intended and unintended actions, to discuss actions by one agent which enable another agent to act, to discuss collaborative efforts in which different agents perform actions which serve as means towards a mutual end, etc. In short, we now have a much more subtle and sophisticated tool with which to approach problems in the logic of action and deontic logic.

Still, we haven't yet achieved all our original goals. Although we now have a present progressive tense operator capable of expressing the claim that I am opening the door, this will not convey any claim that the process will continue to its intended conclusion: that the door be (fully) open. So in effect we now have

[17] Presumably no action is intended, in the sense taken up here, at moments which are not choice points for the agent. The intention function given here is designed only to handle immediate intentions, not (for example) intentions about what I'll do tomorrow.

present tense process verbs, but still don't have a satisfactory array of present tense achievement verbs. This in turn means that our discussion of means and ends is still not fully satisfactory, since it cannot yet relate achievements as means to achievement of an end. And although we can now express the claim that one agent enables another to act (or prevents another from acting) in a certain way, we still have no means of expressing the more subtle claim that one agent influences another to act. So although we have made palpable and promising progress, there is clearly a large domain for future investigation. No doubt it should be reassuring that we have not yet run out of work to do.

References

1. Belnap, Nuel D., Jr. "Backwards and forwards in the modal logic of agency", *Philosophy and Phenomenological Research* 1991, **51** 777–807.
2. Belnap, Nuel D., Jr. "Before refraining: Concepts for agency", *Erkenntnis* 1991, **34** 137–169.
3. Belnap, Nuel D., Jr., Michael Perloff and Ming Xu. *Facing the Future: Agents and Choices in our Indeterminist World.* Oxford University Press, Oxford, 2001.
4. Brown, Mark A. "On the logic of ability", *Journal of Philosophical Logic* 1988, **17**, 1–26.
5. Elgesem, Dag. "The modal logic of agency", *The Nordic Journal of Philosophical Logic* 1997, **2**.2, 1–46.
6. Governatori, Guido, and Antonino Rotolo. "On the axiomatisation of Elgesem's logic of agency and ability", *Journal of Philosophical Logic* 2005, **34**.4, 403–431.
7. Horty, John F. *Agency and Deontic Logic.* Oxford University Press, Oxford, 2001.
8. Horty, John F., and Nuel Belnap, Jr. "The deliberative stit: A study of action, omission, ability, and obligation", *Journal of Philosophical Logic* 1995, **24**, 583–644.
9. Pörn, Ingmar. "Some basic concepts of action", in S. Stenlund (ed.) *Logical Theory and Semantic Analysis.* D. Reidel, Dordrecht, 1977.
10. Thomason, Richmond H. "Deontic logic and the role of freedom in moral deliberation", in Risto Hilpinen (ed.) *New Studies in Deontic Logic.* D. Reidel, Dordrecht, 1981, 177–186. (Synthese Library volume 152).
11. Xu, Ming. "Causation in branching time (I): Transitions, events, and causes", *Synthese* 1997, **112**, 137–192.

A State/Event Temporal Deontic Logic

Julien Brunel, Jean-Paul Bodeveix, and Mamoun Filali

Institut de Recherche en Informatique de Toulouse
Université Paul Sabatier
118 route de Narbonne, 31062 Toulouse, France
{brunel, bodeveix, filali}@irit.fr

Abstract. This paper studies a logic that combines deontic and temporal aspects. We first present a state/event temporal formalism and define a deontic extension of it. Then, we study the interaction between the temporal dimension and the deontic dimension. We present some logical properties, concerning formulas where deontic and temporal operators are nested, and discuss their intuitive meaning. We focus more particularly on the properties of obligation with deadline and define a specific operator to express this notion.

1 Introduction

Deontic logic is useful for specifying normative systems, i.e., systems which involve obligations, prohibitions, and permissions. Applications can be found in computer security, electronic commerce, or legal expert systems [21].

In this paper, we are interested in applications where both temporal and deontic notions appear. For instance, it may be interesting to express an access control policy in which the permissions depend on time, or events. Such a policy can be called an availability policy [8].

Consider a simple resource monitoring problem. An availability policy consists in giving a set of obligations and permissions for users to use resources.

- *$user_i$ has the permission to use the resource r for 5 time units continuously, and he must be able to access it 15 time units after asking, at the latest*
- *$user_i$ has always the permission to use the resource r, and he has to release it after 5 time units of utilization*
- *If $user_i$ is asking for the resource and he has the permission to use it, then the system has the obligation to give it to him before 5 time units*
- *If $user_i$ uses the resource without the permission, he must not ask for it during 10 time units*

The cases where permissions (idem with obligations and prohibitions) are granted may depend on the temporal events of the system, as in the two first sentences. But in the two last sentences, we see that permissions and prohibitions are granted according to other deontic notions. The last sentence gives for instance a prohibition if an obligation is violated.

L. Goble and J.-J.C. Meyer (Eds.): DEON 2006, LNAI 4048, pp. 85–100, 2006.

Cuppens and Saurel have used in [8] predicate logic. They have exhibited four dedicated predicates to express a policy - which gives a limited expressive power - and about ten formulas to check that the system does not violate the policy. Our goal is to define a language which allows to specify easily availability policies, or other systems in which time and norms play an important role.

We will first present a state/event extension of the temporal logic LTL[18]. Section 3 defines a deontic extension of this formalism. Then, we will discuss in section 4 about properties of formulas where temporal and deontic operators are nested. In section 4, we will particularly focus on obligation with deadline.

2 State/Event Extensions of LTL

We present here an extension of the state based *Linear Temporal Logic* (LTL)[18]. We will first discuss about the notions of event and action. Then we will present a state/event extension of Kripke structures. We will also present the logic which is interpreted on such structures : *State/Event Linear Temporal Logic*[5] (SE-LTL), a state/event extension of LTL. Note that these extensions do not increase the expressiveness, but allow to express behaviours in a much more succinct way (see [5] for more details).

2.1 Events or Actions?

The notion of event is close to an action in dynamic logic [12]. But an event is atomic, and has no duration. The actions can be composed with several combinators: sequence, choice, iteration, converse. Some propositions to combine dynamic and temporal logics [13] strengthen the temporal operator until U. $\varphi_1 U^\alpha \varphi_2$ means that $\varphi_1 U \varphi_2$ is satisfied along some path which corresponds to the execution of the action α. This gives a good expressive power, but the specification of properties can be much harder. For instance, consider the sentence "If $user_i$ uses the resource, he will release it". in_use_i is a proposition, and $release_i$ an event (or an atomic action in dynamic logic). In a state/event logic, we express it naturally: $in\_use_i \Rightarrow F\ release_i$.

But in a dynamic temporal logic, we cannot put actions and propositions at the same level. We have to use specific operators to introduce actions: $in\_use_i \Rightarrow \top\ U^{\Sigma*;release_i}\ \top$ where Σ represents any atomic action.

Moreover, in many cases the composition of events, can be expressed without using the combinators of dynamic logic. For instance, the formula $e_1 \wedge X e_2$ expresses that the event e_1 happens, followed by e_2, which corresponds to the action $e_1; e_2$ in a dynamic logic. The formula $e_1 U e_2$ means that there is an arbitrary number of executions of e_1, followed by an execution of e_2, and corresponds to the execution of the composed action $e_1*\ ; e_2$.

Besides, the semantics of events is much simpler, and there exists efficient tools for state/event logics [5, 4]. In the rest of this paper, we have preferred events to actions.

2.2 Labeled Kripke Structures

We define here a labeled Kripke structure. Although it has the same expressive power than event free Kripke structure, it is an interesting implementation formalism because it allows to express behaviours in a much more succinct way.

Definition 1 (LKS). *A labeled Kripke structure (LKS) over* $(P, \mathcal{E})$ *is a tuple* $(S, I, T, \alpha, \beta, \lambda, \nu)$ *where*

- P *is a countable set of atomic propositions that label the states.*
- $\mathcal{E}$ *is a countable set of events that label the transitions.*
- S *is a countable set of states.*
- $I \subseteq S$ *is a set of the initial states.*
- T *is a set of transition identifiers, hereafter simply called transitions.*
- $\alpha : T \to S$ *is a function which associates each transition with its source state.*
- $\beta : T \to S$ *is a function which associates each transition with its destination state.*
- $\lambda : T \to \mathcal{E}$ *is a function which associates each transition with the event performed during the transition.*
  We often write $t : s \xrightarrow{e} s'$*, where* $t \in T, s, s' \in S$*, and* $e \in \mathcal{E}$*, to mean that* $\alpha(t) = s$, $\beta(t) = s'$*, and* $\lambda(t) = e$
  We suppose that every state has an outgoing transition:
  $\forall s \in S\ \exists(t, e, s') \in T \times \mathcal{E} \times S\ \ t : s \xrightarrow{e} s'$[1]
- $\nu : S \to 2^P$ *is a valuation function which associates each state with the set of the atomic propositions it satisfies.*

Definition 2 (Run and trace). *A run* $\rho = (s_0, t_0, s_1, t_1 \ldots)$ *of an LKS is an infinite alternating sequence of states and transitions such that* $s_0 \in I$ *and* $\forall i \in \mathbb{N}\ \ \exists e \in \mathcal{E}\ \ t_i : s_i \xrightarrow{e} s_{i+1}$. *We can talk about infinite runs because we have supposed that there is a starting transition from every state.*

A trace $\tau = (\tau_{pr}, \tau_{ev})$ *over a run* $\rho = (s_0, t_0, s_1, t_1 \ldots)$ *is defined as follows*

- $\tau_{pr} : seq(2^P)$ *is the sequence*[2] *of the atomic proposition sets associated with the states of the run*
  $\forall i\ \tau_{pr}(i) = \nu(s_i)$
- $\tau_{ev} : seq(\mathcal{E})$ *is the sequence of the events associated with the transitions of the run*
  $\forall i\ \tau_{ev}(i) = e_i$ *where* $t_i : s_i \xrightarrow{e_i} s_{i+1}$

2.3 *SE-LTL*

We present now the syntax and the semantics of the specification formalism *SE-LTL* [5]. It is an extension of the state based logic *LTL* which takes into account both events and propositions.

[1] This is equivalent to $range(\alpha) = S$.

[2] $seq(E) \stackrel{def}{=} \mathbb{N} \to E$ is the set of sequences of elements of E.

Definition 3 (Syntax of *SE-LTL*). *Given a countable set P of atomic propositions, and a countable set $\mathcal{E}$ of events, a well-formed formula of SE-LTL is defined by:*

$$\varphi ::= p \in P \mid e \in \mathcal{E} \mid \bot \mid \varphi \Rightarrow \varphi \mid \varphi U^{+} \varphi$$

$\varphi_1 U^{+} \varphi_2$ means that φ_2 will hold at some point m in the strict future, and φ_1 will hold from the next moment until the moment before m.

We can define some usual abbreviations:

$\neg\varphi \stackrel{def}{=} \varphi \Rightarrow \bot$	$\top \stackrel{def}{=} \neg\bot$
$\varphi_1 \vee \varphi_2 \stackrel{def}{=} \neg\varphi_1 \Rightarrow \varphi_2$	$\varphi_1 \wedge \varphi_2 \stackrel{def}{=} \neg(\varphi_1 \Rightarrow \neg\varphi_2)$
$X\ \varphi \stackrel{def}{=} \bot\ U^{+}\ \varphi$	$\varphi_1\ U\ \varphi_2 \stackrel{def}{=} \varphi_2 \vee (\varphi_1\ \wedge\ \varphi_1\ U^{+}\ \varphi_2)$
$F\ \varphi \stackrel{def}{=} \top U \varphi$	$G\ \varphi \stackrel{def}{=} \neg F\ \neg\varphi$

The timed operators (with discrete time) are defined as follows:

$$\varphi_1\ \mathcal{U}_{\leqslant k}\varphi_2 \stackrel{def}{=} \begin{cases} \varphi_2 & \text{if } k = 0 \\ \varphi_2 \vee (\varphi_1 \wedge X\ (\varphi_1\ \mathcal{U}_{\leqslant k-1}\ \varphi_2)) & \text{else} \end{cases}$$

$$\varphi_1\ \mathcal{U}_{=k}\varphi_2 \stackrel{def}{=} \begin{cases} \varphi_2 & \text{if } k = 0 \\ \varphi_1 \wedge X\ (\varphi_1\ \mathcal{U}_{=k-1}\ \varphi_2) & \text{else} \end{cases}$$

We can now define $F_{\leqslant k}\varphi \stackrel{def}{=} \top\ U_{\leqslant k}\ \varphi$, $F_{=k}\ \varphi \stackrel{def}{=} \top\ U_{=k}\ \varphi$, and $G_{\leqslant k}\varphi \stackrel{def}{=} \neg F_{\leqslant k}(\neg\varphi)$.

Definition 4 (Satisfaction). *A formula φ of SE-LTL is interpreted on a trace of an LKS. Given a trace $\tau = (\tau_{pr}, \tau_{ev})$, a natural i, and a formula φ, we can define the satisfaction relation $\models$ by induction on φ:*

$(\tau, i) \models p$	*iff*	$p \in \tau_{pr}(i)$	*where* $p \in P$
$(\tau, i) \models e$	*iff*	$e = \tau_{ev}(i)$ (*e will be performed next*)	*where* $e \in \mathcal{E}$
$(\tau, i) \not\models \bot$			
$(\tau, i) \models \varphi_1 \Rightarrow \varphi_2$	*iff*	$(\tau, i) \models \varphi_1$ *implies* $(\tau, i) \models \varphi_2$	
$(\tau, i) \models \varphi_1\ U^{+}\ \varphi_2$	*iff*	$\exists j > i\ ((\tau, j) \models \varphi_2$ *and* $\forall\ i < k < j\ \ (\tau, k) \models \varphi_1)$	

We say that a trace τ satisfies a formula φ ($\tau \models \varphi$) if the first state of τ satisfies φ.

$$\tau \models \varphi \quad \textit{iff} \quad (\tau, 0) \models \varphi$$

We can easily extend the satisfaction relation to labeled Kripke structures, which are preferred to sequences in order to model programs.

Definition 5 (Satisfaction by a model and validity). *A labeled Kripke structure $\mathcal{M}$ is called a model of a formula φ, and we write $\mathcal{M} \models \varphi$, if all the traces τ of $\mathcal{M}$ satisfy φ.*

A formula φ is said to be valid if every LKS satisfy it.

Remark 1 (Extension to concurent events). We have chosen a state/event point of view, as *SE-LTL* to build our temporal and deontic language. However, we have considered that several events may happen simultaneously. It follows that from each state of a sequence, a set of events can be performed. This reveals to be interesting to model true concurrency (also called non-interleaved concurrency). For instance , if the model of the system contains connected events, they may happen simultaneously. It may be the case with the events *start_sound* and *start_video* in a multimedia context. In this case, the formula *start_sound* $\wedge$ *start_video* has to be satisfiable.

From now, each transition of an *LKS* is labeled with a set of events. ($\lambda : T \rightarrow 2^{\mathcal{E}}$.) A trace over a run $\rho = (s_0, t_0, s_1, t_1, \ldots)$ is a pair $\tau = (\tau_{pr}, \tau_{ev})$ where $\tau_{pr} \in seq(2^P)$ is defined as for *SE-LTL*, and $\tau_{ev} \in seq(2^{\mathcal{E}})$ is now a sequence of sets of events. We call the extension of *SE-LTL* to concurrent events *State/Concurrent Events LTL* (*SCE-LTL*).

Definition 6 (Syntax of *SCE-LTL*). *The syntax of SCE-LTL is the same as the syntax of SE-LTL. Only the semantics of the events differs.*

Definition 7 (Semantics of *SCE-LTL*). *The semantics of an event in SCE-LTL is defined by*

$$\tau, i \models e \quad \textit{iff} \quad e \in \tau_{ev}(i) \qquad \textit{where } e \in \mathcal{E}$$

e is satisfied if it is one of the events that are going to be performed simultaneously.

The other formulas have the same semantics as in SE-LTL.

3 Deontic Extension

We define here a deontic extension of the state/event formalism described in section 2. In *Standard Deontic Logic* (*SDL*) [23, 17], the semantics of deontic modalities is given by a relation on states (also called worlds). We extend this relation to combine both deontic and temporal aspects.

We first present deontic labeled Kripke structures (DLKS). Then, we define the logic *State/Event Deontic Linear Temporal Logic* (*SED-LTL*), extension of *SCE-LTL* with a deontic modality. And in the last part, we discuss about some logic properties in *SED-LTL*.

3.1 Deontic Labeled Kripke Structures

We present here a deontic extension of a labeled Kripke structure. In our framework, which describes temporal behaviours, a world is a LKS. We call these worlds the alternatives, because they represent different possible behaviours.

They all have the same states and transitions, but the labels (on both states and transitions) differ from one alternative to another. Thus, we extend the deontic relation to be a relation on alternatives.

Definition 8 (Deontic labeled Kripke structure). *A deontic labeled Kripke structure over* $(P, \mathcal{E})$ *is a tuple* $(A, a_0, S, I, T, \alpha, \beta, \lambda, \nu, R_o)$ *where*

- *A is a (countable) set of alternative names (hereafter called alternatives).*
- $a_0 \in A$ *is the alternative that corresponds to the real behaviour. The other alternatives are needed to model the deontic aspects.*
- S, I, T, α, *and* β *are defined as in part 2.2.*
- $\lambda : A \times T \to 2^{\mathcal{E}}$ *is the valuation function which associates each transition of an alternative with its label (a set of events that occur simultaneously during the transition). If* $t \in T$, $s, s' \in S$, $E \subseteq \mathcal{E}$, $a \in A$, *we often write* $t : s \to s'$ *to mean that* $\alpha(t) = s$ *and* $\beta(t) = s'$, *and* $t : s \xrightarrow[a]{E} s'$ *to mean that* $\alpha(t) = s$, $\beta(t) = s'$, *and* $\lambda(a, t) = E$.
- $\nu : A \times S \to 2^P$ *is the valuation function that associates each state of an alternative with a set of atomic propositions.* $\nu(a, s)$ *represents the set of the atomic propositions satisfied by* $s \in S$ *in the alternative* $a \in A$.
- $R_o \subseteq A \times A$ *is the deontic relation which associates each alternative with the set of its good alternatives.* R_o *is supposed to be serial.*

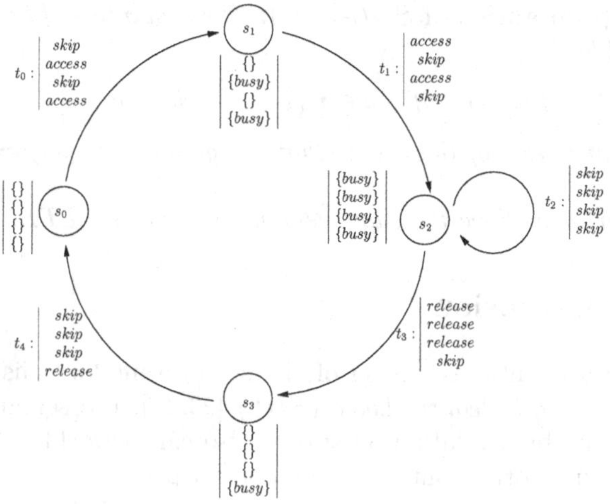

Fig. 1. Example of a DLKS

The figure 1 shows an example of a DLKS in which there are four alternatives. Thus, each transition and each state has four labels. We have not shown the relation R_o.

Definition 9 (Run and trace). *As in part 2.3, we define a run* $\rho = (s_0, t_0, s_1, t_1, \ldots)$ *of a DLKS as an alternating sequence of states and transitions, such that*

$s_o \in I$, and $\forall i \in \mathbb{N}\ t_i : s_i \rightarrow s_{i+1}$. A trace τ of a run $\rho = (s_0, t_0, s_1, t_1, \ldots)$ is a pair (τ_{pr}, τ_{ev}) where the sequences τ_{pr} and τ_{ev} are now indexed by the alternative. Indeed, through one run, the labels of states and transitions may differ from one alternative to another.

- *$\tau_{pr} : A \rightarrow seq(2^P)$ associates each alternative with a sequence of proposition sets.*
  $\forall (i,a) \in \mathbb{N} \times A \quad \tau_{pr}(a)(i) = \nu(a, s_i)$
- *$\tau_{ev} : A \rightarrow seq(2^E)$ associates each alternative with a sequence of event sets.*
  $\forall (i,a) \in \mathbb{N} \times A \quad \tau_{ev}(a)(i) = \lambda(a, t_i)$

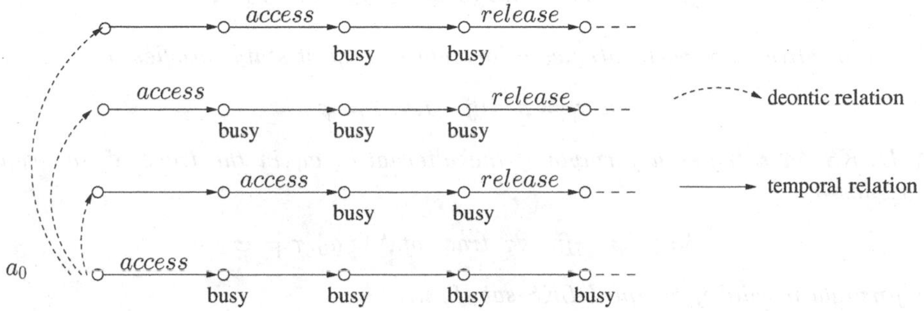

Fig. 2. Example of a DLKS trace

The figure 2 shows an illustration of the trace over the run $(s_0, t_0, s_1, t_1, s_2, t_2, s_2, t_3, s_3, \ldots)$ of the DLKS of the figure 1.

3.2 Syntax and Semantics

We present here the syntax of our specification formalism *State/Event Deontic Linear Temporal Logic* (*SED-LTL*), and its semantics.

Definition 10 (Syntax of *SED-LTL*). *Given a countable set P of atomic propositions, and a countable set $\mathcal{E}$ of events, a well-formed formula of SED-LTL is defined by:*

$$\varphi ::= p \in P \mid e \in \mathcal{E} \mid \bot \mid \varphi \Rightarrow \varphi \mid \varphi\ U^+\ \varphi \mid \mathbf{O}(\varphi)$$

The deontic modality $\boldsymbol{O}$ represents the obligation. We can define the permission ($\boldsymbol{P}$) and the prohibition ($\boldsymbol{F}$) as the following abbreviations:

$$\boldsymbol{P}(\varphi) \stackrel{def}{=} \neg \boldsymbol{O}(\neg\varphi) \quad \boldsymbol{F}(\varphi) \stackrel{def}{=} \boldsymbol{O}(\neg\varphi)$$

We define some usual operators, and timed operators (with discrete time) as for *SE-LTL* (cf part (2.3)).

We define here the semantics of the formulas in *SED-LTL*.

Definition 11 (Satisfaction). *A formula φ is interpreted on a trace $\tau = (\tau_{pr}, \tau_{ev})$ of a DLKS. Given a DLKS $\mathcal{M}$, an alternative a, a trace τ, an integer i, and a formula φ, we can define the satisfaction relation $\models$ by induction on φ:*

$$
\begin{array}{lll}
a,\tau,i \models p & \textit{iff} \quad p \in \tau_{pr}(a)(i) & \textit{where } p \in P \\
a,\tau,i \models e & \textit{iff} \quad e \in \tau_{ev}(a)(i) & \textit{where } e \in \mathcal{E} \\
a,\tau,i \not\models \bot & & \\
a,\tau,i \models \varphi_1 \Rightarrow \varphi_2 & \textit{iff} \quad a,\tau,i \models \varphi_1 \textit{ implies } a,\tau,i \models \varphi_2 & \\
a,\tau,i \models \varphi_1\ U^+\ \varphi_2 & \textit{iff} \quad \exists i'' > i \textit{ such that } a,\tau,i'' \models \varphi_2 \textit{ and} & \\
 & \qquad \forall i'\ i < i' < i'' \ \Rightarrow\ a,\tau,i' \models \varphi_1 & \\
a,\tau,i \models \mathbf{O}\varphi & \textit{iff} \quad \forall a' \textit{ such that } (a,a') \in R_o\ \ a',\tau,i \models \varphi &
\end{array}
$$

An alternative in a trace satisfies a formula if its first state satisfies it.

$$a,\tau \models \varphi \quad \textit{iff} \quad a,\tau,0 \models \varphi$$

A DLKS $\mathcal{M}$ satisfies a formula if the alternative a_0 in the trace of any run satisfies it.

$$\mathcal{M} \models \varphi \quad \textit{iff} \quad \forall\tau \textit{ trace of } \mathcal{M}\ \ a_0,\tau \models \varphi$$

A formula is valid if all the DLKS satisfy it.

$$\models \varphi \quad \textit{iff} \quad \forall\mathcal{M}\ \mathcal{M} \models \varphi$$

Remark 2 (Product). This formalism is very closed to a product of a temporal and a deontic logic. However, in this case, the definition of a product [11] does not match because of the use of events that makes the temporal semantics more complex than in *LTL*. We plan to study the adaptation of decidability results to *SED-LTL*.

Let us consider the trace of figure 2. The behaviour of the alternative a_0 accesses to the resource while it is permitted, and does not release it after three time units although it is obliged. This deontic trace satisfies the following formulas, which express the violation of some obligations.

$a_0,\tau \models F_{=3}(\boldsymbol{O}\ release\ \wedge\ \neg release)$

$a_0,\tau \models \boldsymbol{O}\ (access \Rightarrow F_{\leqslant 3} release)\ \wedge\ (access \wedge \neg F_{\leqslant 3} release)$

$a_0,\tau \models G\boldsymbol{O}\ (busy \Rightarrow F_{\leqslant 3}\neg busy)\ \wedge\ X(busy \wedge \neg F_{\leqslant 3}\neg busy)$

3.3 Expression of Availability Policies

We express here in our formalism the policies given in the introduction.

- *$user_i$ has the permission to use the resource r for 5 time units continuously, and he must be able to access it 15 time units after asking, at the latest* The first part of this sentence has an ambiguity. It is not clear whether it is forbidden to use the resource after 5 time units of utilization. The first formula does not consider it is the case, whereas the second one does.

$$G\ ((access \Rightarrow G_{\leqslant 5}(\boldsymbol{P}use)) \wedge (request \Rightarrow \boldsymbol{O}(F_{\leqslant 15}\ access)))$$

$$G\ ((access \Rightarrow F_{\leqslant 5}\boldsymbol{O}(\neg use)) \wedge (request \Rightarrow \boldsymbol{O}(F_{\leqslant 15}\ access)))$$

- *$user_i$ has always the permission to use the resource r, and he has to release it after 5 time units of utilization*

$$G(\boldsymbol{P}use) \wedge G(access \Rightarrow \boldsymbol{O}(F_{\leqslant 5}release))$$

- *If $user_i$ is asking for the resource and he has the permission to use it, then the system has the obligation to give it to him before 5 time units*

$$G((request \wedge \boldsymbol{P}use)\ \Rightarrow\ \boldsymbol{O}(F_{\leqslant 5}access))$$

- *If $user_i$ uses the resource without the permission, he must not ask for it during 10 time units*

$$G((use \wedge \boldsymbol{O}(\neg use))\ \Rightarrow\ \boldsymbol{O}(G_{\leqslant 10}\neg request))$$

We have shown that the translation from natural language to our formalism is easy for such sentences, that describe availability policies. We now focus on the logical properties of $SED\text{-}LTL$, and discuss their intuitiveness.

3.4 Finiteness of the Set of Alternatives

We have not discussed yet about the nature of the set A of alternatives. Our first idea was to consider A countable, by analogy with a usual Kripke structure. But in order to develop a decision procedure, or to reason on some examples, as the figure 2, it would be simpler to have a finite set of alternatives.

Let us exhibit a formula which may be satisfiable or unsatisfiable, whether we consider A finite or countable. Consider the following statement *"p is always permitted, and p ought to happen at most once"*. It corresponds to the formula $\varphi \stackrel{def}{=} G\boldsymbol{P}p \wedge \boldsymbol{O}(\text{AtMostOne}(p))$, where $p \in P$ is an atomic proposition, and AtMostOne(p) is the abbreviation of $G(p \Rightarrow XG\neg p)$, which means *p happens at most once.* If we consider the class of models such that A is finite, φ is unsatisfiable. Indeed a model of φ necessarily contains a infinite number of good alternatives, since each point of the alternative a_0 satisfies $\boldsymbol{P}p$. For each of these future points, there is an alternative satisfying p different from the other alternatives. It follows that there are as many alternatives as points in the future. The figure 3 shows the construction of a model of φ.

In the same way, φ is clearly satisfiable if we consider that A can be infinite. This shows that our formalism has not the finite model property.

At a first sight, there is no direct contradiction in φ. Thus, it is in the favour of an infinite set A of alternatives.

3.5 Properties

In this part, we will focus on the properties of formulas in which temporal and deontic modalities are nested. For instance, $\boldsymbol{P}X\varphi$, or $\boldsymbol{O}G\varphi$.

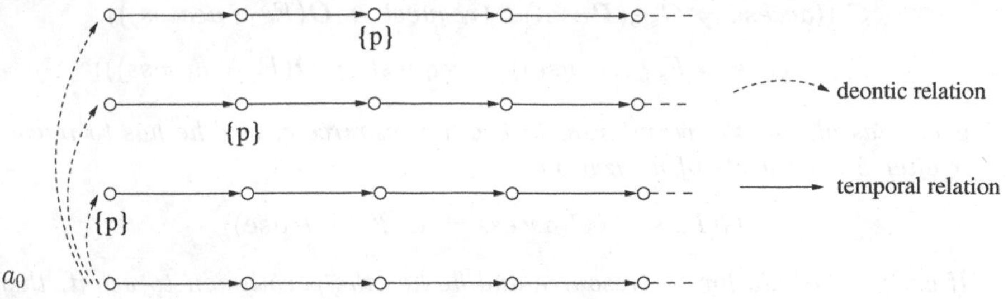

Fig. 3. A model for $GPp \wedge O(\text{AtMostOne}(p)$

The fact that the deontic relation R_o is a relation on sequences, and not on states, has consequences on such formulas. Indeed, the good alternatives of a_0 can be seen as the rules to be obeyed. What is obligatory is then what is true in all the rules, and what is permitted is what is true in at least one rule. Each of these rules is consistent from a temporal point of view. For instance, if $X\varphi$ is permitted in one state (a_0, τ, i), then there exists a good state (a_1, τ, i) which satisfies $X\varphi$. Since $(a_1, \tau, i+1)$ is a good state of $(a_0, \tau, i+1)$, (a_0, τ, i) also satisfies $X\boldsymbol{P}\varphi$. Then, for any alternative a, trace τ, and natural i, we have $(a, \tau, i) \models \boldsymbol{P}X\varphi \Rightarrow X\boldsymbol{P}\varphi$

In the same way, we prove the converse property $(a, \tau, i) \models X\boldsymbol{P}\varphi \Rightarrow \boldsymbol{P}X\varphi$. Therefore,

$$\models \boldsymbol{P}X\varphi \Leftrightarrow X\boldsymbol{P}\varphi$$

This is in accordance with one intuition. Indeed, the natural language gives some ambiguities to the expression "I have the permission to check in tomorrow". One can understand that this permission can be withdrawn before tomorrow. That is not our point of view. We consider the permissions globally, and not from an operational point of view. In other words, we do not model changing norms. If the permission to check in does not exist tomorrow, then we cannot say today that we have the permission to check in tomorrow.

The same reasoning shows that the two modalities $\boldsymbol{O}$ and X commute.

$$\models \boldsymbol{O}X\varphi \Leftrightarrow X\boldsymbol{O}\varphi$$

As a consequence, the deontic modalities commute with the timed operator $F_{=k}$ (cf. part 2.3).

$$\models \boldsymbol{O}F_{=k}\,\varphi \Leftrightarrow F_{=k}\,\boldsymbol{O}\varphi$$

$$\models \boldsymbol{P}F_{=k}\,\varphi \Leftrightarrow F_{=k}\,\boldsymbol{P}\varphi$$

This is not the case with the operators F, i.e., the property $\boldsymbol{O}F\varphi \Leftrightarrow F\boldsymbol{O}\varphi$ is not valid, which is intuitive. Indeed, having the obligation to meet φ in the future does not imply that there will be an immediate obligation to satisfy φ. On the other hand, we would like the obligation to continue while φ is not satisfied.

The general expression of this propagation property is :

$$\boldsymbol{O}(\varphi \vee X\psi) \wedge \neg \boldsymbol{O}\varphi \wedge \neg\varphi \;\Rightarrow\; X\boldsymbol{O}\psi \tag{1}$$

We need $\boldsymbol{O}(\varphi \vee X\psi) \wedge \neg \boldsymbol{O}\phi$ in the hypothesis of (1) rather than only $\boldsymbol{O}(\varphi \vee X\psi)$. Indeed, we want to express the fact that the disjunction $\varphi \vee X\psi$ is obligatory, but not φ.

A consequence of the simpler formula $\boldsymbol{O}(\varphi \vee \psi) \wedge \neg\varphi \;\Rightarrow\; X\boldsymbol{O}\psi$ would be $\boldsymbol{O}(\varphi) \wedge \neg\varphi \;\Rightarrow\; X\boldsymbol{O}\psi$, where ψ can be any formula, which is of course not desirable.

However, the property (1) is not valid because of the semantics of $\boldsymbol{O}$ given by the deontic relation R_o. The problem does not rely on the addition of the temporal dimension, but on the interaction between obligation and disjunction. The problem raised by propagation comes from the fact the following property is not valid in SDL:

$$\boldsymbol{O}(\varphi_1 \vee \varphi_2) \wedge \neg \boldsymbol{O}\varphi_1 \wedge \neg\varphi_1 \;\Rightarrow\; \boldsymbol{O}\varphi_2$$

The intuition suggests a strong interaction between what happens ($\neg\varphi_1$ in this formula) and what is obliged ($\varphi_1 \vee \varphi_2$). This interaction does not exist in SDL, and thus does not exist either in our formalism. The obligation we model is a norm that does not depend on the facts.

We have the same problem with the obligation with deadline ($\boldsymbol{O}F_{\leqslant k}\varphi$), because $F_{\leqslant k}\varphi$ is equivalent to $\varphi \vee XF_{k-1}\varphi$ (if $k \geqslant 1$). We will focus on the notion of obligation with deadline, which is closer to real cases than the obligation to satisfy φ in the future without deadline.

So, we would like the following property to be satisfied

$$\boldsymbol{O}\, F_{\leqslant k}\varphi \wedge \neg \boldsymbol{O}\varphi \wedge \neg\varphi \;\Rightarrow\; X\boldsymbol{O}F_{\leqslant k-1}\,\varphi$$

As explained, this property is not satisfied. To overcome this problem, we introduce a new operator in the next section.

4 Obligation with Deadline

In this part, we will focus on the notion of obligation with deadline. We want to specify that it is obliged to satisfy φ before k time units. We will use the notation $\mathcal{O}_k(\varphi)$.

4.1 A First Definition

The natural way to specify it is : $\mathcal{O}_k(\varphi) \overset{def}{=} \boldsymbol{O}(F_{\leqslant k}\,\varphi)$.

Do we have properties that are in accordance with our intuition? First of all, we have the property of monotonicity with respect to the deadline:

$$\models \mathcal{O}_k(\varphi) \Rightarrow \mathcal{O}_{k'}(\varphi) \quad \text{where } k \leqslant k'$$

It is in accordance with the first intuition we have of obligation with deadline. Indeed, if it is obliged to meet φ before k time units, then it is also obliged to meet φ before a greater deadline k'.

Another property is the property of monotonicity with respect to φ:

$$\models \mathcal{O}_k(\varphi_1 \wedge \varphi_2) \Rightarrow \mathcal{O}_k(\varphi_1) \quad \text{and} \quad \models \mathcal{O}_k(\varphi_1) \;\Rightarrow\; \mathcal{O}_k(\varphi_1 \vee \varphi_2)$$

However, as we said in part (3.5), the property of propagation of an obligation with deadline is not satisfied. It expresses that if it is obliged to satisfy φ before k time units, and if φ is not satisfied now, then in one time unit, it will be obliged to satisfy φ before $k-1$ time units. It corresponds to the formula:

$$\mathcal{O}_k(\varphi) \wedge \neg \boldsymbol{O}\varphi \wedge \neg\varphi \;\Rightarrow\; X\mathcal{O}_{k-1}(\varphi)$$

This is not a valid formula in *SED-LTL*. Indeed, a model satisfies $\mathcal{O}_k\varphi$ if all the good alternatives of a_0 satisfy φ before k time units. But one of these good alternatives may satisfy φ now, and $\neg\varphi$ thereafter. In this case, the obligation does not hold in one time unit, even if φ has not been satisfied.

4.2 A New Operator for Obligation with Deadline

The fact obligation with deadline is not propagated while it is not complied with implies that it can be violated without having an immediate obligation at any moment. In other words, $\boldsymbol{O}(F_{\leqslant k}\varphi) \wedge \neg F_{\leqslant k}\varphi \;\wedge\; G\neg\boldsymbol{O}\varphi$ is satisfiable. A solution is to have a "counter" which is set to k when $\boldsymbol{O}F_{\leqslant k}\;\varphi$ holds, and decremented while φ is not satisfied. So, we have to define a new operator dedicated to the obligation with deadline, $\mathcal{O}_k(\varphi)$, which means that there is an obligation to meet φ before k which has not been fulfilled yet.

$$\begin{array}{lll} a,\tau,i \models \mathcal{O}_k(\varphi) & \text{iff} \quad \exists k' \in \mathbb{N} & a,\tau,i-k' \models \boldsymbol{O}F_{\leqslant k+k'}\;\varphi \;\wedge\; \neg\varphi U_{=k'} \top \\ & \quad\;\; \wedge\; \nexists k'' < k+k' & a,\tau,i-k' \models \boldsymbol{O}F_{\leqslant k''}\;\varphi \end{array}$$

More precisely, $\mathcal{O}_k(\varphi)$ may be read as *"k' time units ago, there were an obligation to satisfy φ before k+k', and the obligation has not been fulfilled yet"*. The second part of the definition (on the second line) means that k' time units ago, there was no obligation to satisfy φ before a shorter deadline. Otherwise, an immediate obligation $\boldsymbol{O}\varphi$ would imply $\mathcal{O}_k\varphi$ (for any k) while φ is not satisfied.

Because of this last point, the new operator cannot be deduced from $\boldsymbol{O}(F_{\leqslant k}\varphi)$. Indeed, if $\boldsymbol{O}\varphi$ is satisfied at some moment, then $\boldsymbol{O}(F_{\leqslant k}\varphi)$ is also true, whereas $\mathcal{O}_k(\varphi)$ may not be true. There exists a,τ,i such that

$$a,\tau,i \not\models \boldsymbol{O}F_{\leqslant k}(\varphi) \Rightarrow \mathcal{O}_k(\varphi)$$

In the same way, the property of monotonicity with respect to the deadline does not hold either: there exists a,τ,i such that

$$a,\tau,i \not\models \mathcal{O}_k(\varphi) \;\Rightarrow\; \mathcal{O}_{k'}(\varphi) \quad \text{where } k \leqslant k'$$

Indeed, unlike the first definition (cf section 4.1), the new definition of $\mathcal{O}_k(\varphi)$ considers an exact deadline k: what is obligatory is to meet φ before k time units, not to meet φ before $k+1$, or $k-1$ time units. Of course, the fact that $\mathcal{O}_k(\varphi)$ is complied with, which corresponds to $F_{\leqslant k}\varphi$, has the monotonicity property with respect to k.

With a deadline 0, the new operator can be seen as a generalization of the obligation $\boldsymbol{O}$. Indeed, when φ is obligatory, in the sense of $\boldsymbol{O}$, then $\mathcal{O}_0(\varphi)$ holds, but the converse property is not true. For any a, τ, i

$$a, \tau, i \models \boldsymbol{O}(\varphi) \Rightarrow \mathcal{O}_0(\varphi)$$

But there exists a, τ, i such that $a, \tau, i \not\models \mathcal{O}_0\varphi \Rightarrow \boldsymbol{O}\varphi$.

Thus, $\mathcal{O}_0$ can be seen as a generalised immediate obligation. Indeed, it holds if

- there is an explicit immediate obligation $\boldsymbol{O}$
- or there is an obligation with deadline which has not been fulfilled, and we have reached the deadline

For instance, consider the case where we check that a violation condition for a formula φ does not hold. $\mathcal{O}_0(\varphi) \wedge \neg\varphi$ is much stronger that the usual violation condition $\boldsymbol{O}(\varphi) \wedge \neg\varphi$.

Note that since the semantics of the operator $\mathcal{O}_k$ involves the past, we give properties for any step i, given a trace τ and an alternative a. Until this point, the temporal operators only involved the future of the current state, so the properties was given at step 0, and thus expressed as validity properties.

Figure 4 shows an illustration of obligation with deadline. In the first state of a_0, there is an obligation to release the resource before three time units. Indeed, each of the three good alternatives of a_0 do so. In the second state, the resource has not been released, but the obligation does not hold anymore because one of the good alternatives has already released the resource. But our new operator $\mathcal{O}_k$ holds until the deadline is reached. And in the fourth state, there is a generalised immediate obligation to release the resource.

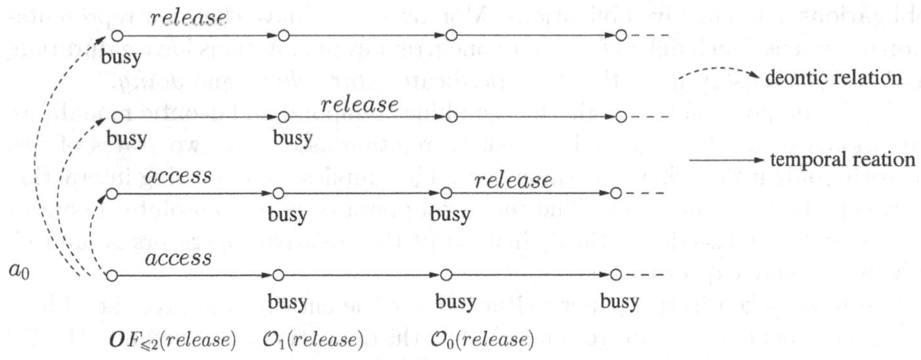

Fig. 4. Illustration of the obligation with deadline operator

5 Related Work

In this section, we compare our formalism with some existing formalisms. We have proposed to define both deontic and temporal concepts using a Kripke model. In [16, 9], Meyer proposes a definition of deontic notions in dynamic logic, based on the reduction of deontic logic to alethic modal logic by Anderson [1]. The idea is to introduce a new proposition V that represents the violation of an obligation, and to define obligation in terms of this violation proposition. This differs from our formalism since we have defined deontic concepts by a relation on states of a Kripke model, and then the violation of some obligation $\boldsymbol{O}(\varphi)$ as $\boldsymbol{O}(\varphi) \wedge \neg\varphi$. A consequence is that reductionist approaches cannot distinguish the violation of a formula φ_1 from the violation of another formula φ_2.

Broersen et al. also suggests in [3] a reductionist approach. The goal is to model obligations with deadline in the branching time logic *CTL* (*Computation Tree Logic* [6, 19]), using a violation constant and an ideality constant. The deadline is a formula and not a concrete duration as suggested in this paper. This is an interesting point of view which can be considered as a more abstract point of view as ours. Dignum et al. [10] have also proposed a reductionist approach to study the specification of deadlines.

Some other formalisms represent both time and deontic notions using relations on states in Kripke models. For instance, in [22, 20] there is no constraint that link the two relations (deontic and temporal). This is another possible interpretation of the natural language, which corresponds to the case where the obligations can change over time. We have chosen here to model obligations that cannot be withdrawn, or added if they are in contradiction with previous obligations. For instance, $\boldsymbol{O}Gp \wedge X\neg\boldsymbol{O}p$ and $\boldsymbol{O}Gp \wedge X\,\boldsymbol{O}(\neg p)$ are both unsatisfiable in *SED-LTL*. If one gets the obligation to always establish p, this obligation cannot be withdrawn tomorrow, and a new obligation cannot be added if it is in contradiction with the previous one.

In [7], Cuppens et al. present the formalism *NOMAD* which can be considered as a starting point of our study. Nevertheless, the only temporal operator is "next", and there is a distinction between two kinds of obligations: contextual obligations, and effective obligations. Moreover, they have chosen a representation of actions which differs from our concurrent events. Actions have a duration, and they are observed via the three predicates *start*, *done*, and *doing*.

Aqvist proposes in [2] a logic that combines temporal and deontic modalities. His model differs from ours. The deontic relation associates two states of two histories only if they share the same past. This implies an interesting interaction between the two dimensions. The temporal operator is the "absolute" operator R_t ("it holds at the time t that") instead of the "relative" operators X and U, which are more expressive.

We have talked in this paper neither about the entities who give the obligations, nor about those who are concerned by the obligations. For instance, [15, 20] present a formalism where the deontic modalities are indexed by agents. There are in the model as many deontic relations as there are agents. It would be interesting to integrate these aspects which would give a greater expressiveness.

6 Conclusion and Future Work

We have proposed a deontic and temporal logic that takes into account states and events, following the idea of *SE-LTL*[5]. We have focused on the interaction between the temporal dimension and the deontic dimension. Thus, we have discussed the intuitiveness of some properties of formulas where deontic and temporal modalities are nested, and we have studied more particularly obligation with deadline.

We plan to study a modification of the semantics of obligation that would take into account an interaction between the facts and the rules, in order to have the propagation property. We also want to study a (sound and complete) axiomatization of *SED-LTL*, and its decidability. Indeed, a decision procedure would allow to check the internal coherency of an availability policy expressed in this formalism. Another direction for future work is to check if a system does not violate an obligation within a policy. This last problem, which requires to reason on both a pure temporal behaviour (the system) and a deontic temporal formula (the policy), need to be formalised.

As we said in section 5, it would be interesting to integrate the notion of agents. We have begun to model access control policies which can be described in the OrBAC[14] model, where the notion of agent appear. Following [20] in which roles and groups of agents are represented, we are studying an agent based extension of *SED-LTL* in order to express such policies.

References

1. A. R. Anderson. A reduction of deontic logic to alethic modal logic. *Mind*, pages 100–103, 1958.
2. L. Aqvist. Combinations of tense and deontic logic. *Journal of Applied Logic*, 3:421–460, 2005.
3. J. Broersen, F. Dignum, V. Dignum, and J.-J. C. Meyer. Designing a deontic logic of deadlines. In *7th International Workshop on Deontic Logic in Computer Science (DEON'04)*, Madeira, Portugal, 26-28 May 2004.
4. S. Chaki, E. Clarke, O. Grumberg, J. Ouaknine, N. Sharygina, T. Touili, and H. Veith. State/event software verification for branching-time specifications. In *Fifth International Conference on Integrated Formal Methods (IFM 05)*, volume 3771 of *Lecture Notes in Computer Science*, pages 53–69, 2005.
5. S. Chaki, E. M. Clarke, J. Ouaknine, N. Sharygina, and N. Sinha. State/event-based software model checking. In E. A. Boiten, J. Derrick, and G. Smith, editors, *Proceedings of the 4th International Conference on Integrated Formal Methods (IFM '04)*, volume 2999 of *Lecture Notes in Computer Science*, pages 128–147. Springer-Verlag, April 2004.
6. E. Clarke and E. Emerson. Design and synthesis of synchronization skeletons using branching-time temporal logic. In *Proceedings of the 3rd Workshop of Logic of Programs (LOP'81)*, volume 131 of *Lecture Notes in Computer Science*, pages 52–71, 1981.
7. F. Cuppens, N. Cuppens-Boulahia, and T. Sans. Nomad: a security model with non atomic actions and deadlines. In *Proceedings of the 18th IEEE Computer Security Foundations Workshop*, June 2005.

8. F. Cuppens and C. Saurel. Towards a formalization of availability and denial of service. In *Information Systems Technology Panel Symposium on Protecting Nato Information Systems in the 21st Century*, Washington, 1999.
9. P. d'Altan, J.-J. C. Meyer, and R. Wieringa. An integrated framework for ought-to-be and ought-to-do constraints. *Artif. Intell. Law*, 4(2):77–111, 1996.
10. F. Dignum and R. Kuiper. Obligations and dense time for specifying deadlines. In *Thirty-First Annual Hawaii International Conference on System Sciences (HICSS)-Volume 5*, 1998.
11. D. Gabbay, A. Kurucz, F. Wolter, and M. Zakharyachev. *Many-Dimensional Modal Logics: Theory and Applications.* Elsevier, 2003.
12. D. Harel, D. Kozen, and J. Tiuryn. Dynamic logic. In D. Gabbay and F. Guenther, editors, *Handbook of Philosophical Logic Volume II — Extensions of Classical Logic*, pages 497–604. D. Reidel Publishing Company: Dordrecht, The Netherlands, 1984.
13. J. G. Henriksen and P. S. Thiagarajan. Dynamic linear time temporal logic. *Annals of Pure and Applied Logic*, 96(1-3):187–207, 1999.
14. A. A. E. Kalam, R. E. Baida, P. Balbiani, S. Benferhat, F. Cuppens, Y. Deswarte, A. Miège, C. Saurel, and G. Trouessin. Organization based access control. In *IEEE 4th International Workshop on Policies for Distributed Systems and Networks (Policy 2003)*, Lake Come, Italy, June 2003.
15. A. Lomuscio and M. Sergot. On multi-agent systems specification via deontic logic. In *Agent Theories Languages, and Architectures*, volume 2333 of *Lecture Notes in Artificial Intelligence*, Seattle, Springer 2001. Springer Verlag.
16. J.-J. C. Meyer. A different approach to deontic logic: Deontic logic viewed as a variant of dynamic logic. *Notre Dame Journal of Formal Logic*, 1988.
17. J.-J. C. Meyer, R. Wieringa, and F. Dignum. The role of deontic logic in the specification of information systems. In *Logics for Databases and Information Systems*, pages 71–115, 1998.
18. A. Pnueli. The temporal semantics of concurrent programs. *Theoretical Computer Science*, 13:45–60, 1981.
19. J.-P. Queille and J. Sifakis. Specification and verification of concurrent systems in cesar. In *Proceedings of the 5thInternational Symposium on Programming (SOP'82)*, volume 137 of *Lecture Notes in Computer Science*, pages 337–351, 1982.
20. L. van der Torre, J. Hulstijn, M. Dastani, and J. Broersen. Specifying multiagent organizations. In *Seventh International Workshop on Deontic Logic in Computer Science (DEON'04)*, volume 3065 of *Lecture Notes in Computer Science*, pages 243–257, 2004.
21. R. J. Wieringa and J.-J. C. Meyer. *Applications of Deontic Logic in Computer Science: A Concise Overview*, pages 17–40. John Wiley & Sons, 1993.
22. B. Wożna, A. Lomuscio, and W. Penczek. Bounded model checking for deontic interpreted systems. In *Electronic Notes in Theoretical Computer Science*, volume 126, pages 93–114, march 2005.
23. G. H. V. Wright. Deontic logic. *Mind*, 1951.

Speech Acts with Institutional Effects in Agent Societies

Robert Demolombe[1] and Vincent Louis[2]

[1] ONERA Toulouse
France
Robert.Demolombe@cert.fr

[2] France Telecom Research & Development
Lannion, France
vincent.louis@francetelecom.com

Abstract. A general logical framework is presented to represent speech acts that have institutional effects. It is based on the concepts of the Speech Act Theory and takes the form of the FIPA Agent Communication Language.

The most important feature is that the illocutionary force of all of these speech acts is declarative. The formal language that is proposed to represent the propositional content has a large expressive power and therefore allows to represent a large variety of speech acts such as: to empower, to appoint, to order, to declare,...etc.

The same formal language is also used to express the feasibility preconditions, the illocutionary effects and the perlocutionary effects.

1 Introduction

Agent communication languages play an important role for interactions between electronic institutions, in particular for electronic commerce [8, 9]. These languages must have a well defined semantics, and they have to be based on concepts as close as possible to those which are used to define communication in natural language, in order to have an intuitive semantics. That is why Speech Act Theory [26] and the concept of institutional fact [5, 22, 21] are recognized as a good framework for this purpose.

In this paper we investigate the formalization of speech acts that have institutional effects in agent societies such as: to create an obligation, to assign a role to an agent, to accept an offer or to declare that auctions are open.

The context of our work is the formalization of interactions between electronic agents, and we concentrate on actions that are communicative acts.

The Speech Act Theory, that has been defined by Searle in [26] and formalized by Searle and Vanderveken in [27], has been already applied to electronic agents, and the semantics of actions such as *inform* and *request* has been formalized in modal logic in the definition of the FIPA Action Communication Language [16]. This formalization was initiated by Sadek's work presented in [25].

It does not seem to be an over simplification to apply concepts of the Speech Act Theory to electronic agents when beliefs are assigned to electronic agents.

L. Goble and J.-J.C. Meyer (Eds.): DEON 2006, LNAI 4048, pp. 101–114, 2006.

However, it is more problematic to assign to them intentions, and it is not clear whether it is sensible to talk about obligations and institutional powers for electronic agents. Indeed, obligations, and norms in general, are intended to influence the behaviour of agents who have some free will.

Some authors, such as McCarthy in [24], do not reject the idea that electronic agents may have a free will, like human agents, and that they can really choose their intentions or that they can choose to violate, or to fulfil, an obligation.

We do not pretend here to give an answer to the philosophical question of free will for electronic agents, and we accept, as an assumption, that, with respect to the rules of a given institution, electronic agents can be considered like human agents.

A justification for this assumption is that electronic agents can be seen as representative of human agents, in a similar way as human agents can represent institutional agents, as it is proposed by Carmo and Pacheco in [2]. Then, we can assume that the actions performed by electronic agents are determined and chosen, explicitly or implicitly, by human agents. So, in our approach electronic agents' actions count as human agents' actions.[1]

Now, if we ask the question: *what will happen if an electronic agent has violated an obligation?*, the answer is that the human agent, who is represented by the electronic agent, will have to repair the violation. Indeed, if an agent has to pay a penalty, to repair a violation we can imagine that it will be possible to decrease the amount of the electronic agent's account (forget the question of what it means that an electronic agent *owns* an account. However if an agent has to go to jail to repair a violation, it is obvious that the electronic agent will not be able to repair the violation himself.

We are perfectly aware that there are difficult issues about the relationships between electronic agents and human agents in terms of responsibility. For instance, assume an electronic agent violates an obligation because he does not what he is supposed to do according to his specification. Such a case occurs in particular when there is an error in the software. Which human agent is then responsible?

However, in this paper we leave open the problems related to these issues, and we just reason about electronic agents like about human agents.

The structure of the paper is as follows. We start with an informal analysis of the components of speech acts with institutional effects. Then, in section 3, we present a formalization of each component within a logical system. In section 4, we compare our proposal to some similar work. Finally, the conclusion summarizes the main results and sketches future work.

2 Informal Analysis

In this work we do not consider all the subtleties of a speech act definition as presented in [26] and we will restrict ourselves, like in the FIPA Agent Communication Language, to the following features:

[1] We take "count as" in the same sense as Searle in [26] or Jones and Sergot in [20].

- the illocutionary force,
- the propositional content,
- the feasibility preconditions,
- the illocutionary effects,
- the perlocutionary effects.

The agent who plays the role of the speaker is called in the following the "sender", and the agent who plays the role of the hearer is called the "receiver". In general the sender and the receiver are called i and j, respectively.

2.1 Illocutionary Force

The illocutionary force is determined by the direction of fit between the words and the world. The kind of speech acts we want to consider here are those that create institutional facts. In other words, their performance "*have the function [...] of affecting institutional state of affairs*", as K. Bach writes in the entry "speech act" of the Routledge Encyclopedia of Philosophy Online (version 2.0). Such speech acts satisfy the double direction of fit, and, for that reason, their illocutionary force is **declarative**.

The kind of institutional facts we have in mind can be made more concrete with a series of examples. However, we feel quite important to distinguish facts that are represented by descriptive sentences from facts that are represented by normative sentences.
Examples of "descriptive institutional facts" are:

1. the auctions are open.
2. agent j holds the role of salesman.
3. agent j has the institutional power to open the auctions.

Examples of "normative institutional facts" are:

4. agent j has the obligation to pay the bill of the hotel.
5. it is obligatory to have a credit card.
6. agent j has the permission to sell wine.
7. agent j is prohibited to sell cocaine.

It seems clear that the illocutionary force of a speech act which would create institutional facts that do not refer to the performance of some action by the receiver (namely 1, 2, 3 and 5, in these examples), is declarative.

The creation of institutional facts that refer to the performance of an action by the receiver, like 4, raises the question: *Is the illocutionary force of the corresponding speech acts really declarative, or directive?*

Indeed, one could argue that in example 4, i's intention is that j pay the bill. That is the case, for example, if i is the cashier of an hotel, and i gives the bill to a client j and says: "*you have to pay the bill!*".

But we can also argue that i's intention is that it be obligatory for j to pay the bill. For instance, the cashier gives the bill, which is an official document, to the client because his intention is that the client know that his statement is not just a request but rather an order, and that by doing so it is obligatory for the client to pay.

Of course, it is true that i's intention is not just to create the obligation to pay, he also has the intention to be paid. But i believes that this obligation is a more efficient means of being paid than just asking j to pay.

Indeed, if j refuses to pay, he knows that he violates an obligation and that he shall have to pay a penalty. i knows that j knows that, and this gives comfort to i about the payment. Moreover, if the threat of the penalty is not enough to influence j's behaviour, then i can ask to a lawyer, or a policeman, to force j to pay, and i expects that such representatives of the institution will be more successful than he would have been if he had only requested j to pay.

Our proposal is that, in cases like 4, the speech act has two perlocutionary effects: (1) the establishment of an institutional fact (in that example, the obligation for j to pay), and (2) the performance of some action by the receiver (in that example, the payment of the bill), which we call a "secondary perlocutionary effect".

In example 6, i's intention is to give j permission to perform the action (to sale wine), but it is not i's intention that j perform the action. In example 7, it is obvious that i's intention is not that j perform the action (to sale cocaine). Thus, in cases 6 and 7, it is clear that the illocutionary force of the act is declarative.

2.2 Propositional Content

The propositional content represents the institutional fact to create by performing the speech act. More precisely, this representation can be decomposed into a reference to an institution, the propositional content itself, and possibly some particular circumstances required for the speech act to obtain its institutional effect.

We have considered several types of propositional contents that are relevant in the context of applications such as electronic commerce, but, depending on the application domain, the list below may be extended.

For propositional contents that represent descriptive institutional facts, we have the following types of propositional contents:

- Propositional contents that represent situations where some actions count, or do not count, as institutional actions. A typical example is a situation where auctions are open. In this situation, bids have an institutional status. Another example is a situation where a given service is proposed. In this situation, under certain conditions, a request to the server creates some obligations. In natural language, the speech acts that create these situations can be called: "**to open**" or "**to close**".
- Propositional contents that represent situations where an agent holds, or does not hold, a role. For example, agent j holds, or does not hold, the role of salesman. In natural language, the speech acts that create these situations can be called: "**to appoint**" or "**to dismiss**".

- Propositional contents that represent situations where an agent has, or does not have, some institutional power. For example, agent j has, or does not have, the institutional power to open the auctions. In natural language, the speech acts that create these situations can be called: **"to empower"** or **"to remove some power"**.

Propositional contents that represent normative institutional facts may actually represent obligations, permissions or prohibitions. When considering norms about actions, like "obligations to do", the speech acts can respectively be called in natural language: **"to order"**, **"to permit"** or **"to prohibit"**. Similar verbs can be used when considering "obligations to be".

In addition to the propositional content itself, the circumstances under which the institutional facts to create are acknowledged by the institution as a regular consequence of the speech act performance, have to be mentioned. In the previous example of the cashier and the client, the fact that the client has actually stayed for a night at the hotel and that the rates are officially displayed, are implicit circumstances or conditions, which make the cashier's order valid with respect to the law. This order can therefore be put into words as: "whereas you stayed for a night and whereas the official rate is such and such an amount, I order you to pay this bill". If such conditions are not satisfied, for instance if the client has not stayed at the hotel, the speech act makes no sense.

Finally, the intuitive meaning of our proposed speech act with institutional effects may be expressed in a more complete form as: "the sender declares to the receiver his willing to change the institutional state of affairs, given the fact that some conditions, which empower him to create this state of affairs from the institution point of view, are satisfied".

2.3 Feasibility Preconditions

The sincerity precondition is that i believes that he has the institutional power to create the institutional fact represented by the propositional content of the speech act, and also believes that the conditions required to exercise this power hold.

Note that there is a significant difference here between to order to do an action (which is considered as a declarative), and to request to do an action (which is considered as a directive). Indeed, if i requests to j to do α, a sincerity precondition is that i believes that j has the capacity to do α, while, if i orders to j to do α, there is no such precondition, because, as we have mentioned before, i's intention in performing this action is to create the obligation to do α.

For instance, if we consider again the example of the cashier and the client, the cashier's intention to be paid is independent of the fact that the client is able to pay. That is why the fact that the client is able to pay is not a sincerity precondition.

The relevance precondition, like for other speech acts, is that i does not believe that the perlocutionary effect already holds.

2.4 Illocutionary Effect

In a first approach, we can say that the illocutionary effect is that j (the receiver) believes that i'intention is that the propositional content holds.

However, it is a bit more complicated if we consider another agent k who observes (receives) the speech act. In that case, the illocutionary effect on k is that k believes that i's intention is that j believes what we have mentioned just before.

2.5 Perlocutionary Effect

One of the perlocutionary effects is that the institutional fact represented by the propositional content holds. Another perlocutionary effect is that the receiver j believes that this fact holds.

For instance, in the example of the cashier and the client, the fact that it be obligatory for the client to pay is not enough. Another significant effect is that the client is informed about this obligation. It is the same, for example, if the perlocutionary effect is to appoint someone to some position.

According to the previous discussion about the illocutionary force, we will distinguish the "*direct perlocutionary effect*" from the "*indirect perlocutionary effect*".

There is an indirect perlocutionary effect only when the meaning of the speech act is to order to do an action. In that case, the indirect effect is that this action has been done.

3 Formalization

We adopt the FIPA Agent Communication Language structure[2] for speech act definitions. A speech act a with institutional effects is formally defined by the following components:

$a \;=\; < i, Declare(j, D_s n, cond) >$
FP $= p$
DRE $= q_1$
IRE $= q_2$

where:

- i is the sender,
- j is the receiver,
- s is the institution,
- n is a formula that represents the propositional content,
- *cond* is a formula that represents a condition,
- p is a formula that represents the feasibility preconditions,

[2] The only difference is that we have two perlocutionary effects.

- q_1 is a formula that represents the direct perlocutionary effects,[3]
- q_2 is a formula that represents the indirect perlocutionary effects.

Such a speech act means that the sender i declares to the receiver j he intends, by performing this act, to create the institutional fact n with respect to the institution s, given the fact that this institution empowers him to do so in the context where the condition *cond* holds.

3.1 Formal Language and Its Semantics

The syntax of the language used to express the formulas n, p, q_1 and q_2 is defined as follows.

Language L_0. L_0 is a language of a classical first order logic.
Language L. If i is the name of an agent, s is the name of an institution, α is the name of an action and p and q are formulas in L_0 or L, then $B_i p$, $E_i p$, $done_i(\alpha, p)$, Op, $Obg_i(\alpha < p)$, $Perm_i(\alpha < p)$, $Proh_i(\alpha < p)$, $D_s p$, $(\neg p)$, $(p \vee q)$ and $(p \Rightarrow_s q)$ are in L.

The reason why L is built upon L_0 is to avoid complications due to quantifiers outside the scope of the modal operators (see [14]).

The intuitive meaning of the modal operators of language L, including the non standard connective $\Rightarrow_s$, are:

$B_i p$: agent i believes p.
$E_i p$: agent i has brought it about that p.
$done_i(\alpha, p)$: agent i has just done the action α, and p was true just before the performance of α.
Op: it is obligatory that p.
$Obg_i(\alpha < p)$: it is obligatory that i performs α before p becomes true.
$Perm_i(\alpha < p)$: it is permitted that i performs α before p becomes true.
$Proh_i(\alpha < p)$: it is prohibited that i performs α before p becomes true.
$D_s p$: in the context of the institution s, we have p.
$p \Rightarrow_s q$: in the context of the institution s, p counts as q.

The other logical connectives: $\wedge$, $\rightarrow$ and $\leftrightarrow$, are defined in function of $\neg$ and $\vee$ as usual. The permission and prohibition to have p are defined in function of Op as usual.

We have introduced the operators $Obg_i(\alpha < p)$, $Perm_i(\alpha < p)$ and $Proh_i(\alpha < p)$ because obligations to do make sense only if there is an explicit deadline (here expressed as "when p becomes true") to check if they are violated.

We leave open the possibility to consider actions as atomic actions or as complex actions, structured with the standard constructors: sequence, non deterministic choice, test,...etc.

[3] In FIPA definitions, the perlocutionary effect of a speech act is called "rational effect" in order to highlight its understanding as the formal reason for which a speech act is selected in a planning process. In this paper, we maintain this appellation by choosing the notations DRE and IRE to refer respectively to direct and indirect perlocutionary effects.

We use the following notations:

$done_i(\alpha) \stackrel{\text{def}}{=} done_i(\alpha, true)$

$power(i, s, cond, \alpha, f) \stackrel{\text{def}}{=} (cond \wedge done_i(\alpha)) \Rightarrow_s f$, where $cond$ and f are formulas in L.

The meaning of $power(i, s, cond, \alpha, f)$ is that, in the context of the institution s, the agent i has the power to create a situation where f holds by doing the action α in circumstances where we have $cond$.

Speech acts whose intuitive meaning are: *to open* or *to close*, are respectively represented by a propositional content n of the form: p or $\neg p$, where p is a formula of L_0.

If $holds(i, r)$ is a predicate that means that the agent i holds the role r, the speech acts *to appoint* or *to dismiss* are respectively represented by a propositional content n of the form: $holds(i, r)$ or $\neg holds(i, r)$.

The speech acts: *to empower* or *to remove some power*, are respectively represented by a propositional content n of the form: $power(i, s, cond, \alpha, f)$ or $\neg power(i, s, cond, \alpha, f)$.

The speech acts: *to order*, *to permit* or *to prohibit* to do an action α before a deadline d, are respectively represented by a propositional content of the form: $Obg_i(\alpha < d)$, $Perm_i(\alpha < d)$ and $Proh_i(\alpha < d)$.

In general, the language L allows to define speech acts that have more complex meaning than those expressed by usual verbs of the natural language.

It is not the main topic of this work to define a formal semantics for the modal operators involved in the language L. Then, we only give brief indications about their semantics and we adopt, when it is possible, quite simple definitions.

For the epistemic operator B_i, we adopt a KD system, according to Chellas terminology [3]. The dynamic operator $done_i$ is defined as a variant and a restriction (see [23]) of the Dynamic Propositional Logic defined by Harel in [19]. The dynamic operator E_i is defined by a system with RE, C, ¬N and T.

For the "obligation to be" operator O, we adopt the Standard Deontic Logic, that is a KD system. For the "obligation to do" operators Obg_i, $Perm_i$ and $Proh_i$, we adopt the semantics defined in [7], which is an extension of the Dynamic Deontic Logic defined by Segerberg in [28].

Finally, to reason about institutional facts, we adopt, for the operator D_s and for the connective $\Rightarrow_s$, the semantics defined by Jones and Sergot in [20].

3.2 Components of a Speech Act with Institutional Effects

Now, we can formally define the components of a speech act with institutional effects.

Propositional content

The propositional content is represented by the two parameters $D_s n$ and $cond$ where n and $cond$ are formulas in L.

Feasibility preconditions
The sincerity precondition expresses that (1) i believes that he has the institutional power to create the institutional fact represented by $D_s n$ by doing the speech act a in some circumstances described by the formula *cond*, and that (2) i believes that these circumstances hold in the current situation. This is represented by the formula: $B_i(power(i, s, cond, a, D_s n) \wedge cond)$.

The relevance precondition is represented by the formula: $\neg B_i D_s n$. Then we have:

$$\text{FP} = B_i(power(i, s, cond, a, D_s n) \wedge cond) \wedge \neg B_i D_s n$$

Illocutionary effect
The fact that j believes that i's intention is that $D_s n$ holds is represented by: $B_j I_i D_s n$. And the fact that an observer k believes that this is i's intention is represented by: $B_k I_i B_j I_i D_s n$.

Then the illocutionary effect E is:

$$\text{E} = B_k I_i B_j I_i D_s n$$

Perlocutionary effects
The direct perlocutionary effect is that $D_s n$ holds, and that j believes that $D_s n$ holds. Then, we have:

$$\text{DRE} = D_s n \wedge B_j D_s n$$

The indirect perlocutionary effect depends on the propositional content n. For example, if n is of the type $Obg_k(\alpha < d)$, where k can be either the sender i or the receiver j, the indirect perlocutionary effect is represented by $done_k(\alpha < d)$. Note that if k is the sender the meaning of the speech act is a commitment. In general we have:

IRE =
- $done_k(\alpha < d)$, if $n = Obg_k(\alpha < d)$,
- $\neg done_k(\alpha < d)$, if $n = Forb_k(\alpha < d)$,
- $true$, in other cases.

The direct perlocutionary effect $D_s n$ is obtained if i has the relevant institutional power $power(i, s, cond, a, D_s n)$, and if we are in circumstances where *cond* holds and the speech act a has been performed, that is when we have:

$$power(i, s, cond, a, D_s n) \wedge cond \wedge done_i(a)$$

In a similar way, the direct perlocutionary effect $B_j D_s n$ is obtained if we have:

$$B_j(power(i, s, cond, a, D_s n) \wedge cond \wedge done_i(a))$$

The indirect perlocutionary effect $done_k(\alpha < d)$ is obtained if k has adopted the intention to do α before d, and if he has the ability to do α. We have not expressed these conditions in formal terms here because the formal representation of ability by itself is a non trivial problem (see [11]).

Note that, even in the case of a commitment, that is when k is the sender i, it may happen that the conditions to reach the perlocutionary effect are not satisfied. For instance, in the example of the cashier and the client, if the locutionary act performed by the client is, for example, to sign an official document where he declares that he will pay the bill before the end of the week, it may happen that, even so, he has not really the intention to pay or that he has not the ability to pay.

The indirect perlocutionary effect $\neg done_k(\alpha < d)$ is obtained if k has adopted the intention to refrain to do α until d, and if he has the ability to do so.

4 Comparison with Other Work

There is a very limited number of papers that have proposed a formalization of speech acts with institutional effects.

In [10], Dignum and Weigand consider speech acts that have the effect to create obligations, permissions and prohibitions. Their analysis is also based on the concepts of Speech Acts Theory.

A significant difference with our work is that, in their approach, the illocutionary force of the speech acts is directive. Another difference is that the perlocutionary effects are obtained if the sender has the power to order to the receiver to perform some action, or if the receiver has authorized the sender to order to do this action. This second type of relationship between the sender and the receiver is quite different of the first one, which is close to an institutional power. Then, in our view, the status of the obligations created in the second context is not clear because we do not know if this obligation counts as an obligation with regard to some institution.

From a technical point of view this work has some weaknesses. In particular, there are two distinct operators (DIR_a and DIR_p) to represent speech acts that have the same illocutionary force. Also, the authors consider assertive speech acts, but the distinction between directives and assertives is not perfectly clear. Finally, axioms of the form: $[DIR(i,j,\alpha)]I(i,\alpha)$, which mean that after i has requested j to do α, necessarily i intends α to be done, say that $I(i,\alpha)$ is an effect of $DIR(i,j,\alpha)$, while it is a feasibility precondition.

We can also see that the expressive power of their logic is more limited than ours. For instance, the action $DIR_p(i,j,\alpha)$ is represented in our framework as a special case of speech act with institutional effects of the form: $< i, Declare(j, D_s(Obg_j(\alpha < true), true) >$. In addition, in their framework there is no way to specify the institution s.

In [13], Firozabadi and Sergot have introduced the operator $Declares_i n$, whose meaning is that the agent i declares that n, and where n is supposed to be an institutional fact. They also have defined the operator $Pow_i n$, which means that the agent i is empowered to create the institutional fact n. The relationship between the two operators is defined by the property:

$$[DECL] \vdash Declares_i n \wedge Pow_i n \rightarrow n$$

where $[DECL]$ "*expresses the exercise of a power to create* n *by designated agent* i". There is a deep analogy between this relationship and the following property that holds in our framework:

$$\vdash cond \wedge done_i(a) \wedge power(i, s, cond, a, n) \rightarrow D_s n$$

where the speech act a is $< i, Declare(j, n) >$.

There are some minor technical differences. In $Declares_i n$ there is no reference to the addressee of the speech act. And the institutional power $Pow_i n$ is independent of the context (there is no condition *cond*).

A more significant difference with our work is that there is no distinction between what we have called the primary and the secondary perlocutionary effects. Maybe this distinction is ignored because the authors consider a particular application domain where n only represents either the permission or the prohibition to do an action (for instance, the permission or prohibition for an agent to read a file). Then, they can assume that the sender's intention is just to create a new normative situation. Another difference is that feasibility preconditions are not mentioned.

In [4], Cohen and Levesque show how performatives can be used as requests or assertions, but they do not consider the creation of institutional facts.

In [18], Fornara, Vigan and Colombetti claim that all the communicative acts can be defined in terms of declarations. They have defined a formal syntax for an Agent Communication Language that refers to the concepts of the Speech Act Theory and of institutions. For each type of communicative act are defined the preconditions and postconditions. But these conditions are different of the feasibility preconditions and perlocutionary effects. Moreover, there is no formal logic to define the semantics of this language.

In [12], El Fallah-Segrouchni and Lemaitre informally analyse the different types of communicative interactions between electronic agents, or groups of electronic agents, who represent companies. However, the formal contribution of their work is limited to the formal definitions of obligations to do for groups of agents.

In this paper, we have presented an extension of the FIPA ACL. It was not our purpose to compare the FIPA approach, which refers to agents' mental states, to other approaches of agent communication languages, in particular the ones proposed by authors like Singh [29, 30], Colombetti et al. [6, 17, 18] and Chaib-draa and Pasquier [15], which refer to the notion of social commitment.

In [18] the authors write: "*the main advantage of this approach* [social commitment] *is that commitments are objective and independent of agent's internal structure, and that it is possible to verify whether an agent is behaving according to the given semantics*".

We would like to point out that in our proposal agents can create commitments, and many other normative positions of agents, like prohibitions and permissions. Also, it is possible to check whether a speech act has actually created such normative positions. Indeed, that depends on the fact that the speaker has the corresponding institutional power, and this can be checked in the context of a given institution.

For instance, in the context of the institution of e-commerce, the fact that an agent has paid for a given service counts as the fact that he has the right to use this service. Then, by asking to use this service (this can be expressed in our framework by a declare act with the appropriate propositional content) the agent can create the obligation for the server to provide him with this service.

In that example, we see that in order to check whether the obligation has been created, an observer has to check whether the agent has paid for this service, and that raises no specific problem. Moreover, after the obligation has been created, it is possible to check whether the obligation has been fulfilled by the server, i.e. whether the service has been delivered.

However, there is no means to check, for example, the agent's sincerity, or to check whether the agent's intention was to create the rational effect of a given speech act. But, even if there is some degree of uncertainty about these mental states, they can be quite useful in the perspective of plan generation and intention recognition.

5 Conclusion

We have presented a general formal definition of speech acts whose intended effects are to create institutional facts. The original aspect of our work is that all of them, including orders, are considered as declaratives. Another significant aspect is that the formalization is perfectly compatible and homogeneous with the formalization of assertives and directives in the FIPA Agent Communication Language framework. Then, the results can be seen as a proposal for an extension of this language.

In another context (not discussed within this paper), we have checked the practical usability of our approach with the Letter Credit procedure presented in [1]. This procedure is supposed to guarantee that a customer that has bought some goods will receive the goods, and that the supplier will be paid for that. The procedure is a bit complex and involves three other agents: the carrier, the issuing bank and the corresponding bank. We did not find any difficulty to represent the procedure in terms of speech acts. For example, the procedure involves an action of the type "notification", to officially inform the customer that the carrier has carried on the good at its destination. This can be easily expressed with a propositional content of the form: $D_s B_j(goods.are.arrived)$.

In further work, we will investigate how the axioms that determine the planning of speech acts by a rational agent have to be adapted to this type of speech act.

References

1. G. Boella, J. Hulstin, Y-H. Tan, and L. van der Torre. Transaction trust in normative multi agent systems. In *AAMAS Workshop on Trust in Agent Societies*, 2005.
2. J. Carmo and O. Pacheco. Deontic and action logics for collective agency and roles. In R. Demolombe and R. Hilpinen, editors, *Proceedings of the 5th International workshop on Deontic Logic in Computer Science*. ONERA, 2000.

3. B. F. Chellas. *Modal Logic: An introduction.* Cambridge University Press, 1988.
4. P. R. Cohen and H. Levesque. Performatives in a Rationally Based Speech Act Theory. In R. C. Berwick, editor, *Proc. of 28th Annual meeting of Association of Computational Linguistics.* Association of Computational Linguistics, 1990.
5. R. M. Colomb. Information systems technology grounded on institutional facts. In *Workshop on Information Systems Foundations: Constructing and Criticising.* The Australian National University, Canberra, 2004.
6. M. Colombetti and M. Verdicchio. An analysis of agent speech acts as institutional actions. In C. Castelfranchi and W. L. Johnson, editors, *Proceedings of the first international joint conference on Autonomous Agents and Multiagent Systems*, pages 1157–1166. ACM Press, 2002.
7. R. Demolombe, P. Bretier, and V. Louis. Formalisation de l'obligation de faire avec dlais. In *Troisimes Journes francophones Modles Formels de l'Interaction*, 2005.
8. F. Dignum. Software agents and e-business, Hype and Reality. In R. Wieringa and R. Feenstra, editors, *Enterprise Information Systems III.* Kluwer, 2002.
9. F. Dignum. *Advances in Agent Communication.* Springer verlag LNAI 2922, 2003.
10. F. Dignum and H. Weigand. Communication and Deontic Logic. In R. Wieringa and R. Feenstra, editors, *Information Systems, Correctness and Reusability.* World Scientific, 1995.
11. D. Elgesem. *Action Theory and Modal Logic.* PhD thesis, University of Oslo, Department of Philosophy, 1992.
12. A. El Fallah-Seghrouchni and C. Lemaitre. A framework for social agents' interaction based on communicative action theory and dynamic deontic logic. In *Proceedings of MICAI 2002, LNAI 2313.* Springer Verlag, 2002.
13. B. S. Firozabadi and M. Sergot. Power and Permission in Security Systems. In B. Christianson, B. Crispo, and J. A. Malcolm, editors, *Proc. 7th International Workshop on Security Protocols.* Springer Verlag, LNCS 1796, 1999.
14. M. Fitting and R. L. Mendelsohn. *First-Order Modal Logic.* Kluwer, 1998.
15. R. Flores, P. Pasquier, and B. Chaib-draa. Conversational semantics with social commitments. In M-P. Huget R. van Eijk and F. Dignum, editors, *International Workshop on Agent Communication (AAMAS'04)*, 2004.
16. Foundation for Intelligent Physical Agents. FIPA Communicative Act Library Specification. Technical report, http://www.fipa.org/specs/fipa00037/, 2002.
17. N. Fornara and M. Colombetti. Defining interaction protocols using a commitment-based agent communication language. In *Proceedings of the second international joint conference on Autonomous Agents and Multi Agent Systems*, pages 520–527. ACM Press, 2003.
18. N. Fornara, F. Vigan, and M. Colombetti. Agent communication and institutional reality. In R. van Eijk, M. Huget, and F. Dignum, editors, *Developments in Agent Communication.* Springer Verlag LNAI 3396, 2005.
19. D. Harel. Dynamic logic. In D. Gabbay and F. Guenthner, editors, *Handbook of Philosophical Logic*, volume 2. Reidel, 1984.
20. A. J. Jones and M. Sergot. A formal characterisation of institutionalised power. *Journal of the Interest Group in Pure and Applied Logics*, 4(3), 1996.
21. S. O. Kimbrough and S. A. Moore. On automated message processing in Electronic Commerce and Work Support Systems: Speech Act Theory and Expressive Felicity. *ACM Transactions on Information Systems*, 15(4), 1997.
22. S. O. Kimbrough and Y-H. Tan. On lean messaging with unfolding and unwrapping for Electronic Commerce. *International Journal of Electronic Commerce*, 5(1), 2000.

23. V. Louis. *Conception et mise en oeuvre de modèles formels du calcul et du suivi de plans d'actions complexes par un agent rationnel dialoguant.* PhD thesis, Université de Caen, France, 2002.
24. J. McCarthy. Free will - even for robots. *Journal of Experimental and Theoretical Artificial Intelligence*, (to appear).
25. D. Sadek. A study in the logic of intention. In *Proc. of the 3rd Conference on Principles of Knowledge Representation and Reasoning (KR'92)*, 1992.
26. J. R. Searle. *Speech Acts: An essay in the philosophy of language.* Cambridge University Press, New-York, 1969.
27. J. R. Searle and D. Vanderveken. *Foundations of Illocutionary Logic.* Cambridge University Press, Cambridge, 1984.
28. K. Segerberg. Some Meinong/Chisholm thesis. In K. Segerberg and K. Sliwinski, editors, *Logic, Law, Morality. A festrichft in honor of Lennart Aqvist*, volume 51, pages 67–77. Uppsala Philosophical Studies, 2003.
29. M. P. Singh. Social and psychological commitments in multiagent systems. In *AAAI Fall Symposium on Knowledge and Action at Social and Organizational Levels*, 1991.
30. M. P. Singh. A social semantics for agent communication languages. In F. Dignum and M. Greaves, editors, *Issues in Agent Communication*, pages 31–45. Springer Verlag, 2000.

Counts-as: Classification or Constitution? An Answer Using Modal Logic

Davide Grossi, John-Jules Ch. Meyer, and Frank Dignum

Utrecht University,
The Netherlands
{davide, jj, dignum}@cs.uu.nl

Abstract. By making use of modal logic techniques, the paper disentangles two semantically different readings of statements of the type X counts as Y in context C (the classificatory and the constitutive readings) showing that, in fact, 'counts-as is said in many ways'.

1 Introduction

The term "counts-as" derives from the paradigmatic formulation that in [19] and [20] is attributed to the non-regulative component of institutions, i.e., constitutive rules:

> [...] "institutions" are systems of constitutive rules. Every institutional fact is underlain by a (system of) rule(s) of the form "X counts as Y in context C" ([19], pp.51-52).

In legal theory the non-regulative component of normative systems has been labeled in ways that emphasize a classificatory, as opposed to a normative or regulative, character: *conceptual rules* ([2]), *qualification norms* ([18]), *definitional norms* ([13]). Constitutive rules are definitional in character:

> The rules for checkmate or touchdown must 'define' *checkmate in chess* or *touchdown in American Football* [...] ([19], p.43).

With respect to this feature, a first reading of counts-as is thus readily available: it is plain that counts-as statements express classifications. For example, they express what *is classified* to be a checkmate in chess, or a touchdown in American Football. However, is this all that is involved in the meaning of counts-as statements?

The interpretation of counts-as in merely classificatory terms does not do justice to the notion which is stressed in the label "constitutive rule", that is, the notion of *constitution*. Aim of the paper is to show that this notion, as it is presented in some work in legal and social theory, is amenable to formal characterization (in modal logic) and that the theory we developed in [11] provides a ground for its understanding. These investigations stem therefore from the acknowledgment that what is commonly studied under the label "counts-as", hides

L. Goble and J.-J.C. Meyer (Eds.): DEON 2006, LNAI 4048, pp. 115–130, 2006.

in fact (at least) two different, though related, phenomena. Aim of the work is to disentangle these two different meanings and to do it formally. The result will be a formal characterization of both of them together with their logical relations.

The work is structured as follows. In Section 2 we provide an informal analysis of the differences between the classificatory and the constitutive readings and we isolate some constraints concerning in particular the semantics of the constitutive reading. In Section 3 we expose a modal logic of contextual classifications and by means of it we make explicit what a classificatory view of counts-as implies. In Sections 4 and Section 5 a formal characterization of the constitutive view is instead provided and some logical interdependencies between the two readings are derived. Section 6 is devoted to a brief discussion of what we deem to be the most characteristic features of our approach with respect to the literature available on the topic. Conclusions follow in Section 7.

2 Counts-as Between Classification and Constitution

Consider the following inference: it is a rule of normative system Γ that self-propelled conveyances count as vehicles; it is always the case that cars count as self-propelled conveyances; therefore, according to normative system Γ, cars count as vehicles. This is an instance of a typical normative reasoning pattern: from the rule of a given normative system and a common-sense fact, another fact is inferred which holds with respect to that normative system. The count-as locution occurs three times. However, the first premise states a constitutive rule, the second one states a generally acknowledged classification, and the conclusion states a classification which is considered to hold with respect to the given normative system. The formal analysis proposed in this paper moves from the observation that these different occurrences of the same term counts-as denote, in effect, different concepts. Counts-as can be said in many ways, and the aim of the paper is to show that these 'many ways' all have a precise formal semantics.

The distinction we are going to focus on can be distilled in the following observation: counts-as statements used to express constitutive rules have a different meaning from the counts-as statements which are instead used for expressing what follows from the existence of a constitutive rule. We call the first ones *constitutive counts-as statements*, and the second ones *classificatory counts-as statements*. We will see (Proposition 5) that the formal counterpart of the above reasoning pattern is a validity of the modal logic framework that we are going to introduce for capturing and analyzing this distinction.

2.1 The Classificatory Reading of Counts-as

According to a classificatory perspective on the semantics of counts-as, such as the one we investigated in [9,10,11], the fact that A counts-as B in context c is interpreted as "A is a subconcept of B in context c". In other words, counts-as statements are read as contextual classifications.

A notion of context is necessary because classifications holding for a normative system are not of a universal kind, they do not hold in general. What does this mean? The set of classifications stated as constitutive rules by a normative system (for instance, "self-propelled conveyances count as vehicles") can be seen as exactly the set of situations (possible worlds) which make that set of classifications true. Hence, the set of constitutive rules of any normative system can be seen as a set of situations. And a set of situations is what is called a context in much literature on context theory (see for instance [22, 7]). To put it in a nutshell, a context is a set of situations, and if the constitutive rules of a given normative system Γ are satisfied by all and only the situations in a given set, then that set of situations is the context defined by Γ[1]. This simple observation allows us to think of contexts as "systems of constitutive rules" ([19], p.51). Getting back to the above example: the statement "according to Γ, cars count as vehicles" is read as "in the set of situations defined by the rules of system Γ, `car` is a subconcept of `vechicle`". These features, here just informally introduced, will be fully captured by the formalism we are going to present in the following sections.

Understanding counts-as statements in classificatory terms sheds light on an essential function of constitutive rules within normative systems, namely the function of specifying the ontology that each normative system presupposes in order to be able to carry out its regulative task ([8]). Counts-as statements describe the ontology that normative systems use in order to distribute obligations, rights, prohibitions, permissions: vehicles are not admitted to public parks (general norm), but then, if cars count as vehicles (classification), cars are not admitted to public parks (specific norm). An analysis in terms of contextual classification captures exactly this type of reasoning patterns enabled by counts-as statements.

2.2 The Constitutive Reading of Counts-as

The classificatory perspective does not exhaust, though, all aspects involved in the meaning of counts-as statements. As the analysis developed in [14] shows, there is something more. According to that work, the fact that A counts-as B in context c is read as a statement to the effect that A represents "conditions for guaranteeing the applicability of particular classificatory categories" [14], in this case the category B in context c. One is not generally entitled to infer B from A, but it is the counts-as itself which guarantees the soundness of that inference. Indeed, if we can say that cars count as vehicles in c, is just because there is a constitutive rule of normative system Γ defining c, which states that self-propelled conveyances are vehicles. Without this constitutive rule the conclusion could not be drawn. As said in [14], the constitutive rule "guarantees" the soundness of the inference. The constitutive reading of counts-as stresses exactly this aspect.

In this view, counts-as statements do not only state contextual classifications, but they state new classifications which would not otherwise hold. This is perfectly in line with what maintained in [19]:

[1] The definition of sets of situations via sets of formulae is a well-known formal phenomenon, which has been object of deep investigations especially in epistemic logic. We will come back to this in more detail in Section 4 and 5.

> [...] where the rule (or systems of rules) is constitutive, behaviour which is in accordance with the rule can receive specifications or descriptions which it could not receive if the rules did not exist [p.35] ([19]).

In other words, A counts as B in context c because, in general, it does not hold that A is classified as B. Otherwise such a constitutive rule would be futile.

Remark 1. Constitutive counts-as statements are classifications which hold with respect to a context (set of situations) but which do not hold in general (i.e., with respect to all situations).

There is yet another feature characterizing the constitutive reading of counts-as. Let us go back to the first premise of our example: it is a rule of normative system Γ that self-propelled conveyances count as vehicles. Being normative systems sets of rules, this means that "self-propelled conveyances are classified as vehicles" is one of the rules specifying Γ. We know that any set of rules defines a context, namely, the context of all and only the situations which satisfy that set of rules, so:

Remark 2. A constitutive counts-as statement is a classification such that:
a) it is an element of the set of rules specifying a given normative system Γ;
b) the set of rules of Γ define the context (set of situations) to which the counts-as statement pertains.

In other words when statements "A counts as B in the context c of normative system Γ" are read as constitutive rules, what is meant is that the classification of A under B is a promulgation of the normative system Γ defining context c.

Finally, let us spend some preliminary words about the relation between the classificatory and the constitutive reading which we should expect to be enabled in a framework capturing both those meanings. As the informal analysis above points out, the classificatory view is somehow implied by the constitutive one: a constitutive counts-as does not only express that A is classified as B in c, but it expresses also that this classification is not a universally valid one, and that it is an element of the the system Γ defining c. A clear logical relation between the two views should therefore be expected.

Remark 3. Constitution implies classification: if A counts as B in a constitutive sense, then A counts as B in a classificatory sense.

Such a logical relation is precisely the ground for the type of reasoning typically involved in the manipulation of constitutive rules. The presence of a constitutive rule entitles the reasoner to apply reasoning patterns which are typical of reasoning with concepts and classifications. This aspect is thoroughly investigated in Section 5.5.

Along the lines just sketched, the work presents a proposal for developing the formal analysis we presented in [11] in order to deliver a unified modal logic framework able to capture both the constitutive and the classificatory views of counts-as.

3 Modal Logic of Classificatory Counts-as

This section summarizes the results presented in [11]. We first introduce the languages we are going to work with: propositional n-modal languages $\mathcal{ML}_n$ ([5]). The alphabet of $\mathcal{ML}_n$ contains: a finite set $\mathbb{P}$ of propositional atoms p; the set of boolean connectives $\{\neg, \wedge, \vee, \rightarrow\}$; a finite non-empty set of n (context) indexes C, and the operator $[\ ]$. Metavariables $i, j, ...$ are used for denoting elements of C. The set of well formed formulas ϕ of $\mathcal{ML}_n$ is then defined by the following BNF:

$$\phi ::= \bot \mid p \mid \neg\phi \mid \phi_1 \wedge \phi_2 \mid \phi_1 \vee \phi_2 \mid \phi_1 \rightarrow \phi_2 \mid [i]\phi.$$

We will refer to formulae ϕ in which at least one modal operator occurs as modalized formulae. We call instead objective formulae in which no modal operator occur and we denote them using the metavariables $\gamma_1, \gamma_2, \ldots$.

3.1 Semantics

Semantics for these languages is given via structures $\mathcal{M} = \langle \mathcal{F}, \mathcal{I} \rangle$, where:

- $\mathcal{F}$ is a CXT multi-frame, i.e., a structure $\mathcal{F} = \langle W, \{W_i\}_{i \in C} \rangle$, where W is a finite set of states (possible worlds) and $\{W_i\}_{i \in C}$ is a family of subsets of W.
- $\mathcal{I}$ is an evaluation function $\mathcal{I} : \mathbb{P} \longrightarrow \mathcal{P}(W)$ associating to each atom the set of states which make it true.

Some observations are in order here. To put it another way, CXT multi-frames can be seen as Kripke frames in which, instead of the family of sets $\{W_i\}_{i \in C}$, a family of accessibility relations $\{R_i\}_{i \in C}$ is given which defines for each world w the same set of accessible worlds W_i. Relations enjoying such a property are called *locally universal*[2]. Such multi-frames model thus n different contexts i which might be inconsistent, if the corresponding set W_i is empty, or global if W_i coincides with W itself. This implements in a straightforward way the thesis developed in context modeling according to which contexts can be soundly represented as sets of possible worlds ([22]).

Satisfaction for modal formulae of these languages is then defined as follows:

$$\mathcal{M}, w \models [i]\phi \text{ iff } \forall\ w' \in W_i : \mathcal{M}, w' \models \phi.$$

Satisfaction of atoms and boolean formulae is omitted and runs as usual. A formula ϕ is said to be valid in a model $\mathcal{M}$, in symbols $\mathcal{M} \models \phi$, iff for all w in W, $\mathcal{M}, w \models \phi$. It is said to be valid in a frame $\mathcal{F}$ ($\mathcal{F} \models \phi$) if it is valid in all models based on that frame. Finally, it is said to be valid on a class of frames F (F $\models \phi$) if it is valid in every frame $\mathcal{F}$ in F.

It is instructive to make a remark about the $[i]$-operator clause, which can be seen as the characterizing feature of the modeling of contexts as sets of worlds[3].

[2] See [11] for a detailed discussion of these frames.

[3] Propositional logics of context without this clause are investigated in [4,3].

It states that the truth of a modalized formula abstracts from the point of evaluation of the formula. In other words, the notion of "truth in a context i" is a *global* notion: $[i]$-formulae are either true in every state in the model or in none. This reflects the idea that what is true or false in a context does not depend on the world of evaluation, and this is what we would intuitively expect especially for contexts interpreted as normative systems: what holds in the context of a given normative system is not determined by the point of evaluation but just by the system in itself, i.e., by its rules.

3.2 Axiomatics

The multi-modal logic that corresponds, i.e., that is sound and complete with respect to the class of CXT multi-frames, is a system we call here $\mathbf{K45^{ij}_n}$. It consists of a logic weaker than the logic $\mathbf{KD45^{ij}_n}$ investigated in [11] in that the semantic constraint has been dropped which required the sets in family $\{W_i\}_{i\in C}$ to be non-empty. As a consequence the D axiom is eliminated. To put it in a nutshell, the system is the very same logic for contextual classification developed in [11] except for the fact the we want to allow here the representation of empty contexts as well. In the knowledge representation setting we are working in, where contexts can be identified with the normative systems defining them, this amounts to accept the possibility of normative systems issuing inconsistent constitutive rules.

Logic $\mathbf{K45^{ij}_n}$ is axiomatized via the following axioms and rules schemata:

(P) all tautologies of propositional calculus
(K) $[i](\phi_1 \rightarrow \phi_2) \rightarrow ([i]\phi_1 \rightarrow [i]\phi_2)$
(4^{ij}) $[i]\phi \rightarrow [j][i]\phi$
(5^{ij}) $\neg[i]\phi \rightarrow [j]\neg[i]\phi$
(MP) $\phi_1,\ \phi_1 \rightarrow \phi_2\ /\ \phi_2$
(N) $\phi\ /\ [i]\phi$

where i, j denote elements of the set of indexes C. The system is a multi-modal homogeneous **K45** with the two interaction axioms 4^{ij} and 5^{ij}. The system is a subsystem of the **EDL** system studied in [17]. The proof of the soundness and completeness of the system with respect to CXT multi-frames can be derived by the proof of the completeness of **EDL** ([17]).

A remark is in order especially with respect to axiomata 4^{ij} and 5^{ij}. In fact, what the two schemata do, consists in making the nesting of the operators reducible which, leaving technicalities aside, means that truth and falsehood in contexts ($[i]\phi$ and $\neg[i]\phi$) are somehow absolute because they remain invariant even if evaluated from another context ($[j][i]\phi$ and $[j]\neg[i]\phi$). In other words, they express the fact that whether something holds in a context i is not something that a context j can influence. This is indeed the kind of property to be expected given the semantics presented in the previous section.

3.3 Classificatory Counts-as Formalized

Using a multi-modal logic $\mathbf{K45^{ij}_n}$ on a language $\mathcal{ML}_n$, the formal characterization of the classificatory view on counts-as statements runs as follows.

Definition 1. (Classificatory counts-as: $\Rightarrow^{cl}_c$)
"γ_1 counts as γ_2 in context c" is formalized in a multi-modal language $\mathcal{ML}_n$ as the strict implication between two objective sentences γ_1 and γ_2 in logic $\mathbf{K45^{ij}_n}$:

$$\gamma_1 \Rightarrow^{cl}_c \gamma_2 \ := \ [c](\gamma_1 \rightarrow \gamma_2)$$

These properties for $\Rightarrow^{cl}_c$ follow.

Proposition 1. (Properties of $\Rightarrow^{cl}_c$)
In logic $\mathbf{K45^{ij}_n}$, the following formulas and rules are valid:

$$\gamma_2 \leftrightarrow \gamma_3 \ / \ (\gamma_1 \Rightarrow^{cl}_c \gamma_2) \leftrightarrow (\gamma_1 \Rightarrow^{cl}_c \gamma_3) \quad (1)$$
$$\gamma_1 \leftrightarrow \gamma_3 \ / \ (\gamma_1 \Rightarrow^{cl}_c \gamma_2) \leftrightarrow (\gamma_3 \Rightarrow^{cl}_c \gamma_2) \quad (2)$$
$$((\gamma_1 \Rightarrow^{cl}_c \gamma_2) \wedge (\gamma_1 \Rightarrow^{cl}_c \gamma_3)) \rightarrow (\gamma_1 \Rightarrow^{cl}_c (\gamma_2 \wedge \gamma_3)) \quad (3)$$
$$((\gamma_1 \Rightarrow^{cl}_c \gamma_2) \wedge (\gamma_3 \Rightarrow^{cl}_c \gamma_2)) \rightarrow ((\gamma_1 \vee \gamma_3) \Rightarrow^{cl}_c \gamma_2) \quad (4)$$
$$\gamma \Rightarrow^{cl}_c \gamma \quad (5)$$
$$(\gamma_1 \Rightarrow^{cl}_c \gamma_2) \wedge (\gamma_2 \Rightarrow^{cl}_c \gamma_3) \rightarrow (\gamma_1 \Rightarrow^{cl}_c \gamma_3) \quad (6)$$
$$(\gamma_1 \Rightarrow^{cl}_c \gamma_2) \wedge (\gamma_2 \Rightarrow^{cl}_c \gamma_1) \rightarrow [c](\gamma_1 \leftrightarrow \gamma_2) \quad (7)$$
$$(\gamma_1 \Rightarrow^{cl}_c \gamma_2) \rightarrow (\gamma_1 \wedge \gamma_3 \Rightarrow^{cl}_c \gamma_2) \quad (8)$$
$$(\gamma_1 \Rightarrow^{cl}_c \gamma_2) \rightarrow (\gamma_1 \Rightarrow^{cl}_c \gamma_2 \vee \gamma_3) \quad (9)$$

We omit the proofs, which are straightforward via application of Definition 1. This system validates all the intuitive syntactic constraints isolated in [14] (validities 1-5). In addition, this semantic-oriented approach to classificatory counts-as enables the four validities 6-9. Besides, this analysis shows that counts-as conditionals, once they are viewed as conditionals of a classificatory nature, naturally satisfy reflexivity (5), transitivity (6), and a form of "contextualized" antisymmetry (7), strengthening of the antecedent (8) and weakening of the consequent (9).

The property of transitivity, in particular, deserves a special comment. In [14] the transitivity of counts-as is accepted, but not with strong conviction: "we have been unable to produce any convincing counter-instances and are inclined to accept it" ([14], p.436). What our approach shows is that once we first proceed to the isolation of the exact meaning we are aiming at formalizing, no room for uncertainty is then left about the syntactic properties enjoyed by the formalized notion: if we intend counts-as statements as contextual classifications, then transitivity must be accepted on the ground of pure logical reasons.

4 Counts-as Beyond Contextual Classification

The previous section has provided a formal analysis of the classificatory view of counts-as (Definition 1), explicating what logical properties are to be expected once such an analytical option on the semantics of counts-as is assumed (Proposition 1). In this section, on the basis of Remark 1 and 2, we develop a formal semantics of the constitutive reading of counts-as.

4.1 From Classification to Constitution

What has to be done is just to give formal clothes to Remarks 1 and 2 stated in Section 2. Let us define the set $\mathbb{T}(X)$ of all formulae which are satisfied by all worlds in a finite set of worlds X in a CXT model $\mathcal{M}$:

$$\mathbb{T}(X) = \{\phi \mid \forall w \in X : \mathcal{M}, w \models \phi\}.$$

and the set $\mathbb{T}^{\rightarrow}(X)$ of all implications between objective formulae which are satisfied by all worlds in a finite set of worlds X:

$$\mathbb{T}^{\rightarrow}(X) = \{\gamma_1 \rightarrow \gamma_2 \mid \forall w \in X : \mathcal{M}, w \models \gamma_1 \rightarrow \gamma_2\}.$$

Obviously, for every X: $\mathbb{T}^{\rightarrow}(X) \subseteq \mathbb{T}(X)$. In the classificatory reading, given a model $\mathcal{M}$ where the set of worlds $W_c \subseteq W$ models context c, the set of all classificatory counts-as statements holding in c, i.e., $\mathbb{CL}(W_c)$, is nothing but the set $\mathbb{T}^{\rightarrow}(W_c)$:

$$\mathbb{CL}(W_c) \equiv \mathbb{T}^{\rightarrow}(W_c).$$

Obviously, $\mathbb{CL}(W_c)$ is a superset of all conditional truths of W, that is, of the "universal" context of model $\mathcal{M}$. This is to say that interpreting counts-as as a mere classification makes it inherit all trivial classifications which hold globally in the model, and in this consists precisely the crux of the failure of contextual classifications in capturing a notion of constitution. The notion of contextual classification is indifferent to what the context *adds* to standard classifications.

This suggests, though, a readily available strategy to give a formal specification of Remark 1: the set of constitutive counts-as statements holding in a context c should be contained in the set $\mathbb{T}^{\rightarrow}(W_c)$ from which all the global classifications are eliminated:

$$\mathbb{CO}(W_c) \subseteq \mathbb{T}^{\rightarrow}(W_c) \setminus \mathbb{T}(W). \tag{10}$$

Intuitively, the set of constitutive counts-as holding in c corresponds to the set of implications holding in c, i.e. the set of classificatory counts-as statements of c, minus those classifications which hold globally.

As to Remark 2, what comes to play a role is the notion of a *definition* of the context of a counts-as statement. A definition of a context c is a set of objective formulae Γ such that, $\forall w \in W$:

$$\mathcal{M}, w \models \Gamma \text{ iff } w \in W_c.$$

that is, the set of forumlae Γ such that all and only the worlds in W_c satisfy Γ.

In practice, we are making use, in a different setting but with exactly analogous purposes, of a well-known technique developed in the modal logic of knowledge, i.e., the interpretation of modal operators on "inaccessible worlds" typical, for instance, of the "all that I know" epistemic logics ([15]). Consequently, when it comes to Remark 2 we do not just refer to a context, but also (even if often implicitly) to the set of formulae (or rules) that define it. From a formal point of view constitutive counts-as should thus be indexed not only by the context they pertain to, but also by the definition of the context:

$$\mathbb{CO}_{\Gamma}(W_c) \subseteq \{\gamma_1 \rightarrow \gamma_2 \mid \gamma_1 \rightarrow \gamma_2 \in \Gamma \text{ and } \forall w(\mathcal{M}, w \models \Gamma \text{ iff } w \in W_c)\} \quad (11)$$

That is to say, the set of constitutive counts-as statements of a context c w.r.t. a definition Γ should be a subset of the set of implications which belong to the set of formulae Γ defining c.

If we take the constraints expressed in formulae 10 and 11 to exhaust the meaning of a constitutive view of counts-as, then the set of constitutive counts-as statements of a given set c w.r.t. a definition Γ can be defined as follows.

Definition 2. (Set of constitutive counts-as in c w.r.t. definition Γ)
The set $\mathbb{CO}_{\Gamma}(W_c)$ of constitutive counts-as statements of a context c defined by Γ is:

$$\mathbb{CO}_{\Gamma}(W_c) := \begin{cases} (\mathbb{T}^{\rightarrow}(W_c) \setminus \mathbb{T}(W)) \cap \Gamma, & \textit{if: } \forall w(\mathcal{M}, w \models \Gamma \textit{ iff } w \in W_c) \\ \emptyset, & \textit{otherwise.} \end{cases} \quad (12)$$

Section 5 is devoted to the development of a modal logic based on this definition and to a detailed analysis of this interpretation of counts-as statements especially in relation with the analytical option of viewing counts-as as mere contextual classificatory statements.

5 Counts-as as Constitution

In the following section a logic is developed which implements the above definition 2. By doing this, all the following requirements will be met at the same time: first, capture the intuitions discussed in Section 2 concerning the intuitive reading of counts-as statements in constitutive terms (Remark 1 and 2); second, maintain the possible worlds semantics of context exposed in Section 3 and developed in order to account for the classificatory view of counts-as; third, account for the logical relation between classificatory and constitutive counts-as (Remark 3).

5.1 Expanding $\mathcal{ML}_n$

Language $\mathcal{ML}_n$ is expanded as follows. The set of context indexes C consists of a finite non-empty set K of m atomic indexes c among which the special context index u denoting the universal context, and their negations $-c$. The following morphological clause is thus needed:

$$k ::= c \mid -c$$

The cardinality n of K is obviously equal to $2m$. We call this language $\mathcal{ML}_n^{u,-}$. Metavariables $i, j, \ldots$ for context identifiers range on the elements of C.

5.2 Semantics

Languages $\mathcal{ML}_n^{u,-}$ are given a semantics via a special class of CXT multi-frames, namely the class of CXT multi-frames $\mathcal{F} = \langle W, \{W_c\}_{c \in C}\rangle$ such that there always exists a $W_u \in \{W_c\}_{c \in C}$ s.t. $W_u = W$. That is, the frames in this class, which we call CXT$^\top$, always contain the global context among their contexts.

The semantics for $\mathcal{ML}_n^{u,-}$ is thus easily obtained interpreting the formulae on models built on CXT$^\top$ frames. The new clauses needed to be added to the definition of the satisfaction relation in Section 3 are the following ones:

$$\mathcal{M}, w \models [u]\phi \text{ iff } \forall\, w' \in W_u : \mathcal{M}, w' \models \phi$$
$$\mathcal{M}, w \models [-c]\phi \text{ iff } \forall\, w' \in W \backslash W_c : \mathcal{M}, w' \models \phi.$$

Intuitively, the first clause states that the $[u]$ operator is interpreted on the universal 1-frame contained in each CXT$^\top$ multi-frame, and the second states that the $[-c]$ operator ranges over the complement of the set W_c on which $[c]$ instead ranges.

In fact, the $[c]$ operator specifies a lower bound on what holds in context c ('something more may hold in i'), that is, a formula $[c]\phi$ means that ϕ *at least* holds in context c. The $[-c]$ operator, instead, specifies an upper bound on what holds in c ('nothing more holds in c'), and a $[-c]\phi$ formula means therefore that ϕ *at most* does hold in c, i.e., ϕ *at least* does hold in the complementary context of c[4].

5.3 Axiomatics

To axiomatize the above semantics an extension of logic $\mathbf{K45_n^{ij}}$ is needed which can characterize also atomic context complementation. The extension, which we call logic $\mathbf{Cxt^{u,\backslash}}$, results from the union $\mathbf{K45_n^{ij}} \cup \mathbf{S5_u}$, that is, from the union of $\mathbf{K45_n^{ij}}$ with the $\mathbf{S5_u}$ logic for the $[u]$ operator together with the interaction axioms $(\subseteq .ui)$ $(\subseteq .uc)$ and $(-.\neg)$ below. The axiomatics is thus as follows:

(P) all tautologies of propositional calculus
$(\mathtt{K}^i)$ $[i](\phi_1 \to \phi_2) \to ([i]\phi_1 \to [i]\phi_2)$
(4^{ij}) $[i]\phi \to [j][i]\phi$
(5^{ij}) $\neg[i]\phi \to [j]\neg[i]\phi$
$(\mathtt{T}^u)$ $[u]\phi \to \phi$
$(\subseteq .ui)$ $[u]\phi \to [i]\phi$
$(\subseteq .uc)$ $[c]\phi \wedge [-c]\phi \to [u]\phi$
$(-.\neg)$ $[-c]v \to \neg[c]v$
(MP) $\phi_1,\ \phi_1 \to \phi_2\ /\ \phi_2$
$(\mathtt{N}^i)$ $\phi\ /\ [i]\phi$

[4] For an extensive discussion of this technique we refer the reader to [17].

where i, j are metavariables for the elements of K, c denotes elements of the set of atomic context indexes C, and v ranges over uniquely satisfiable objective formulae, i.e., "objective" formulae which are true in at most one world.

The interaction axioms deserve some comments. Axiom ($\subseteq .ui$) is quite intuitive. It just says that what holds in the global context, holds in every context. Axiom ($\subseteq .uc$) is needed in order to axiomatize the interplay between atomic contexts and their complements: if some formula holds in both a context and its complement, than it holds globally. Axiom ($-.\neg$) states instead that the contexts denoted by c and $-c$ are strongly disjoint, in the sense that they do not contain the same valuations.

The system is nothing but a multi-modal version of a fragment of the **S5O** system investigated in [17] (that system contained also a "strongly universal context", that is the context of all logically possible valuations), which is in turn an extension of the propositional fragment of the "all I know" logic studied in [15]. We conjecture that the completeness proof can be obtained extending the completeness result concerning **S5O** provided in [17].

5.4 Constitutive Counts-as Formalized

Using a multi-modal logic $\mathbf{Cxt}^{\mathbf{u},\backslash}$ on a language $\mathcal{ML}_n^{u,-}$, the constitutive reading of counts-as statements can be formalized as follows.

Definition 3. (Constitutive counts-as: $\Rightarrow^{co}_{c,\Gamma}$)
Given a set of formulae Γ the conjunction on which is $\Gamma = (\phi_1 \wedge \ldots \wedge (\gamma_1 \rightarrow \gamma_2) \wedge \ldots \wedge \phi_n)$, the constitutive counts-as statement "γ_1 counts as γ_2 in context c" is formalized in a multi-modal logic $\mathbf{Cxt}^{\mathbf{u},\backslash}$ on language $\mathcal{ML}_n^{u,-}$ as:

$$\gamma_1 \Rightarrow^{co}_{c,\Gamma} \gamma_2 := [c]\Gamma \wedge [-c]\neg\Gamma \wedge \neg[u](\gamma_1 \rightarrow \gamma_2)$$

with γ_1 and γ_2 objective formulae.

A detailed comment of the definition is in order. The definition implements in modal logic the intuition summarized in Remark 1 and 2, and formalized in Definition 2: constitutive counts-as correspond to those classifications which are stated by the definition Γ of the context c. Of course, for a given context, we can have a number of different equivalent definitions. The linguistic aspect comes thus into play and that is why $\Rightarrow^{co}_{c,\Gamma}$ statements need to be indexed also with the chosen definition of the context, i.e. with the set of explicit promulgations of the normative system at issue. The warning "no logic of norms without attention to a system of which they form part" ([16], pag. 29) is therefore taken seriously. As a result, constitutive counts-as statements can also be viewed as forms of speech acts creating a context: given that $\gamma_1 \rightarrow \gamma_2$ is a formula of Γ, $\gamma_1 \Rightarrow^{co}_{c,\Gamma} \gamma_2$ could be read as "let it be that $\gamma_1 \rightarrow \gamma_2$ with all the statements of Γ and only of Γ or, using the terminology of [21], "fiat Γ and only Γ".

On the other hand, notice that because of this linguistic component of Definition 3 there is no logic, in a classical sense, of constitutive statements pertaining to one unique context description. That is to say, given a set of $\Rightarrow^{co}_{c,\Gamma}$ statements, nothing can be inferred about $\Rightarrow^{co}_{c,\Gamma}$ statements which are not already contained in the set Γ. How awkward this might sound it is perfectly aligned with the intuitions on the notion of constitution which backed Definition 3: constitutive counts-as are those classifications which are explicitly stated in the specification of the normative system. In a sense, constitutive statements are just given, and that is it. This does not mean, however, that constitutive statements cannot be used to perform reasoning. The following proposition touches upon the logical link relating constitutive statements pertaining to different but equivalent specifications of the same context.

Proposition 2. (Equivalent sets of $\Rightarrow^{co}_{c,\Gamma}$)
Let $\mathcal{M}$ be a $\textsc{Cxt}^{\top}$ *model and Γ and Γ' be context definitions containing no ϕ s.a. $\mathcal{M} \models \phi$. Let Γ contain $(\gamma_1 \to \gamma_2) \wedge (\gamma_3 \to \gamma_4)$, and let Γ' be obtained by Γ substituting $(\gamma_1 \to \gamma_2) \wedge (\gamma_3 \to \gamma_4)$ with a formula $(\gamma_5 \to \gamma_6)$. Then, it is valid that:*

$$\mathcal{M} \models [u](\Gamma \leftrightarrow \Gamma') \to ((\gamma_1 \Rightarrow^{co}_{c,\Gamma} \gamma_2 \wedge \gamma_3 \Rightarrow^{co}_{c,\Gamma} \gamma_4) \leftrightarrow (\gamma_5 \Rightarrow^{co}_{c,\Gamma'} \gamma_6)).$$

This states nothing but that the substitutions of equivalents in the description Γ of a context always yields a set of constitutive $\Rightarrow^{co}_{c,\Gamma'}$ statements which is equivalent with the first set, provided that the the two definitions do not contain globally true statements.

Proof. Given that $[u](\Gamma \leftrightarrow \Gamma')$ and the conditions on Γ and Γ', it follows from the construction of Γ' that $[u](((\gamma_1 \to \gamma_2) \wedge (\gamma_3 \to \gamma_4)) \leftrightarrow (\gamma_5 \to \gamma_6))$. □

Precise logical relations hold also between the two characterizations of Definition 1 and 3 and are investigated in the next section.

5.5 Classification vs Constitution

Since $\mathbf{Cxt}^{\mathbf{u},\backslash}$ is an extension of $\mathbf{K45}^{\mathbf{ij}}_{\mathbf{n}}$ it can be used to represent and reason about, at the same time, both $\Rightarrow^{cl}_{c}$ and $\Rightarrow^{co}_{c,\Gamma}$ formulae. The logical relations between the two are formally explicated and studied in the following three propositions.

Proposition 3. ($\Rightarrow^{cl}_{c}$ vs $\Rightarrow^{co}_{c,\Gamma}$)
In logic $\mathbf{Cxt}^{\mathbf{u},\backslash}$, the following formula is valid for every context definition Γ containing $\gamma_1 \to \gamma_2$:

$$(\gamma_1 \Rightarrow^{co}_{c,\Gamma} \gamma_2) \to (\gamma_1 \Rightarrow^{cl}_{c} \gamma_2) \tag{13}$$

Proof. It follows from Definitions 1 and 3: given that $\gamma_1 \to \gamma_2$ is in Γ, then $\gamma_1 \Rightarrow^{co}_{c,\Gamma} \gamma_2$ implies that $[c]\Gamma$ holds. From which it follows that $[c](\gamma_1 \to \gamma_2)$. □

The proposition translates the following intuitive fact: the statement of a constitutive rule *guarantees*, to say it with [14], the possibility of applying specific classificatory rules. If it is a rule of Γ that self-propelled conveyances count as vehicles (constitutive sense) then self-propelled conveyances count as vehicles in the context c defined by Γ (classificatory sense).

Furthermore, it is noteworthy that formula 13 has a striking resemblance with the syntactic constraint that in [14] was imposed on the conditional minimal model semantics of counts-as conditionals: $A \Rightarrow_c B \rightarrow D_c(A \rightarrow B)$. This constraint was used to relate the logic of counts-as and the multi-modal $\mathbf{KD_n}$ system chosen there as a logic of "general institutional constraints". Such a constraint, though intuitively motivated from a syntactic point of view, appears in that work a bit ad hoc since the choice for the $\mathbf{KD_n}$ logic is not thoroughly investigated and it is even explicitly considered to be a "provisional proposal" ([14], p.437). In our characterization, instead, the constraint emerges as a validity of the system, finding thus in our semantics a strong grounding and a clear formal motivation. It seems then safe to say that what authors in [14] meant under the label "constraint operative in institution c" was nothing but the notion of truth in a context (defined by a given normative system), i.e., what we have here represented via the $[i]$ operator. As a consequence, their strict implication under a $\mathbf{KD_n}$ logic ($D_c(\gamma_1 \rightarrow \gamma_2)$) was in our view an attempt to capture the kind of statements that we investigated in [11] and that we have here called classificatory counts-as statements ($\gamma_1 \Rightarrow^{cl}_c \gamma_2$). A more detailed comparison of the two approaches can be found in [12].

The following result about constitutive counts-as statements also follows.

Proposition 4. (Impossibility of $\Rightarrow^{co}_{u,\Gamma}$)
Constitutive counts-as are impossible in the universal context u. In fact, the following formula is valid for every Γ containing $\gamma_1 \rightarrow \gamma_2$:

$$(\gamma_1 \Rightarrow^{co}_{u,\Gamma} \gamma_2) \rightarrow \bot \tag{14}$$

Proof. The proposition is proved considering Definition 3: if $\gamma_1 \Rightarrow^{co}_{u,\Gamma} \gamma_2$ then $[u](\gamma_1 \rightarrow \gamma_2)$ and $\neg[u](\gamma_1 \rightarrow \gamma_2)$, from which contradiction follows. □

In other words, what holds in general is never a product of constitution. This is indeed a very intuitive property strictly related to what discussed about Remark 1 in Section 2. As a matter of fact, that apples are classified as fruits is not due to any constitutive activity. Asserting that "apple always count as fruits" can only be true in a classificatory sense, while intending it as "apple always constitute fruits" is, logically speaking, nonsense. On the contrary, contextual classification statements are perfectly sound with respect to the universal context since no contradiction follows from sentences such as $\gamma_1 \Rightarrow^{cl}_u \gamma_2$. This is related with the last property we report in this paper:

Proposition 5. (From $\Rightarrow^{co}_{c,\Gamma}$ to $\Rightarrow^{cl}_c$ via $\Rightarrow^{cl}_u$)
The following formula is valid for every Γ containing $\gamma_1 \rightarrow \gamma_2$:

$$(\gamma_2 \Rightarrow^{co}_{c,\Gamma} \gamma_3) \rightarrow ((\gamma_1 \Rightarrow^{cl}_u \gamma_2) \rightarrow (\gamma_1 \Rightarrow^{cl}_c \gamma_3)) \tag{15}$$

Proof. Straightforward from Definition 1, Definition 3, Proposition 3 and the transitivity of classificatory counts-as (Proposition 1). □

Formula 15 represents nothing but the form of the reasoning pattern occurring in the illustrative example with which our informal analysis started in Section 2: if it is a rule of Γ that $\gamma_2 \rightarrow \gamma_3$ ("self-propelled conveyances count as vehicles") and it is always the case that $\gamma_1 \rightarrow \gamma_2$ ("cars count as self-propelled conveyances"), then $\gamma_1 \rightarrow \gamma_3$ ("cars count as vehicles") in the context c defined by normative system Γ. What is remarkable about this property is that it neatly shows how the two senses of counts-as both play a role in the kind of reasoning we perform with constitutive rules: the constitutive sense, though enjoying extremely poor logical properties, enables all the rich reasoning patterns proper of classificatory reasoning.

5.6 Counts-as and the "Transfer Problem"

The two meanings of counts-as behave differently with respect to the so-called transfer problem ([14]). This problem can be exemplified as follows: suppose that somebody brings it about that a priest effectuates a marriage. Does this count as the creation of a state of marriage? In other words, is the possibility to constitute a marriage transferable to anybody who brings it about that the priest effectuates the ceremony?

From the standpoint of classificatory counts-as, the transfer problem is a feature of the formalism because $\Rightarrow_c^{cl}$ enjoys the strengthening of the antecedent (Proposition 1). This is nevertheless what we would intuitively expect when interpreting counts-as statements as contextual classifications: whatever situation in which a priest performs a marriage ceremony, and therefore also the situation in which the priest is, for instance, forced to carry out the ritual, is classified as a situation in which a marriage state comes to be, with respect to the clerical statutes.

This is not the case for the constitutive reading of counts-as statements. In this view, counts-as statements represent the rules specifying a normative system. So, all that it is explicitly stated by the'institution of marriage' is that if the priest performs the ceremony then the couple is married, while no rule belongs to that normative systems which states that the action of a third party bringing it about that the priest performs the ceremony also counts as a marriage. Our formalization fully captures this feature. Let the 'marriage institution' c be represented by the set of rules $\Gamma = \{p \rightarrow m\}$, i.e., by the rule "if the priest performs the ceremony, then the couple is married". Let then t represent the fact that a third party brings it about that p. For Definition 3 the counts-as $(t \wedge p) \Rightarrow_{c,\Gamma}^{co} m$ is just an undefined expression, because $((t \wedge p) \rightarrow m) \notin \Gamma$, that is, because the 'marriage institution' does not state such a classification.

6 Discussion

The main difference between our analysis and the formal approaches available -to our knowledge- in the literature at this moment ([14, 1, 6]) is of a methodological nature. In fact, though using modal logic like also [14, 6], we did not proceed to the isolation of syntactic constraints that seem intuitive on the basis of the common usage of the term "counts-as". Instead, we made some precise interpretations formal of what counts-as statements mean (see Remarks 1-3), following hints from related disciplines such as legal and social theory. We deem the core advantage of this methodology to reside in the fact that the obtained formal characterizations could not be the result of mixing under the the same logical representation (in this case a conditional $\Rightarrow_c$) different semantic flavors that, from an analytical point of view, should be separated. For instance, suppose to establish whether transitivity is a meaningful property for a conditional characterization of counts-as. Tackling this issue syntactically amounts to check, on the basis of the natural usage of the term, the acceptability of the inference: if "A counts as B" and "B counts as C" then "A counts as C". However, it can be case that the sense we naturally attach to the term "counts as" in the first premise is not the same sense we attach to the term "counts as" in the second premise or in the conclusion. This is indeed the case in the example with which our analysis started in Section 2 and which has been formalized in Proposition 5. A syntax-driven approach runs the risk of overlooking these ambiguities risking to import them in the formalization.

A last aspect worth mentioning concerns how our semantic analysis relates to syntax-driven approaches to counts-as which emphasize the defeasibility aspects of counts-as statements such as in particular [6]. Approaches analyzing counts-as as a form of defeasible rule or defeasible conditional deal in fact with something which is *in essence* different from what has been investigated here. In a sense they investigate yet another possible meaning of counts-as statements: "normally (or, unless stated otherwise), A counts as B in context C". Our approach stands to those ones as the classical approach to the formalization of a statement such as "all birds can fly" (`bird` $\subseteq$ `fly`) stands to those approaches formalizing it as a defeasible conditional or rule, that is, as "normally, all birds can fly".

7 Conclusions

The work has aimed at providing a clarification of the different meanings that can be given to counts-as statements. It has shown that there exist at least two senses in which 'counts-as can be said': the classificatory and the constitutive ones (a third one is investigated in [12]), both of which find grounds in legal and social theory literature. The modal logic characterizations of these two senses has made explicit the formal properties yielded by the two semantic options and, more noticeably, the logical relations holding between them.

Acknowledgments

We would like to thank the anonymous reviewers of DEON'06 for their helpful comments, and Prof. Andrew Jones for the inspiring discussions that motivated the research presented in this work.

References

1. G. Boella and L. van der Torre. Attributing mental attitudes to normative systems. In *Proceedings of AAMAS '03*, pages 942–943, New York, NY, USA, 2003. ACM Press.
2. E. Bulygin. On norms of competence. *Law and Philosophy 11*, pages 201–216, 1992.
3. S. Buvač, S. V. Buvač, and I. A. Mason. The semantics of propositional contexts. *Proceedings of the 8th ISMIS. LNAI-869*, pages 468–477, 1994.
4. S. V. Buvač and I. A. Mason. Propositional logic of context. *Proceedings AAAI'93*, pages 412–419, 1993.
5. D.M. Gabbay, A. Kurucz, F. Wolter, and M. Zakharyaschev. *Many-dimensional modal logics. Theory and applications.* Elsevier, 2003.
6. J. Gelati, A. Rotolo, G. Sartor, and G. Governatori. Normative autonomy and normative co-ordination: Declarative power, representation, and mandate. *Artificial Intelligence and Law*, 12(1-2):53–81, 2004.
7. C. Ghidini and F. Giunchiglia. Local models semantics, or contextual reasoning = locality + compatibility. *Artificial Intelligence*, 127(2):221–259, 2001.
8. D. Grossi, H. Aldewereld, J. Vázquez-Salceda, and F. Dignum. Ontological aspects of the implementation of norms in agent-based electronic institutions. In *Proceedings of NorMAS'05.*, pages 104–116, Hatfield, England, April 2005. AISB.
9. D. Grossi, F. Dignum, and J-J. Ch. Meyer. Contextual taxonomies. In J. Leite and P. Toroni, editors, *Post-proceedings of CLIMA V*, LNAI 3487, pages 33–51. Springer-Verlag, 2005.
10. D. Grossi, F. Dignum, and J-J. Ch. Meyer. Contextual terminologies. In F. Toni and P. Torroni, editors, *Post-proceedings of CLIMA VI*, LNAI 3900, pages 284–302. Springer-Verlag, 2006.
11. D. Grossi, J-J. Ch. Meyer, and F. Dignum. Modal logic investigations in the semantics of counts-as. In *Proceedings of the Tenth International Conference on Artificial Intelligence and Law (ICAIL'05)*, pages 1–9. ACM, June 2005.
12. D. Grossi, J-J. Ch. Meyer, and F. Dignum. Classificatory aspects of counts-as: An analysis in modal logic. Under submission, 2006.
13. A. J. I. Jones and M. Sergot. Deontic logic in the representation of law: towards a methodology. *Artificial Intelligence and Law 1*, 1992.
14. A. J. I. Jones and M. Sergot. A formal characterization of institutionalised power. *Journal of the IGPL*, 3:427–443, 1996.
15. H. J. Levesque. All i know: A study in autoepistemic logic. *Artificial Intelligence*, (42):263–309, 1990.
16. D. Makinson. On a fundamental problem of deontic logic. In P. McNamara and H. Prakken, editors, *Norms, Logics and Information Systems. New Studies in Deontic Logic and Computer Science*, pages 29–53. IOS Press, Amsterdam, 1999.
17. J.-J. Ch. Meyer and W. van der Hoek. *Epistemic Logic for AI and Computer Science*, volume 41 of *Cambridge Tracts in Theoretical Computer Science*. Cambridge University Press, 1995.
18. A. Peczenik. *On Law and Reason.* Kluwer, Dordrecht, 1989.
19. J. Searle. *Speech Acts.* Cambridge University Press, Cambridge, 1986.
20. J. Searle. *The Construction of Social Reality.* Free Press, 1995.
21. B. Smith. Fiat objects. *Topoi*, 2(20):131–148, 2001.
22. R. Stalnaker. On the representation of context. In *Journal of Logic, Language, and Information*, volume 7, pages 3–19. Kluwer, 1998.

Don't Ever Do That! Long-Term Duties in PD_eL

Jesse Hughes and Lambèr Royakkers

Eindhoven University of Technology
The Netherlands
{j.hughes, l.m.m.royakkers}@tm.tue.nl

Abstract. This paper studies long-term norms concerning actions. In Meyer's Propositional Deontic Logic (PD_eL), only immediate duties can be expressed, however, often one has duties of longer durations such as: "Never do that", or "Do this someday". In this paper, we will investigate how to amend PD_eL so that such long-term duties can be expressed. This leads to the interesting and suprising consequence that the long-term prohibition and obligation are not interdefinable in our semantics, while there is a duality between these two notions. As a consequence, we have provided a new analysis of the long-term obligation by introducing a new atomic proposition I (indebtedness) to represent the condition that an agent has some unfulfilled obligation.

1 Introduction

The classical deontic logic introduced by von Wright (1951) is based on a set of "ideal" or "perfect" worlds, in which all obligations are fulfilled, and introduces formula-binding deontic operators. Meyer (1988, 1989) instead based deontic logic on dynamic logic by introducing a special violation atom V, indicating that in the state of concern a violation of the deontic constraints has been committed. But there is a deeper difference than this stress of violation over ideal outcomes. Namely, Meyer's PD_eL (Propositional Deontic Logic) is a *dynamic* logic.

Following Anderson's proposal in (1967), Meyer introduced deontic operators to propositional dynamic logic (PDL) as follows: an action α is forbidden in w if doing α in w inevitably leads to violation. Similarly, α is obligatory in w if doing anything *other than* α inevitably leads to violation. In PD_eL, then, duties bind actions rather than conditions: one is obligated to *do* something, rather than bring about some condition.

The benefit from this approach of reducing deontic logic to dynamic logic is twofold. Firstly, in this way we get rid of most of the nasty paradoxes that have plagued traditional deontic logic (cf. Castañeda (1981)), and secondly, we have the additional advantage that by taking this approach to deontic logic and employing it for the specification of integrity constraints for knowledge based systems we can directly integrate deontic constraints with the dynamic ones.

L. Goble and J.-J.C. Meyer (Eds.): DEON 2006, LNAI 4048, pp. 131–148, 2006.

Nonetheless, PD_eL comes with its own limitations, notably in the kinds of ought-to-do statements that can be expressed. In particular, PD_eL's deontic operators express norms about *immediate* rather than *eventual* actions. Certainly, some prohibitions are narrow in scope: "Do not do that *now*." But other prohibitions restrict action more broadly: "Don't *ever* do that" (i.e., at every point in the future, do not do that). Our aim here is to investigate how to amend PD_eL so that such long-term norms can be expressed.

Interestingly, our semantics for long-term obligation is not as closely related to long-term prohibition as one might expect. The essential difference comes in evaluating whether a norm has been violated. A long-term prohibition against α is violated if there is some time at which the agent has done α. Thus, long-term prohibitions can be expressed in terms of reaching worlds in violation. Long-term obligations are different: an obligation to do α is violated just in case the agent *never* does α. But there is no world corresponding to this condition. At each world, the agent may later do α and thus fulfill his obligation. In learning-theoretic terms (Kelly (1996)), violations of prohibitions are verifiable with certainty but not refutable, and dually fulfillment of obligations are verifiable with certainty but not refutable.

Thus, while there *is* a duality between long-term prohibitions and obligations, the two notions are not inter-definable in our possible world semantics. Instead, we must provide a new analysis of obligation that is considerably subtler than Meyer's definition of immediate obligation. We find that the asymmetry between our long-term normative concepts is one of the most interesting and surprising consequences of our investigations.

Our presentation begins with a summary of a somewhat simplified version of PD_eL, introducing Meyer's definitions of (immediate) prohibition and obligation. In Section 3, we introduce our definition of long-term contiguous prohibition, an admonition to never perform a particular sequence of actions one after the other. We also introduce long-term contiguous obligation and explain why inter-definability fails for these two concepts. In Section 4, we briefly discuss non-contiguous variations for prohibitions and obligations. These include prohibitions against doing a sequence of actions in order, but with other actions interspersed (and an analog for obligations).

We close with a few comments about future directions for dynamic deontic logic.

For reasons of space, we have omitted most of the proofs. However, we have given enough properties of the relations and concepts involved so that the missing derivations are simple and straightforward. We have included a few proofs where the reasoning is not obvious and immediate from previous discussion, but our focus here is on semantic appropriateness rather than technical developments.

2 The Basic System PD_eL

We present here a somewhat simplified form of PD_eL. Our presentation is primarily based on Meyer (1988, 1989).

2.1 Actions and Their Interpretations

PD_eL is a dynamic logic aimed at reasoning about duties and prohibitions. It differs from most deontic logics by focusing on actions rather than conditions: Things one ought to do (or not do) rather than conditions one ought to bring about (or avoid). The syntax is similar to other dynamic logics, with complex action terms built from a set A of atomic action terms (or atoms) and complex propositions built from a set of atomic propositions. We use $a, b, \ldots$ to range over A. The semantics, too, are similar to other dynamic logics: models are given by a labeled transition system on a set $\mathcal{W}$ of worlds and actions are interpreted as sets of paths in this transition system.

PD_eL differs from classical PDL (Harel (1984); Meyer (2000)) primarily in the set of action-constructors. In particular, PD_eL includes synchronous composition (doing both α and β at the same time) and negation (doing something *other* than α). Synchronous composition adds a new degree of non-determinism to our semantics, because actions can be specified to greater or lesser degree. If one does α & β, then he has done α, but the converse is not true. On this approach, even atomic action terms are not fully specified: a is interpreted as a set of alternatives, which include doing a and b simultaneously, doing a and c simultaneously, doing a by itself and so on.

Therefore, our semantics comes with an extra step: we interpret actions as sets of sequences of *fully specified atomic actions* (what Meyer calls "synchronicity sets" or "s-sets"). Meyer took his set of fully specified atomic actions to be $\mathcal{P}^+\mathcal{P}^+A$, where $\mathcal{P}^+S$ is the set $\mathcal{P}S \setminus \{\emptyset\}$ of non-empty subsets of S. He maps atoms a to subsets of $\mathcal{P}^+A$ via

$$a \mapsto \{S \subseteq A \mid a \in S\},$$

and hence interpretation of actions is a function $TA \to \mathcal{P}^+\mathcal{P}^+A$. This concrete interpretation is well-motivated, but we prefer a simpler, more flexible and abstract approach. We fix a set X to be our fully specified atomic actions together with a function

$$i : A \to \mathcal{P}^+X,$$

where we recover Meyer's interpretation by choosing $X = \mathcal{P}^+A$ and using the mapping above. Intuitively, the set X provides the alternative ways in which one can do each atomic action term a. An atomic term a may describe the act of whistling, say, but there are many different ways to whistle. One may whistle while walking or whistle while writing a letter; one may also whistle this tune or that. The set X specifies each of these alternatives.

Our semantics is more flexible than Meyer's in the following sense: in Meyer (1989), each pair of atomic action terms can be performed simultaneously, i.e. $[\![a \,\&\, b]\!] \neq \emptyset$, but this is not always reasonable. By choosing X and i appropriately, we allow that some pairs of atomic actions (whistling and chewing crackers, say) cannot be done at the same time. In particular, a and b can be done simultaneously just in case $i(a) \cap i(b) \neq \emptyset$.

To summarize: we fix a set A of atomic action terms, a set X of fully specified atomic actions and a function $i : A \to \mathcal{P}X$. We build a set TA of *action terms* from the elements of A. Each action term will be interpreted as a set of sequences over X, yielding

$$[\![-]\!] : TA \to \mathcal{P}(X^{<\omega}),$$

so $\mathcal{P}(X^{<\omega})$ is our set of *actions*, the interpretations of action terms. The set $[\![\alpha]\!]$ represents the alternative fully specified ways of doing α. Such X-sequences will define a set of paths in our X-labeled transition system on $\mathcal{W}$ and this yields the usual interpretation of the dynamic operator $[\alpha]$, but let us not get ahead of ourselves.

The set TA of action terms is defined by

$$\beta ::= a \mid \underline{\emptyset} \mid \epsilon \mid \textbf{any} \mid \alpha \cup \beta \mid \alpha \,\&\, \beta \mid \alpha;\beta \mid \overline{\beta}$$

The action term $\underline{\emptyset}$ describes the impossible action, ϵ the do-nothing action,[1] **any** the do-any atomic action and **any*** the do-any complex action. As mentioned, $\alpha \,\&\, \beta$ represents simultaneous performance of α and β and $\overline{\beta}$ represents doing *anything but* β. As usual, $\alpha \cup \beta$ represents the non-deterministic choice between α and β and $\alpha;\beta$ for the sequential composition of actions α and β.

Table 1. The interpretation of action terms as sets of X-sequences

Definition of $[\![\alpha]\!]$

$$
\begin{aligned}
[\![a]\!] &= \{\, \langle x \rangle \mid x \in i(a) \,\} \\
[\![\textbf{any}]\!] &= \{\, \langle x \rangle \mid x \in X \,\} \\
[\![\textbf{any}^*]\!] &= X^{<\omega} \\
[\![\underline{\emptyset}]\!] &= \emptyset \\
[\![\epsilon]\!] &= \{\langle\,\rangle\} \\
[\![\alpha \cup \beta]\!] &= [\![\alpha]\!] \cup [\![\beta]\!] \\
[\![\alpha;\beta]\!] &= \{\, r * s \mid r \in [\![\alpha]\!],\, s \in [\![\beta]\!] \,\} \\
[\![\alpha_1 \,\&\, \alpha_2]\!] &= \{\, s \mid s \in [\![\alpha_i]\!] \text{ and } \exists n \,.\, s \restriction n \in [\![\alpha_j]\!],\ j \neq i \,\} \\
[\![\overline{\alpha}]\!] &= \begin{cases} \{\, s * \langle x \rangle \mid \mathsf{cmp}([\![\alpha]\!], s) \wedge \neg\mathsf{cmp}([\![\alpha]\!], s * \langle x \rangle) \,\} & \text{if } [\![\alpha]\!] \neq \emptyset \\ [\![\textbf{any}]\!] & \text{else} \end{cases}
\end{aligned}
$$

Before defining the interpretation $TA \to \mathcal{P}(X^{<\omega})$, we must introduce a bit of terminology for sequences.

Let $|s|$ denote the length of the sequence s. If $n \leq |s|$, the sequence $s \restriction n$ is the prefix of s of length n, i.e.

$$s \restriction n = \langle x_0, x_1, \ldots, x_{n-1} \rangle.$$

If $n \geq |s|$, then $s \restriction n = s$.

[1] This symbol does not occur in other versions of dynamic logic in the literature. It is, however, comparable with the ϵ process in process algebra.

We write $s * t$ for the concatenation of s and t. We say that s is a *prefix* of t if there is some n such that $s = t \upharpoonright n$, equivalently there is some r such that $t = s * r$, and s is a *proper* prefix if the chosen r is not the empty sequence $\langle\rangle$. Two sequences are *comparable* if one is a prefix of the other. If S is a set of sequences, we define

$$\mathsf{cmp}(S, s) \Leftrightarrow \exists t \in S \ . \ t \text{ is comparable to } s.$$

A set S of sequences is *n-uniform* iff every sequence s in S has length n. If S is n-uniform for some n, then it is *uniform*. We will also say that α is (n-)uniform if $[\![\alpha]\!]$ is and that a set $S \subseteq TA$ of action terms is uniform if there is some n such that each $\alpha \in S$ is n-uniform.

Our definition of $[\![-]\!] : TA \rightarrow \mathcal{P}(X^{<\omega})$ is found in Table 1. This definition is a slight simplification of Meyer (1989). In Meyer's system, all the sequences are infinite, but only finite initial segments are *relevant* (specified) by marking the s-sets. We do not deal with marked s-sets, since we admit only finite sequences and all s-sets in the treated sequences we consider "relevant". Furthermore, we do not have the restriction of action terms to be in normal form, i.e., a form in which every subexpression of the form $\alpha \cup \beta$ has the property that $[\![\alpha \mathbin{\&} \beta]\!] = \emptyset$, and dually, every subexpression of the form $\alpha \mathbin{\&} \beta$ has the property that $[\![\overline{\alpha} \mathbin{\&} \overline{\beta}]\!] = \emptyset$. This condition is necessary in Meyer's system, since otherwise some axioms would not be sound.[2] This is a result of his definition of the "$\cup$"-operator, which is not the set-theoretic union as in our language. It gives the union of two sets of sequences but subtracts every sequence s in the union comparable with some sequence t in the union and is not a proper prefix of t. So, $[\![\alpha \cup (\alpha; \beta)]\!] \subseteq [\![\alpha]\!]$, which is not a property in our language.

Consequently, we lose a few properties, such as the desirable property $[\![\overline{\overline{\beta}}]\!] = [\![\beta]\!]$. However, these properties play no significant role in our development of long-term norms. Thus, we prefer to simplify PD_eL and focus on the original work as far as possible.

See Figures 1 and 2 for a pictorial explanation of & and $\overline{-}$.

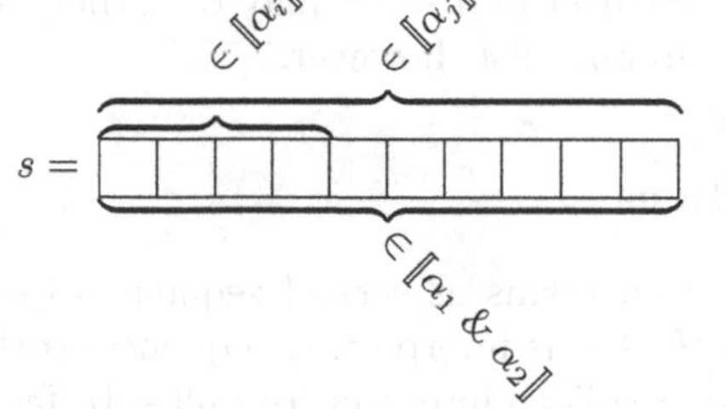

Fig. 1. s is in $[\![\alpha_1 \mathbin{\&} \alpha_2]\!]$ if s is in one of the α_i's and a prefix of s is in the other

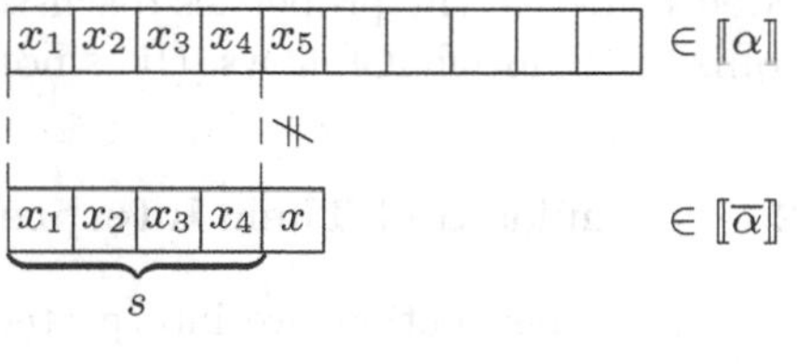

Fig. 2. $s * \langle x \rangle \in [\![\overline{\alpha}]\!]$ if s is comparable to something in $[\![\alpha]\!]$, but $s * \langle x \rangle$ is not

[2] E.g. axiom $[\alpha \cup \beta]\phi \equiv [\alpha]\phi \wedge [\beta]\phi$: In Meyer's system it holds that $[\![a]\!] = [\![a \cup a; b]\!]$, however, $[a \cup a; b]\phi \equiv [a]\phi \not\equiv [a]\phi \wedge [a; b]\phi$.

If every sequence in $[\![\alpha]\!]$ is also in $[\![\beta]\!]$, then consequences of doing β are also consequences of doing α. Because this fact is so basic to our reasoning, we introduce the partial order $\lesssim$, defined by

$$\alpha \lesssim \beta \text{ iff } [\![\alpha]\!] \subseteq [\![\beta]\!].$$

We write $\alpha \approx \beta$ iff $[\![\alpha]\!] = [\![\beta]\!]$.

Table 2. Basic properties of $\lesssim$

Properties of $\lesssim$	
$\overline{\alpha \cup \beta} \approx \overline{\alpha} \mathbin{\&} \overline{\beta}$	$\alpha \mathbin{\&} \beta \lesssim \alpha$ if $\{\alpha, \beta\}$ is uniform
$\overline{\alpha \mathbin{\&} \beta} \approx \overline{\alpha} \cup \overline{\beta}$ if $[\![\alpha \mathbin{\&} \beta]\!] \neq \emptyset$	$\alpha \mathbin{\&} \beta \lesssim \alpha; \mathbf{any}^*$
$\overline{\alpha} \lesssim \overline{\alpha; \beta}$	$\alpha \mathbin{\&} (\beta \cup \gamma) \approx (\alpha \mathbin{\&} \beta) \cup (\alpha \mathbin{\&} \gamma)$
$\alpha \lesssim \mathbf{any}^*$	$\alpha \cup (\beta \mathbin{\&} \gamma) \lesssim (\alpha \cup \beta) \mathbin{\&} (\alpha \cup \gamma)$
$\alpha \lesssim \alpha; \mathbf{any}^*$	$\overline{\mathbf{any}} \approx \overline{\mathbf{any}^*} \approx \underline{\emptyset}$
$\alpha \lesssim \mathbf{any}^*; \alpha$	$\overline{\alpha} \approx \overline{\alpha; \mathbf{any}} \approx \overline{\alpha; \mathbf{any}^*}$
$a \lesssim \mathbf{any}$	$\overline{\mathbf{any}^*; \alpha} \approx \underline{\emptyset}$ if $[\![\alpha]\!] \neq \emptyset$
$\alpha; \underline{\emptyset} \approx \underline{\emptyset}; \alpha \approx \underline{\emptyset}$	$\overline{\underline{\emptyset}} \approx \mathbf{any}$
$\epsilon; \beta \approx \beta; \epsilon \approx \beta$	
$\left.\begin{array}{l} \alpha \cup \gamma \lesssim \beta \cup \delta \\ \alpha; \gamma \lesssim \beta; \delta \end{array}\right\}$ if $\alpha \lesssim \beta$ and $\gamma \lesssim \delta$	

We give some of the basic properties regarding $\lesssim$ and $\approx$ in Table 2. In each case, the derivation is fairly simple. Moreover, Table 2 contains every property needed to derive the properties discussed hereafter and so we may take them as axioms for $\lesssim$ in what follows. (It is not a minimal list, however.)

2.2 Formulas and Their Interpretations

In the previous section, we interpreted action terms as sets of sequences over X. The final step in defining semantics for PD_eL is interpreting sequence-world pairs as sets of paths in our model and using this to interpret formulas. In fact, as with other dynamic logics, we are not interested in the paths per se, but only with the final worlds in each path. This simplifies our definitions a bit.

A PD_eL model consists of a set $\mathcal{W}$ of worlds together with an X-labeled transition system and an interpretation of atomic propositions. We define an interpretation

$$[\![-]\!] : X^{<\omega} \to (\mathcal{P}\mathcal{W})^{\mathcal{W}}$$

taking a sequence $\langle x_1, \ldots, x_n \rangle$ and world w to the set of all w's reachable from w via a path like so:

$$w \xrightarrow{x_1} w_1 \xrightarrow{x_2} \cdots \xrightarrow{x_n} w'.$$

Explicitly,

$$\llbracket \langle\rangle \rrbracket(w) = \{w\},$$
$$\llbracket s * \langle x \rangle \rrbracket(w) = \{\, w' \mid \exists w'' \in \llbracket s \rrbracket(w) \text{ and } w'' \xrightarrow{x} w' \,\}.$$

This induces an interpretation $\llbracket - \rrbracket : TA \to (\mathcal{PW})^{\mathcal{W}}$ defined by

$$\llbracket \alpha \rrbracket(w) = \{\, w' \mid \exists s \in \llbracket \alpha \rrbracket \; . \; w' \in \llbracket s \rrbracket(w) \,\}.$$

Clearly, we are overloading the notation $\llbracket - \rrbracket$ here, but we hope that our meaning is clear from context. When we write $\llbracket \alpha \rrbracket$, we mean a set of X-sequences and when we write $\llbracket \alpha \rrbracket(w)$, we mean a set of worlds.

Assertions in PD_eL are either atomic propositions, logical compositions $\neg\phi$, $\phi_1 \vee \phi_2$, $\phi_1 \wedge \phi_2$, $\phi_1 \to \phi_2$, $\phi_1 \equiv \phi_2$, or expressions of the form $[\alpha]\phi$ with intended meaning that ϕ holds after the performance of action α. The semantics of the formula $[\alpha]\phi$ is defined by

$$w \models [\alpha]\phi \text{ iff } \forall w' \in \llbracket \alpha \rrbracket(w) \; . \; w' \models \phi.$$

We summarize the rules and axioms for our simplified PD_eL in the top half of Table 3. This theory is sound but not complete (see Meyer (1988, 1989)).

Thus far, we have defined a variant of PDL, with no particular relevance for reasoning about prohibitions or obligations. The reduction of deontic operators to dynamic ones uses Anderson's violation atom V (1967) to represent deontic violations. This yields deontic operators f and o, representing that an action is prohibited/obligatory, resp., presented in Table 3 along with some prominent theorems.

Note that these definitions of prohibition and obligation are about *immediate* actions. Let us focus on prohibition for a moment. A world w satisfies $f(\alpha)$ just in case *in* w, the result of doing the action $\llbracket \alpha \rrbracket$ is a world in violation. But if our agent performs some other action first, say $\llbracket \beta \rrbracket$, then he may no longer be in w and so the fact that $w \models f(\alpha)$ is not relevant for him. In other words, the formula $f(\alpha)$ expresses that an agent is prohibited from doing $\llbracket \alpha \rrbracket$ *now*, not that he is prohibited from *ever* doing $\llbracket \alpha \rrbracket$.

We close this section with some comments about one unfortunate consequence of this approach. It seems reasonable that, if α is forbidden, so is any action beginning with α, i.e. $f(\alpha) \to f(\alpha; \beta)$. But this property does not hold in general. Indeed, it is easy to see that

$$\vdash (f(\epsilon) \to f(\epsilon; \mathbf{any}^*)) \equiv (V \to [\mathbf{any}^*]V).$$

Thus, if one wants $f(\alpha) \to f(\alpha; \beta)$ to hold in general, he must either give up the defining axiom for $[\alpha; \beta]$ or require $V \to [\mathbf{any}^*]V$. This is a very strong and usually undesirable condition which we briefly discuss in Section 3.1.

Table 3. The theory PD_eL

Inference rules

$$\frac{\phi}{[\alpha]\phi}(\text{N}) \qquad \frac{\phi \to \psi \quad \phi}{\psi}(\text{MP})$$

Axioms

every propositional tautology	$[\beta]\phi \to [\alpha]\phi$ if $\alpha \lesssim \beta$
$[\beta](\phi_1 \to \phi_2) \to ([\beta]\phi_1 \to [\beta]\phi_2)$	$[\alpha \cup \beta]\phi \equiv [\alpha]\phi \wedge [\beta]\phi$
$[\alpha;\beta]\phi \equiv [\alpha]([\beta]\phi)$	$[\alpha]\phi \vee [\beta]\phi \to [\alpha \mathbin{\&} \beta]\phi$ if $\{\alpha, \beta\}$ is uniform
$[\emptyset]\phi$	$[\mathbf{any}]\phi \to [a]\phi$
$[\epsilon]\phi \equiv \phi$	$[\mathbf{any}^*]\phi \to \phi \wedge [\mathbf{any}][\mathbf{any}^*]\phi$

Deontic definitions

$f(\alpha) \equiv [\alpha]V$	$o(\alpha) \equiv [\overline{\alpha}]V$

PD_eL theorems

$f(\alpha;\beta) \equiv [\alpha]f(\beta)$	$f(\alpha \cup \beta) \equiv f(\alpha) \wedge f(\beta)$
$f(\beta) \to f(\alpha)$ if $\alpha \lesssim \beta$	$f(\alpha \mathbin{\&} \beta) \wedge o(\alpha) \equiv f(\beta) \wedge o(\alpha)$
$o(\alpha \cup \beta) \wedge f(\alpha) \to o(\beta)$ if α, β are both 1-uniform	$o(\alpha \mathbin{\&} \beta) \equiv o(\alpha) \wedge o(\beta)$ if $[\![\alpha \mathbin{\&} \beta]\!] \neq \emptyset$

$$\left.\begin{array}{l} o(\alpha) \vee o(\beta) \to o(\alpha \cup \beta) \\ f(\alpha) \vee f(\beta) \to f(\alpha \mathbin{\&} \beta) \end{array}\right\} \text{if } \{\alpha, \beta\} \text{ is uniform}$$

It may be argued that, in the end, a dynamic deontic logic indeed wants $f(\alpha) \to f(\alpha;\beta)$ and $\neg(V \to [\mathbf{any}^*]V)$. The natural way to satisfy this is to change the semantics to interpret $[\![\alpha]\!](w)$ as a set of paths and define: $w \models [\alpha]V$ just in case for each path

$$w \xrightarrow{x_1} w_1 \xrightarrow{x_2} \cdots \xrightarrow{x_n} w_n$$

in the interpretation of α, there is a world $w_i \models V$. Such an interpretation violates not only $[\alpha;\beta]\phi \equiv [\alpha][\beta]\phi$, but also the axiom K: $[\beta](\phi_1 \to \phi_2) \to ([\beta]\phi_1 \to [\beta]\phi_2)$. We postpone this alternative for later research.

3 The Long-Term Contiguous System

We turn our attention now to long-term norms. The definitions of f and o from the previous section are intended to express immediate duties, but often one has duties of longer duration, such as: "Never do that", or, "Do this someday." In this section, we introduce the machinery to express such long-term norms.

3.1 The Long-Term, Contiguous Prohibition

As previously mentioned, the formula $f(\alpha)$ expresses one important kind of prohibition, namely, that the agent is prohibited from doing $[\![\alpha]\!]$ in *this* world. But many prohibitions are stronger than this. They express that one is *never* allowed to perform a particular act[3], such as "Never point a loaded gun at an innocent man and pull the trigger." But we mean a particular interpretation of this "never". We do not mean "in every world, one should not point a loaded gun...," but rather "in every world *reachable from the current world*, one should not point a loaded gun..."

Such prohibitions are easy to express in the logic at hand. One is *never* allowed to do the action $[\![\alpha]\!]$ just in case, for any action term β, doing $[\![\beta]\!]$ followed by $[\![\alpha]\!]$ results in violation. In other words, in world w, one is forever prohibited from doing $[\![\alpha]\!]$ iff for all β,

$$\forall w' \in [\![\beta;\alpha]\!](w) \;.\; w' \models V.$$

But this is true just in case for every $w' \in [\![\mathbf{any}^*;\alpha]\!](w)$, we have $w' \models V$, i.e. just in case $w \models f(\mathbf{any}^*;\alpha)$. Thus, we define $F(\alpha) \equiv f(\mathbf{any}^*;\alpha)$.

We summarize our definition of F and give a derived rule of inference and several theorems in Table 4. The derivations are straightforward.

Table 4. The definition of F and several Consequences

<u>Defining axiom</u>	<u>Rule of inference</u>
$F(\alpha) \equiv f(\mathbf{any}^*;\alpha)$	$\dfrac{f(\alpha) \rightarrow f(\beta)}{F(\alpha) \rightarrow F(\beta)}$
<u>PD_eL theorems for F</u>	
$F(\beta) \rightarrow F(\alpha)$ if $\alpha \lesssim \mathbf{any}^*;\beta$	$F(\alpha) \rightarrow f(\alpha)$
$F(\beta) \rightarrow F(\alpha)$ if $\alpha \lesssim \beta$	$F(\alpha;\beta) \equiv [\mathbf{any}^*;\alpha]f(\beta)$
	$F(\alpha;\mathbf{any}^*;\beta) \equiv [\mathbf{any}^*;\alpha]F(\beta)$
$F(\alpha) \rightarrow F(\beta;\alpha)$	$F(\alpha) \wedge F(\beta) \equiv F(\alpha \cup \beta)$
$F(\alpha) \equiv F(\mathbf{any}^*;\alpha)$	$F(\alpha) \vee F(\beta) \rightarrow F(\alpha \,\&\, \beta)$ if $\{\alpha,\beta\}$ is uniform

3.2 The Long-Term, Contiguous Obligation

As we have seen, long-term prohibition is easy to express in PD_eL and in fact requires only one small change to Meyer's 1989 presentation: the addition of

3 In practice, such prohibitions are likely to include a conditional, such as, "Always obey local traffic laws unless in an emergency." Admittedly, these conditional prohibitions are not expressible as $F(\alpha)$ in the sense given here, because we have omitted the test action constructor $\phi?$. But we did so only for simplicity's sake. The test constructor presents no particular difficulty for our logic.

the action term **any***. But the situation for long-term obligation is considerably more subtle.

Suppose one has promised to repay a loan, without any deadline regarding repayment. When can we conclude that he has failed to fulfill his duty? What behavior satisfies his obligation? One fulfills his obligation provided there is *some* point at which he actually repays the loan. Until then, he has an outstanding obligation.

One is naturally tempted to define O in terms of F, just as o was defined in terms of f. Unfortunately this yields unreasonable results: suppose $O(a) \equiv F(\overline{a})$. The latter is equivalent $[\mathbf{any}^*]f(\overline{a})$ and hence to $[\mathbf{any}^*]o(a)$. But $[\mathbf{any}^*]o(a) \to [\mathbf{any}^*; a]o(a)$ and so this obligation can never be discharged! After doing $[\![a]\!]$, the agent must do it again and again and .... This may be adequate for perpetual obligations, like honoring one's parents,[4] but not for long-term obligations.

So what properties should O capture? When we assert that α is a long-term obligation, we mean that we cannot discharge our obligation *without* first doing $[\![\alpha]\!]$. This does not mean that we must do $[\![\alpha]\!]$ *now*. It also does not mean that if we do $[\![\alpha]\!]$, our obligation will be discharged: if we are indebted Paul, we have to acquire the necessary funds to repay him. But acquiring the funds is not enough to discharge our debt, of course: we must also repay Paul! Long term obligations are about necessary actions, rather than sufficient actions.

Thus, $O(\alpha)$ should be interpreted as: our obligations will not be discharged *unless* we do some act in $[\![\mathbf{any}^*; \alpha; \mathbf{any}^*]\!]$. That is, whatever fully specified s we perform, if $s \notin [\![\mathbf{any}^*; \alpha; \mathbf{any}^*]\!]$, then the outcome will be a world in which we still have an unfulfilled duty. Consequently, we introduce a new action term, $\widehat{\alpha}$ and interpret it as:

$$[\![\widehat{\alpha}]\!] = X^{<\omega} \setminus [\![\mathbf{any}^*; \alpha; \mathbf{any}^*]\!].$$

We also introduce a defining axiom scheme for $\widehat{\alpha}$ in Table 5 along with a few useful properties. The defining axiom can be understood as follows: suppose that never doing $[\![\alpha]\!]$ guarantees a ϕ-world, i.e. $w \models [\widehat{\alpha}]\phi$. Suppose also that doing $[\![\beta]\!]$ always results in a $\neg\phi$-world. Then it must be the case that by doing $[\![\beta]\!]$, one has also done α, i.e. that $\beta \lesssim \mathbf{any}^*; \alpha; \mathbf{any}^*$. Hence, in this case $w \models [\mathbf{any}^*; \alpha; \mathbf{any}^*]\psi \to [\beta]\psi$ for every ψ.

We provide proofs for the soundness of the axiom scheme and the property $\widehat{\alpha} \lesssim \widehat{\alpha \mathbin{\&} \beta}$, since these proofs are not as obvious as the others.

Proof (of defining axiom). We aim to show that, for every world w, pair of actions α, β and pair of formulas ϕ, ψ,

$$w \models ([\widehat{\alpha}]\phi \land [\beta]\neg\phi) \to ([\mathbf{any}^*; \alpha; \mathbf{any}^*]\psi \to [\beta]\psi).$$

Suppose that $w \models [\widehat{\alpha}]\phi \land [\beta]\neg\phi$. First, we will establish that, for every $s \in [\![\widehat{\alpha}]\!] \cap [\![\beta]\!]$, $[\![s]\!](w) = \emptyset$.[5] Let such s be given. Because $w \models [\widehat{\alpha}]\phi$, we see that

[4] Thanks to one of our reviewers for this observation.

[5] Note that Meyer assumes that for every world w and atomic action x, there is a w' such that $w \xrightarrow{x} w'$. Under this assumption, one can show $[\![\widehat{\alpha}]\!] \cap [\![\beta]\!] = \emptyset$.

Table 5. Properties of the $\widehat{\alpha}$ constructor

Defining axiom

$$([\widehat{\alpha}]\phi \wedge [\beta]\neg\phi) \rightarrow ([\mathbf{any}^*; \alpha; \mathbf{any}^*]\psi \rightarrow [\beta]\psi)$$

Properties for $\widehat{\alpha}$ and $\lesssim$

If $\alpha \lesssim \beta$ then $\widehat{\beta} \lesssim \widehat{\alpha}$	$\widehat{\alpha} \lesssim \widehat{\alpha \mathbin{\&} \beta}$	$\widehat{\beta} \lesssim \widehat{\alpha \mathbin{\&} \beta}$
$\widehat{\alpha \cup \beta} \lesssim \widehat{\alpha} \mathbin{\&} \widehat{\beta}$	$\widehat{\alpha} \lesssim \widehat{\alpha; \beta}$	$\widehat{\beta} \lesssim \widehat{\alpha; \beta}$
	$\widehat{\alpha} \approx \widehat{\mathbf{any}^*; \alpha}$	$\widehat{\alpha} \approx \widehat{\alpha; \mathbf{any}^*}$

$[\![s]\!](w) \subseteq [\![\phi]\!]$. But since $w \models [\beta]\neg\phi$, we also see that $[\![s]\!](w) \subseteq [\![\neg\phi]\!]$. Hence $[\![s]\!](w)$ is empty.

Now suppose that $w \models [\mathbf{any}^*; \alpha; \mathbf{any}^*]\psi$ and we will complete the proof by showing $w \models [\beta]\psi$. Let $s \in [\![\beta]\!]$ be given and we must show $[\![s]\!](w) \subseteq [\![\psi]\!]$. If $s \in [\![\mathbf{any}^*; \alpha; \mathbf{any}^*]\!]$ then $[\![s]\!](w) \subseteq [\![\psi]\!]$ by assumption. Otherwise, $[\![s]\!](w) = \emptyset$ and so is trivially contained in $[\![\psi]\!]$.

Proof (of $\widehat{\alpha} \lesssim \widehat{\alpha \mathbin{\&} \beta}$). We will prove the claim by showing that

$$[\![\mathbf{any}^*; (\alpha \mathbin{\&} \beta); \mathbf{any}^*]\!] \subseteq [\![\mathbf{any}^*; \alpha; \mathbf{any}^*]\!].$$

Let s be an element of the set $[\![\mathbf{any}^*; (\alpha \mathbin{\&} \beta); \mathbf{any}^*]\!]$. Then there are sequences s_1, s_2 and s_3 such that $s_2 \in [\![\alpha \mathbin{\&} \beta]\!]$ and $s = s_1 * s_2 * s_3$.

By definition of $[\![\alpha \mathbin{\&} \beta]\!]$, there is some n such that $s_2 \upharpoonright n \in [\![\alpha]\!]$. Thus, we can find t_1 and t_2 such that $s_2 = t_1 * t_2$ and $t_1 \in [\![\alpha]\!]$ (namely, we take $t_1 = s_2 \upharpoonright n$). Then $s = s_1 * t_1 * (t_2 * s_3)$ and hence $s \in [\![\mathbf{any}^*; \alpha; \mathbf{any}^*]\!]$, as desired.

Finally, there is one more characteristic difference between long-term obligation and prohibition. When we are obligated to pay a debt, say, or perform a promised act, then we have an unsatisfied duty. This is not the same as being in violation nor is there any obvious way of expressing this condition in terms of our violation predicate V. Rather, we should introduce a new atomic proposition I (for *indebtedness*) to represent the condition that an agent has some unfulfilled obligation and define O in terms of I. We investigate some possible relations between I and V below.

Thus $O(\alpha)$ represents that $[\![\alpha]\!]$ is a necessary means to $[\![\neg I]\!]$, i.e. that a $\neg I$-world will not be reached unless we do some sequence in $[\![\alpha]\!]$. Therefore, we propose to define $O(\alpha)$ by $[\widehat{\alpha}]I$.

We give a few simple theorems regarding O in Table 6. The proofs are routine.

There is an unfortunate consequence of this definition: if an agent is in a world in which $\neg I$ is unreachable, he is obligated to do *everything*, which is absurd. Thus, one may be tempted to amend the definition so that $O(\alpha)$ is defined as

$$[\widehat{\alpha}]I \wedge \langle \mathbf{any}^* \rangle \neg I.$$

Table 6. The definition of O and some consequences

Defining axiom

$$O(\alpha) \equiv [\widehat{\alpha}]I$$

PD_eL theorems for O

$O(\beta) \to O(\alpha)$ if $\widehat{\alpha} \lesssim \widehat{\beta}$	$O(\alpha \,\&\, \beta) \to O(\alpha) \wedge O(\beta)$
$O(\beta) \to O(\alpha)$ if $\beta \lesssim \alpha$	$O(\alpha; \beta) \to O(\alpha) \wedge O(\beta)$
$O(\alpha) \equiv O(\mathbf{any}^*; \alpha)$	$O(\alpha) \vee O(\beta) \to O(\alpha \cup \beta)$
$O(\alpha) \equiv O(\alpha; \mathbf{any}^*)$	

We do not use this more complicated definition here, partly for simplicity's sake and partly for consistency with the prior definition of o. Also, the amended definition has its own motivational problem: an agent that cannot reach $\neg I$ is never obligated to do *anything*, which seems similarly absurd.

We have presented only the most basic and useful theorems for O in Table 6, but this list can be extended in many natural directions. For instance, if one is obligated to do $[\![a]\!]$ and also $[\![b]\!]$, then he is obligated to do $[\![a]\!]$ and later $[\![b]\!]$ or $[\![b]\!]$ and later $[\![a]\!]$ or both at once, i.e.

$$O(a) \wedge O(b) \equiv O((a; \mathbf{any}^*; b) \cup (b; \mathbf{any}^*; a) \cup (a \,\&\, b)).$$

This is the long-term analogue of $o(\alpha \,\&\, \beta) \equiv o(\alpha) \wedge o(\beta)$ and applies to any pair of 1-uniform actions.

Another intuitive example: one expects that, if an agent is obliged to eventually do $\alpha; \beta$, then after doing α, he will still be obliged to do β. In fact, this is not quite the case, if α is sufficiently complex, but a similar claim *does* hold. For this, let us introduce a new action term constructor, $\ldots\alpha$.

Let $[\![\ldots\alpha]\!]$ be the set of sequences s such that (i) s ends in an α-sequence, i.e. $s = t * t'$ where $t' \in [\![\alpha]\!]$ and (ii) no proper prefix of s ends in an α-sequence. In other words, $[\![\ldots\alpha]\!] = [\![\mathbf{any}^*; \alpha]\!] \setminus [\![\mathbf{any}^*; \alpha; \mathbf{any}; \mathbf{any}^*]\!]$. Then one can easily show:

$$\models O(\alpha; \beta) \to O(\alpha) \wedge [\ldots\alpha]O(\beta).$$

In other words, the agent obligated to do $\alpha; \beta$ is still obligated to do β at the instant he has first completed α. For atomic actions a, it is easy to see that $\ldots a = \widehat{a}; a$, and thus

$$\models O(a; \beta) \to O(a) \wedge [\widehat{a}; a]O(\beta).$$

This formula is analogous to the formula $o(\alpha; \beta) \to o(\alpha) \wedge [\alpha]o(\beta)$, which is valid if $[\![\beta]\!] \neq \emptyset$.

Because O is defined in terms of a new indebtedness proposition, we have lost the strong connection between prohibition and obligation. In fact, we think this is natural: long term obligations are not a simple conjugate of prohibition. They impose looser restrictions on behavior and cannot be captured in terms of V. Nonetheless, it would be natural to suppose some connection between I and V and we briefly consider a few proposals here.

Loosely, a moral agent aims to reach a world in which his obligation is relieved: he aims to realize $\neg I$. This is not quite accurate, however, since new obligations may be created before paying old ones. Borrowing from Peter to pay Paul relieves the obligation to Paul at the expense of creating a new obligation and thus remaining in a state of indebtedness. Nonetheless, this strategy is not obviously immoral (provided that Peter is eventually repaid, perhaps by borrowing from Paul), regardless of its practical merits. A moral agent may fulfill each obligation without ever reaching a $\neg I$-world!

On the other hand, perhaps one should avoid situations in which he can *never* fulfill his outstanding obligations without first acquiring new ones. In such situations, the agent has exceeded his ability to meet his duties. Thus, one may wish to relate I and V by requiring that our models satisfy the *axiom of eventual repayment*:

$$[\mathbf{any}^*]I \to V \qquad \textbf{(ER)}$$

Alternatively, we may wish to restrict the ways in which an agent discharges his obligations. One should not fulfill obligations by doing prohibited acts: it may be okay to borrow from Peter to pay Paul, but robbing Peter is out of bounds. The natural way to restrict such disreputable strategies is the converse of *eventual repayment*, which we call the *axiom of forbidden means*:

$$V \to [\mathbf{any}^*]I \qquad \textbf{(FM)}$$

Thus, if one *ever* reaches a world in violation, he is thereafter in an I-world. With this axiom, one can prove

$$O(\alpha \cup \beta) \wedge F(\alpha) \to O(\beta).$$

(The proof depends on a proof of the abstruse property $\widehat{\alpha \cup \beta} \cup (\mathbf{any}^*; \alpha; \mathbf{any}^*) \approx \widehat{\beta} \cup (\mathbf{any}^*; \alpha; \mathbf{any}^*)$, but is otherwise straightforward.)

One may adopt *both* of the above axioms. In this case, $[\mathbf{any}^*]I$ is equivalent to V. But this means that, once in a V world, every path leads to a $[\mathbf{any}^*]I$ world, and hence to another V world. Consequently, such models satisfy the *axiom of unforgiveness*:

$$V \equiv [\mathbf{any}^*]V. \qquad \textbf{(UF)}$$

Once an agent is in violation, he remains there. This is an unforgiving model of deontic logic!

4 The Non-contiguous System

In the previous section, we explored long-term prohibitions and obligations of certain kinds of actions, namely *contiguous* actions. The prohibition $F(\alpha; \beta)$

expresses that one is never allowed to do $[\![\alpha]\!]$ *immediately* followed by $[\![\beta]\!]$, but it does not restrict one from doing $[\![\alpha]\!]$, then something else and then $[\![\beta]\!]$.

It seems reasonable that most prohibitions do involve such contiguous actions. One is not allowed to aim the gun at an innocent and pull the trigger, but he can aim the gun at an innocent[6], then point it at the ground and pull the trigger. The effect of aiming the gun can be undone before pulling the trigger.

But in rare situations, the effects of doing $[\![\alpha]\!]$ cannot be undone and thus one should *never* do $[\![\beta]\!]$ thereafter. Suppose that a big yellow button arms a bomb and that, once armed, it cannot be disarmed. Suppose also that a big red button detonates the bomb if it is armed. Then one should never press the yellow button followed eventually by pressing the red button. We denote this kind of prohibition by F^*.

Admittedly, such strong prohibitions tend to be as artificial as our bomb example, but we claim that long-term non-contiguous *obligations* are fairly common. We will discuss these in Section 4.1, but let us first examine prohibitions.

We must introduce a few relations on sequences and actions in order to express non-contiguous prohibitions. The first relation, $r \sqsubseteq s$, expresses that r is a subsequence of s and that the last element of r is the last element of s. Explicitly, $r \sqsubseteq s$ iff there is a $f : |r| \rightarrow |s|$ satisfying the following:

- f is strictly increasing;
- $f(|r| - 1) = |s| - 1$;
- for every $i < |r|$, we have $s(f(i)) = r(i)$.

In other words, $r \sqsubseteq s$ iff r is a subsequence of s such that the last element of r is also the last element of s. We say in this case that r is a *tail-fixed subsequence* of s. See Figure 3 for an illustration.

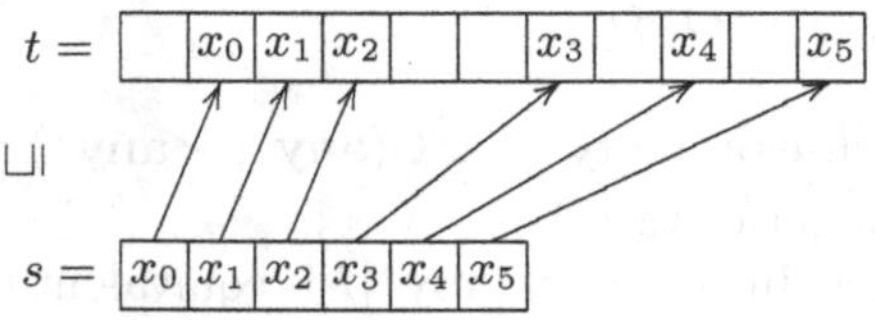

Fig. 3. Illustration of $s \sqsubseteq t$

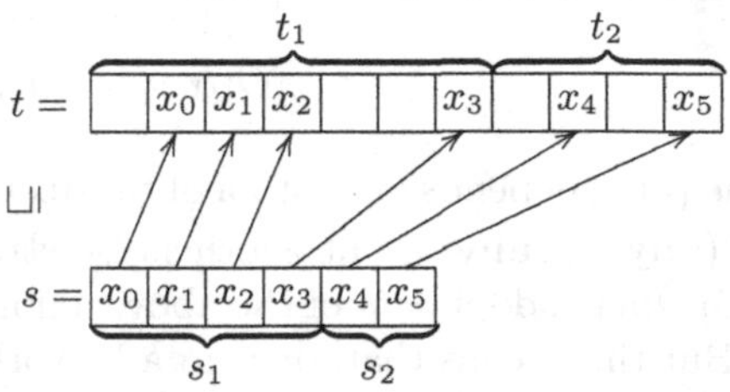

Fig. 4. Illustration for $\sqsubseteq$ properties (2) and (3)

We can express the intended meaning of long-term non-contiguous prohibition in terms of $\sqsubseteq$. Suppose that $F^*(\beta)$ and $s \in [\![\beta]\!]$. Then any fully specified t satisfying $s \sqsubseteq t$ will lead to violation. In order to express this prohibition on the level of action terms, we introduce an action constructor $\widetilde{\beta}$ defined by

[6] Let us briefly ignore very sensible rules regarding gun handling and safety and the very disturbing effect caused by staring down the wrong end of a gun barrel.

Table 7. Properties of $\sqsubseteq$ and $\widetilde{\alpha}$

Properties of $\sqsubseteq$

1. $\sqsubseteq$ is a partial order.
2. If $s_1 \sqsubseteq t_1$ and $s_2 \sqsubseteq t_2$ then $s_1 * s_2 \sqsubseteq t_1 * t_2$.
3. If $s_1 * s_2 \sqsubseteq t$, then there are t_1, t_2 such that $t = t_1 * t_2$, $s_1 \sqsubseteq t_1$ and $s_2 \sqsubseteq t_2$.
4. If $s \sqsubseteq t$ then $s \sqsubseteq r * t$ for any r.
5. If $s_1 * s_2 \sqsubseteq t$ then $s_2 \sqsubseteq t$.
6. $\langle x \rangle \sqsubseteq \langle x_1, \ldots, x_n \rangle$ iff $x = x_n$.

Definition of $\widetilde{\alpha}$

$$[\![\widetilde{\alpha}]\!] = \{ s \in X^{<\omega} \mid \exists r \in [\![\alpha]\!] \,.\, r \sqsubseteq s \}$$

Properties of $\widetilde{\alpha}$ and $\lesssim$

$\alpha \lesssim \widetilde{\alpha}$	$\widetilde{\alpha; \beta} \approx \widetilde{\alpha}; \widetilde{\beta}$
If $\alpha \lesssim \beta$ then $\widetilde{\alpha} \lesssim \widetilde{\beta}$	$\widetilde{\alpha \cup \beta} \approx \widetilde{\alpha} \cup \widetilde{\beta}$
$\widetilde{\alpha; \beta} \lesssim \widetilde{\beta}$	$\widetilde{a} \approx \mathbf{any}^*; a$
$\widetilde{\widetilde{\beta}} \approx \widetilde{\beta}$	$\mathbf{any}^* \approx \widetilde{\mathbf{any}^*}$

$$[\![\widetilde{\beta}]\!] = \{ s \in X^{<\omega} \mid \exists r \in [\![\beta]\!] \,.\, r \sqsubseteq s \}.$$

Hence, $\alpha \lesssim \widetilde{\beta}$ iff each s in $[\![\alpha]\!]$ has some r in $[\![\beta]\!]$ as a tail-fixed subsequence. Thus, if $w \models F^*(\beta)$ and $\alpha \lesssim \widetilde{\beta}$, then $w \models f(\alpha)$.

In case $\alpha \lesssim \widetilde{\beta}$, we say that α *involves* β. In this case, however one does $[\![\alpha]\!]$ (whichever fully specified sequence is chosen), one is doing $[\![\beta]\!]$ "along the way", i.e. doing some $t \in [\![\beta]\!]$ as a tail-fixed subsequence. For example, α involves a just in case each $s \in [\![\alpha]\!]$ has as last element some element of $i(a)$.

We summarize properties of $\sqsubseteq$ and $\widetilde{\alpha}$ in Table 7.

4.1 Long-Term Non-contiguous Prohibition and Obligation

We have already tipped our hand regarding the long-term non-contiguous prohibition F^*. When we say that β is forbidden in this sense, we mean that each fully specified sequence s containing some $t \in [\![\beta]\!]$ as a tail-fixed subsequence leads to violation. In other words, we define $F^*(\beta) \equiv f(\widetilde{\beta})$, equivalently $F^*(\beta) \equiv [\widetilde{\beta}]V$.

With this definition, one can easily derive $F^*(\beta) \rightarrow f(\alpha)$ whenever $\alpha \lesssim \widetilde{\beta}$. In fact we can derive a stronger consequence in this case. If $\alpha \lesssim \widetilde{\beta}$ then $\widetilde{\alpha} \lesssim \widetilde{\widetilde{\beta}} \approx \widetilde{\beta}$, so $\vdash F^*(\beta) \rightarrow F^*(\alpha)$.

This and other properties of F^* are presented in Table 8. Again, the proofs are straightforward from the properties stated previously. We can also see now that the three prohibition operators are comparable, with F^* the strongest and f the weakest, since $F^*(\alpha) \rightarrow F(\alpha)$ and $F(\alpha) \rightarrow f(\alpha)$.

Table 8. The long-term non-contiguous prohibition operator F^*

Defining axioms	
$F^*(\alpha) \equiv [\widetilde{\alpha}]V$	$O^*(\alpha) = [\widehat{\widetilde{\alpha}}]I$
PD_eL **theorems for** F^*	
$[\widetilde{\beta}]\phi \rightarrow [\alpha]\phi$ if $\alpha \lesssim \widetilde{\beta}$	$F^*(\alpha) \rightarrow F(\alpha)$
$F^*(\beta) \rightarrow F^*(\alpha)$ if $\alpha \lesssim \widetilde{\beta}$	$F(a) \equiv F^*(a)$
$F^*(\beta) \rightarrow F^*(\alpha)$ if $\alpha \lesssim \beta$	$F^*(\alpha; a) \equiv [\widetilde{\alpha}]F(a)$
$F^*(a; \alpha) \equiv [\mathbf{any}^*; a]F^*(\alpha)$	$F^*(\alpha; \beta) \equiv [\widetilde{\alpha}]F^*(\beta)$
$F^*(\alpha \cup \beta) \equiv F^*(\alpha) \wedge F^*(\beta)$	$F^*(\alpha) \vee F^*(\beta) \rightarrow F^*(\alpha \,\&\, \beta)$
$F^*(\beta) \rightarrow F^*(\alpha; \beta)$	
PD_eL **theorems for** O^*	
$O^*(\alpha) \rightarrow O^*(\beta)$ if $\widetilde{\alpha} \lesssim \widetilde{\beta}$	$O(\alpha) \rightarrow O^*(\alpha)$
$O^*(\alpha) \rightarrow O^*(\beta)$ if $\alpha \lesssim \beta$	$O^*(\alpha \,\&\, \beta) \rightarrow O^*(\alpha) \wedge O^*(\beta)$
$O^*(\alpha; \beta) \rightarrow O^*(\alpha) \wedge O^*(\beta)$	$O^*(\alpha) \vee O^*(\beta) \rightarrow O^*(\alpha \cup \beta)$
$O^*(\alpha) \equiv O^*(\mathbf{any}^*; \alpha)$	$O^*(a) \equiv O(a)$
$O^*(\alpha) \equiv O^*(\alpha; \mathbf{any}^*)$	

The converse implications are not valid, with a counterexample given in Figure 5 (where $X = A$ and $i : A \rightarrow \mathcal{P}^+A$ is the singleton map $a \mapsto \{a\}$). World w_1 satisfies $f(a)$ but not $F(a)$ since $[\![\langle b, a\rangle]\!](w_1) \not\subseteq [\![V]\!]$. World w_0 satisfies $F(a; a)$ but not $F^*(a; a)$, since $[\![\langle a, b, a\rangle]\!] \not\subseteq [\![V]\!]$.

As we have admitted, long-term non-contiguous prohibitions may be fairly rare, since they involve actions with effects that cannot be undone. However, it seems that non-contiguous *obligations* are fairly common. Suppose that Peter owes Paul five dollars, but does not have five dollars. Then he is obliged to first acquire five dollars (or more) and then repay it. But he does not have to repay the money immediately after acquiring it. Rather, he is free to do other things in between. Of course, if he loses the money in between acquiring and repayment,

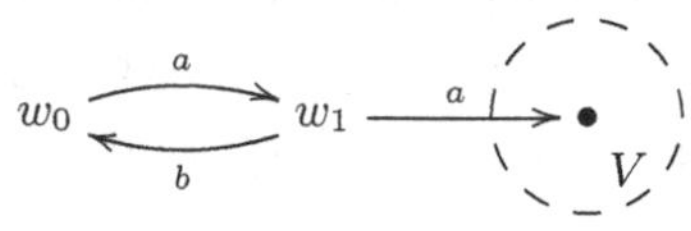

Fig. 5. A counterexample: w_1 satisfies $f(a) \wedge \neg F(a)$ and w_0 satisfies $F(a; a) \wedge \neg F^*(a; a)$

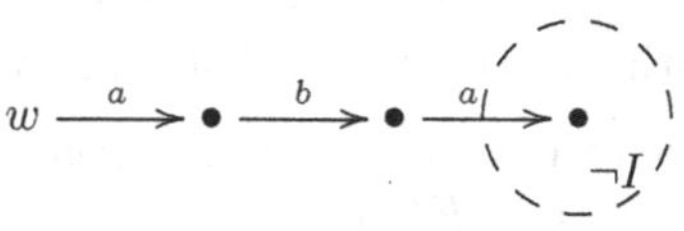

Fig. 6. A counterexample: The world w satisfies $O^*(a; a)$ but not $O(a; a)$

then he cannot discharge his obligation — but we are interested here in necessity rather than sufficiency, and it is necessary that he first acquires and some time later repays.

This suggests the definition $O^*(\alpha) \equiv O(\widetilde{\alpha})$, equivalently $O^*(\alpha) \equiv [\widehat{\widetilde{\alpha}}]I$. Properties for O^* can be found in Table 8. The consequence relation between O and O^* is dual to that between F and F^*, namely $O(\alpha) \rightarrow O^*(\alpha)$. Again, the converse does not hold, as indicated in Figure 6.

Unfortunately, there is no simple relation between O and o. It is clearly not the case that $o(\alpha) \not\rightarrow O(\alpha)$, but this is not too surprising, since the motivation for o (avoiding violation) is different than for O (eventually reaching $\neg I$). It is not hard to show that, for 1-uniform α, the axiom (**FM**) proves $(o(\alpha) \wedge I) \rightarrow O(\alpha)$, but a tighter relationship eludes us.

5 Concluding Remarks

Meyer's work on PD_eL has contributed a formal logic for certain kinds of obligation and prohibition, namely, the *immediate* kind. One of the nice features of his approach is that the two normative concepts are inter-definable: obligation is the same as prohibition from refraining. We aimed to extend his work to include duties of wider scope, duties to never do $[\![\alpha]\!]$ or to eventually do $[\![\alpha]\!]$. As we have seen, however, the natural duality between obligation and prohibition has become obscured by our possible world semantics. Obligations are violated only in the limit, and this is not expressible in terms of worlds reached along the way.

Our work is an extension of an existing framework for deontic logic to include new normative expressions. But we also believe it suggests a new direction for dynamic deontic logic. We would like to recover the duality between prohibitions and obligations that seems so natural in the immediate case. To do so, one needs to evaluate actions in terms of infinite X-sequences and $\mathcal{W}$-paths rather than the worlds encountered at the end of finite paths. In this conceptual setting, it makes sense to discuss failure to meet obligations (i.e., *never* doing what is required) and adherence to long-term prohibitions (i.e., *never* doing what is forbidden). Moreover, we believe that the topological approach of learning theory gives a natural framework for investigating these infinite paths. We hope to return to this topic in future work.

We also believe that some of our considerations provide argument for a hybrid of dynamic and propositional deontic logic. In Section 4.1, we discussed the obligation to obtain funds in order to repay one's debt. But why does a debt impose an obligation to obtain money? Because having money is a necessary precondition for repaying the debt and obtaining money is a means to realize this precondition. It is natural to discuss both ought-to-do and ought-to-be in explaining derivative obligations like the obligation to obtain money. We would like a single framework that includes dynamic operators for both actions and conditions and that allows for reasoning about derived obligations and prohibitions. This would allow for new constructions like, "while ϕ, do $[\![\alpha]\!]$." We expect that existing work on agent planning would be relevant for this project.

References

Anderson, A.R. 1967. Some nasty problems in the formalization of ethics. *Noûs* 1: 345–360.

Castañeda, H.-N. 1981. The paradoxes of deontic logic. In *New studies in deontic logic*, ed. R. Hilpinen, 37–85. Dordrecht: Reidel.

Harel, David. 1984. Dynamic logic. In *Handbook of philosophical logic*, ed. D. Gabbay and F. Guenthner, vol. II, 497–604. D. Reidel Publishing Company.

Kelly, Kevin. 1996. *The logic of reliable inquiry.* Logic and Computation in Philosophy, Oxford University Press.

Meyer, J.-J. Ch. 1988. A different approach to deontic logic: Deontic logic viewed as a variant of dynamic logic. *Notre Dame Journal of Formal Logic* 29:106–136.

———. 1989. Using programming concepts in deontic reasoning. In *Semantics and contextual expression*, ed. R. Bartsch, J.F.A.K. van Benthem, and P. van Emde Boas, 117–145. Dordrecht/Riverton: FORIS publications.

———. 2000. Dynamic logic for reasoning about actions and agents. In *Logic-based artificial intelligence*, 281–311. Norwell, MA, USA: Kluwer Academic Publishers.

von Wright, G.H. 1951. Deontic logic. *Mind* 60:1–15.

On the Normative Aspect of Signalling Conventions

Andrew J.I. Jones[1] and Steven O. Kimbrough[2]

[1] Department of Computer Science
King's College London, London, UK
andrewji.jones@kcl.ac.uk

[2] Operations & Information Management
University of Pennsylvania, Philadelphia, USA
kimbrough@wharton.upenn.edu
http://opim-sky.wharton.upenn.edu/~sok/

Abstract. The paper outlines an approach to the formal representation of signalling conventions, emphasising the prominent role played therein by a particular type of normative modality. It is then argued that, in terms of inferencing related to this modality, a solution can be given to the task J. L. Austin set but failed to resolve: finding a criterion for distinguishing between what Austin called *constatives* and *performatives*. The remainder of the paper indicates the importance of the normative modality in understanding a closely related issue: reasoning about trust in communication scenarios; this, in turn, facilitates a clear formal articulation of the role of a *Trusted Third Party* in trade communication.

1 Introduction

The approach to the analysis of communicative acts taken in this paper differs from those currently most in vogue, in that its focus is neither on the intentions of communicators (FIPA: `http://www.fipa.org/`, and in particular `http://www.fipa.org/repository/bysubject.html`) nor on their supposed commitments [1, 2]. By contrast, the focus here is on the conventions that—as we shall say—*constitute* any given communication system s. These conventions make possible the performance of meaningful communicative acts by the agents, human or electronic, who have adopted s as a means of communicating with each other. We begin by summarising some of the main features of the approach.

2 Signalling Conventions

A convention-based system that defines a framework for agent interaction may appropriately be called an *institution*.[1] In common with other institutions,

[1] This section summarizes the approach to signalling conventions described in [3] and [4], but with somewhat closer attention here to the normative aspect.

L. Goble and J.-J.C. Meyer (Eds.): DEON 2006, LNAI 4048, pp. 149–160, 2006.

communication systems exist to serve a purpose; specifically, their purpose, or point, obviously, is to facilitate the transmission of information of various kinds.

In order to develop these intuitions, and to begin to move towards a formal model, we look first at the communicative act of *asserting* (or *stating*, or *saying*) *that such-and-such is the case.* The key question is this: in the constitution of communication system/institution s, what is it that makes it possible for an agent, if he so wishes, to make an assertion? Our answer is that s contains conventions according to which the performance of particular acts *count as* assertions, and which also specify what those acts mean. Consider, by way of illustration, the institution that was once operative for sea-going vessels, in virtue of which they were able to send signals indicating aspects of the state of a vessel by hoisting sequences of flags. Raising flag sequence *q1* would count (by convention) as a means of saying that the vessel was carrying explosives, raising flag-sequence *q2* would conventionally count as indicating that the vessel carried injured crew members. . . and so on. Note the general form of the conventions themselves: they each associate a particular type of act with a particular state of affairs, and because they are conventions for *asserting* (i.e., for *that* type of communicative act) they each count as a means of *saying that* the associated state of affairs holds.

For present purposes, it matters not at all which sorts of acts are used in a given communication system; the account of communication conventions we offer is entirely neutral on that issue.[2]

Suppose now that in communication system/institution s, the act of bringing it about that A counts as a means of asserting that the state of affairs described by B obtains; (abbreviating: by convention in s, doing A counts as an assertion that B.) And suppose further that agent j, who is an s-user, does A in circumstances in which B does not hold.[3] Then it is appropriate to say that, from the point of view of the institution s, something has gone wrong, *in as much as the purpose or function within institution s of acts of asserting is to facilitate the transmission of reliable information.* The point of asserting, as an institutionalised act, is to be able to show how things stand in a given state of affairs. Given that this *is* the point of asserting, the doing of A in circumstances where B does not hold is a form of *abuse* of the system. Relative to the purpose of asserting, as an institutionalised act, A *ought* to be done *only* when B is the case, and so the doing of A in non-B circumstances amounts to a deviation from the ideal that the system is supposed to achieve.

The conventions for asserting make it possible for acts of assertion to be performed, and they do so by indicating what *would* be the case in circumstances in which the purpose of asserting, qua institutionalised act, is fulfilled. If, by convention in s, doing A counts as an assertion that B, then in ideal circumstances B holds whenever A is done. These observations are the key to understand-

[2] By 'communication system' we here mean the set of conventions that constitute the system, together with the set of agents who make use of those conventions.

[3] It is irrelevant to the present point whether or not j believes that B does not hold.

ing the intuitions on which is grounded the general logical form we assign to communication conventions of the assertoric type.[4]

Following the theory developed in [3], the form of the signalling convention (sc) according to which, in s, agent j's seeing to it that A counts as an assertion that B, is given by

$$\text{(sc-assert)} \qquad E_j A \Rightarrow_s I_s^* B$$

where expressions of the form $E_j A$ are read 'j sees to it that A', $\Rightarrow_s$ is the 'counts as' connective of [8], and I_s^* is a normative operator, intended to capture the sense of 'ought', or ideality, alluded to above. Details of the logics and semantics for the action and 'counts as' modalities are given in [8]. Expressions of the form (sc-assert) say that j's seeing to it that A counts in conventional signalling system s as a means of indicating that, were s to be in an ideal/optimal state with respect to its function of facilitating the transmission of reliable information, B would be true.

The logic of the normative modality is that of a (relativised) normal modality of type K. Closure under logical consequence is a natural assumption, given the intended interpretation of the operator, for if a signalling system would be in an ideal state only if B were true, then it would be in an ideal state only if the logical consequences of B were also true. Note also (cf. [9, p.184]) that the absence of the D. schema reflects the obvious fact that mutually inconsistent assertions can be made: according to one of them, B ought to be true, but according to the other, B ought to be false.

Such other types of communicative acts as *commanding, promising, requesting* and *declaring* (the latter in the sense of [10]) are characterised in terms of signalling conventions of the same basic form as that of (sc-assert) with, crucially, some further elaboration of the scope-formula B falling to the immediate right of the I_s^* operator in the consequent (cf. [3]). This means, of course, that each of these communicative act-types is here treated as a sub-species of the act of asserting, a consequence of the fact that—in stark contrast to Austin [11]—we take all communicative acts to be acts of transmitting information which may, or may not, be true. We shall see in due course how this approach provides the basis for formally articulating the distinction that Austin sought, but failed to capture, between what he called *constatives* and *performatives*.

The form of the signalling convention for *commanding* is

$$\text{(sc-command)} \qquad E_j A \Rightarrow_s I_s^* O E_k B$$

where the O operator is a directive normative modality representing obligation. (We do not here specify a logic of obligation, since it is not the focus of our

[4] [5] is the source from which we take the idea that, in order to understand the communicative act of asserting, one must understand in what sense of 'ought' that which is asserted *ought to be true*. Stenius's much neglected paper is in our opinion one of the most insightful essays written on the analysis of different types of communicative acts. The idea that the 'counts as' notion figures crucially in the convention constituting asserting appears for the first time, to our knowledge, in [6]. For further discussion of the philosophical roots of our approach, see [7].

concern. For present purposes, SDL (standard deontic logic) would suffice.) According to (sc-command), if j sees to it that A, s would then be in an ideal state (things would then be as they ought to be), relative to s's function of facilitating the transmission of reliable information, if there were then an obligation on k (the agent to whom the command is addressed) to see to it that B (where B is the state of affairs that k is commanded to bring about).

The form of the signalling convention for *promising* is

$$\text{(sc-promise)} \qquad E_j A \Rightarrow_s I_s^* O E_j B$$

According to (sc-promise), if j sees to it that A, s would then be in an ideal state (things would then be as they ought to be), relative to s's function of facilitating the transmission of reliable information, if there were then an obligation on j (the agent making the promise) to see to it that B (where B is the state of affairs that j promises to bring about).[5]

The form of the signalling convention for *requesting* is

$$\text{(sc-request)} \qquad E_j A \Rightarrow_s I_s^* H_j E_k B$$

where expressions of the form $H_j A$ are read 'j attempts to see to it that A', and the logic of the *attempts* operator is essentially that of the action operator *minus* the 'success' condition (the T. schema). According to (sc-request), if j sees to it that A, s would then be in an ideal state (things would then be as they ought to be), relative to s's function of facilitating the transmission of reliable information, if j were attempting to get k to see to it that B.

The point of declaratives is to create a new state of affairs, as when, for instance, a couple are declared married, or a meeting is declared open. Let j be the agent issuing the declarative, and let B describe the state of affairs to be created by the performance of the declarative. Then the form of the governing convention is

$$\text{(sc-declare)} \qquad E_j A \Rightarrow_s I_s^* E_j B$$

According to (sc-declare), if j sees to it that A, s would then be in an ideal state (things would then be as they ought to be), relative to s's function of facilitating the transmission of reliable information, if it were then the case that j has indeed seen to it that B.

3 Distinguishing Constatives from Performatives

Austin sought a grammatical criterion for distinguishing between constative sentences (characteristically used in communicative acts the point of which is

[5] We accept that a case can be made for inserting the operator E_j immediately to the left of the obligation operator in the consequent of (sc-command) and (sc-promise), since it is the agent j who, by performing the communicative act $E_j A$, sees to it that the obligation is created. A move of that sort would then make commanding and promising sub-species of declaring (see below), which is perhaps a very natural way of viewing these matters. A change of this kind could be made without necessitating revision of the main points addressed in this paper.

essentially to state or assert that such-and-such is the case) and performative sentences, which are characteristically employed—as he saw it—in doing the *other* kinds of things that one does with words, i.e., other than stating/asserting, such as giving orders, accepting offers, making promises, opening meetings and naming ships. The first seven lectures recorded in the posthumously published *How to Do Things with Words* [11] describe his ultimately unsuccessful attempt to define an appropriate distinguishing criterion—a criterion compatible with the basic assumption he made to the effect that performative sentences, unlike constatives, lack truth values. On our view it was in part that very assumption that prevented him from finding what he sought.[6]

We have characterised four types of performatives (*commanding, promising, requesting* and *declaring*) in terms of conventions that are all special cases of the convention for asserting, (sc-assert). So we are maintaining that the general form of all of these conventions is expressed by (sc-assert). Suppose now that agents j and k are users of communication system s, and that they are mutually aware of the content of the various instances of (sc-assert), each of which shows what the communicative acts performable in s mean. (j's seeing to it that $A1$ counts as an assertion that $B1$, j's seeing to it that $A2$ counts as an assertion that $B2$, j's seeing to it that $A3$ counts as a command to do $B3$, j's seeing to it that $A4$ counts as a request to do $B4$ ... and so on. The particular instances of (sc-assert) are, we may say, the code that constitutes s.)

In terms of the general form of communicative conventions, as expressed by (sc-assert), we may say that k, on witnessing j's performance of the act E_jA, forms a belief[7] the content of which is the consequent of (sc-assert):[8]

$$B_k I_s^* B \tag{1}$$

The key question now is this: under what conditions would k, as a rational agent, be prepared to trust the reliability of j's communicative act, and move from the belief expressed by (1) to (2)?

$$B_k B \tag{2}$$

Crucially, the answer to this question will depend on whether j's act is a performative or a constative, in Austin's sense. If it is a performative (for instance,

[6] What follows in due course below has its roots, in part at least, in an old idea. A number of early contributors to the literature on performatives (Lemmon, Åqvist and Lewis among them) suggested that the characteristic feature of performatives, in contrast to constatives, was 'verifiability by use', or the fact that 'saying makes it so'. See [7] for references and discussion.

[7] For present purposes we shall assume that the belief modality is assigned the logic of a relativised normal modality of type KD. We leave open the question as to whether the positive and negative introspection axioms (4. and 5.) should also be adopted for the logic of belief—the BDI framework ordinarily does adopt them—since this appears to have no bearing on the issues we are here primarily concerned to address.

[8] This is the default conclusion k will draw, on the assumption that j's act is a serious communicative act, i.e., a literal implementation of the governing (sc-assert) convention. For more detail on this, see [3].

one of the four types mentioned above) then k will be justified in making the inference from (1) to (2) provided merely that j is relevantly *empowered*—i.e., empowered to give commands, or empowered to make requests, or empowered to make promises, or empowered to make declarations. Consider commanding: if j is empowered/authorised to give commands, then his performance of the communicative act of commanding will indeed create an obligation on the addressee, k, to see to it that B. The scope formula to the right of the I_s^* operator in the convention is made true by j's performance of the communicative act. If he is empowered, then 'saying makes it so'.

The situation with respect to constatives, however, is quite different, for here there is no notion of empowerment or authorisation which would *itself* license the inference of B from $I_s^* B$. The closest one could get to such a notion would arise in cases in which j is deemed to be an authority on the subject about which he is making an assertion. But even then, his *saying* that B does not in itself *make it the case* that B. The signal he transmits is not 'verifiable by its use', but by appeal to the facts on which he is deemed to have expert, or authoritative, knowledge.

Does this analysis do justice to a distinction—considered by Austin to be important—between *fully performative* and *merely descriptive* usage of performative sentences? To explain the question, consider the utterance by the officer-in-command of the performative sentence 'I command you to open fire' in two different contexts: in the first, he is using the utterance itself to give the command (the fully performative usage), but in the second he is giving the command by signing a written order, and uttering the sentence 'I command you to open fire' so as to describe what (by signing) he is doing. The answer to the question is surely affirmative, for the difference between the two cases lies precisely in the evidence that would be required in order to justify inferring that an obligation (to open fire) had been created. For the fully performative case, the inference is justified if the communicator is indeed empowered to issue commands. But in the descriptive case more evidence is needed, for there the inference is justified only if it is the case *both* that the communicator is empowered to command *and* that he is performing another action by means of which he is exercising that authority (signing the written order). The descriptive case falls then in the category of constatives, according to our criterion, and this is surely in line with the point Austin had in mind regarding these different usages of performative sentences.

Towards the end of Lecture VII in [11], Austin gives up the pursuit of a distinguishing criterion. He says this:

> Now we failed to find a grammatical criterion for performatives, but we thought that perhaps we could insist that every performative *could* be in principle put into the form of an explicit performative, and then we could make a list of performative verbs. Since then we have found, however, that it is often not easy to be sure that, even when it is apparently in explicit form, an utterance is performative or that it is not; and typically anyway, we still have utterances beginning 'I state that...' which seem

> to satisfy the requirements of being performative, yet which surely are the making of statements, and surely are essentially true or false.
>
> It is time to make a fresh start on the problem. We want to reconsider more generally the senses in which to say something may be to do something, or in saying something we do something (and also perhaps to consider the different case in which *by* saying something we do something). [11, p. 91]

And then it is in the remaining lectures that Austin developed the now familiar distinction between *locutionary, illocutionary* and *perlocutionary*—indeed the latter two are already hinted at in the last bit of the passage just quoted. On our view, by contrast, there is no need to despair of finding a means of distinguishing constative from performative, but one should look not for a grammatical criterion, as Austin did, but at the grounds upon which one may justifiably infer a belief of form (2), above, from a belief of form (1).

There is also no need to resort to the theory of illocutionary acts; for we can supply a formal characterisation of different types of communicative acts—as outlined above—that makes no explicit use of the notion of illocutionarity, and which, in contrast to the approach taken by FIPA (`http://www.fipa.org/`, and in particular `http://www.fipa.org/repository/bysubject.html`), does not focus on the intended perlocutionary effects (what FIPA call the 'rational effects') of communication.

As indicated above, we give the analysis in terms of conventions that specify what ought to hold true when, for instance, an order is given or a request or promise is made. The normative, ideality operator is the key element, marking what will be the case if the governing convention is exploited in a way that conforms to the function that the communication/signalling system is designed to fulfil: the transmission of reliable information.

This, in turn, enables us to represent in a very straightforward way the belief of an agent who is aware of what a particular transmitted signal means (see above, formula (1)). The content of that belief is a normative expression, of form $I^*_s B$, where s is the communication system used in transmitting the signal. To be aware of what the signal means, on our view, is just to be aware of what, by convention, ought to be true given that the signal has been sent—it is to be aware of what would be the case if the reliability of the communicator could be trusted.[9] In contrast to some other approaches to the analysis of Agent Communication Languages (ACLs), we do not need to require the recipient to believe that the communicator is intending to produce in him the belief that B or that the communicator believes that B, or the belief that the communicator intends to get him to recognise that it is the communicator's intention to get him to believe that B... or indeed any other part of the convoluted Gricean mechanism.[10] Our

[9] The authors are considering a further paper which would address CTD issues arising in situations in which ideality conditions are not met, i.e., in which the communicator cannot be trusted.

[10] A considerably more detailed critique of the Gricean approach—in which of course the FIPA approach has its roots—is to be found in [7, chapter 4].

approach is very much simpler, and is made possible, essentially, by the role played by the normative operator.

We note in passing two additional advantages of our approach. First, it facilitates third party determination of what is said. Conventions, unlike intentions, beliefs, and desires, are quite public and open to objective assessment by disinterested parties. This is a key property if disputes are to be resolved in a manner that discourages cheating and reneging. Intentions, beliefs, desires and other mental states are, perhaps, not entirely inaccessible to neutral third parties. Even so, they are quite problematic in comparison to established conventions. This is apparent in the case of commercial transactions and electronic commerce in particular, but the point applies in the large, to all forms of communication for which it is valuable to be able to ascertain what was said in a fair and objective way. A second additional advantage of our approach is that it fits well with naturalistic accounts of the evolutionary emergence of communication and signalling systems. These appear in organisms—such as plants and bacteria—for which beliefs, intentions, and even desires are not plausibly ascribable. Philosophical work (e.g., [12, chapter 5] [13, chapter 4]) and scientific work (e.g., [14]) is underway, with results that, we believe, accord well with the theory on offer here. These are, however, matters that must await future research.

4 Reasoning About Messages Received

As we have seen, the formal characterisation of the belief state of a message recipient k enables us to represent what it would be for k to trust the reliability of the message sent: k would make the transition from a belief of type (1) to a belief of type (2). The formalism also facilitates the representation of the reasoning of k in a situation prior to that in which he has decided whether or not to trust messages he has received. This is important at least for the reason that, in trying to determine whether trust is justified, k—as a rational agent—will want to evaluate the consistency of the messages he has received with other beliefs he already holds.

To illustrate, suppose that k has received a message asserting that B, and a message asserting the conditional 'if B then C'. Then

$$B_k I_s^* B \wedge B_k I_s^* (B \rightarrow C) \tag{3}$$

Since the belief modality is normal it follows that

$$B_k (I_s^* B \wedge I_s^* (B \rightarrow C)) \tag{4}$$

Since the I_s^* modality is also normal, we also have

$$\vdash (I_s^* B \wedge I_s^* (B \rightarrow C)) \rightarrow I_s^* C \tag{5}$$

Since the belief modality, as a normal modality, is closed under logical consequence, it now follows from (4) and (5) that

$$B_k I_s^* C \tag{6}$$

Suppose now that, prior to receiving the two assertions, k already had the belief that C is false, i.e., $B_k \neg C$. Since the D. schema holds for the belief modality, it now follows that $\neg B_k C$, from which it follows by the normality of the belief modality that

$$\neg(B_k B \wedge B_k(B \to C)) \tag{7}$$

From this it now follows that k cannot trust both of the messages he has received, so long as he retains his belief (which he might, of course, choose to revise) that C is false. This is a rather simple example, but it nevertheless serves to exhibit how the combination of the logics of the belief and ideality operators may be used to represent aspects of the recipient k's reasoning, as he tries to work out which messages he can trust.

5 Business Communication and the Trusted Third Party

In [15] and [16] we develop a synthesis of Jones's convention-based analysis of communicative acts and Kimbrough's FLBC (Formal Language for Business Communication, see [17, 18, 19, 20, 21, 22, 23]), together with a detailed look at how the resulting combined formal models might be applied to the description of a trading scenario, involving, essentially, a buyer, a seller and a TTP (Trusted Third Party). Both [15] and [24] also discuss design of a Prolog implementation of the combined model. This combined model affords the prospect of deep and, we believe, plausibly complete formal integration of the theory described in this paper with the mundane, but complex, requirements of modern transaction processing. Moreover, we believe that the combined model will facilitate, in an entirely practicable and deployable manner, automated reasoning about communicated messages. These claims are under development and investigation. We content ourselves here with a brief indication of how the notions of conventional signalling systems, discussed in this paper, may be extended to support reasoning with additional sources of information.

Consider, then, a scenario in which a seller of goods and a prospective buyer communicate with each other not directly, but via a TTP. The seller, v, and the buyer, b, send via TTP messages of various kinds, which will typically include (among others) messages that serve to state facts about available goods and their mode of delivery, to request information, and—if a deal is initiated—to create obligations. As the recipient of these messages, the TTP (agent t) forms a set of beliefs of the type exhibited by

$$B_t I_s^* B \qquad \text{(cf. (1), above)} \tag{8}$$

where, as we have earlier emphasised, the scope formula to the right of the I_s^* operator may take a number of different forms, depending on the nature of the communicative act performed.

In terms of our formal theory, the role of the Third Party, qua *Trusted* Third Party, is easily articulated. The key task for which TTP is responsible is to determine which inferences may be accepted from schemas of type (8) to schemas of type

$$B_t B \quad \text{(cf. (2), above)} \tag{9}$$

and then to communicate to buyer and seller the result of his deliberations. Since t is assumed by v and b to be trusted, they will accept what he says as true (they may even be obligated to do so by the contractual agreements they made in order to participate in the system). In other words, for v and b the task of making inferences from schemas of type (1) to schemas of type (2) has been delegated to t: they trust him to do that job for them.

Read schemas of the form $Says_t B/A$ as 't says that B by seeing to it that A', where it is understood that 'says' is a generic term, referring to any type of communicative act. We define $Says_t B/A$ as follows

$$\text{(Df.says)} \qquad Says_t B/A \stackrel{\text{def}}{=} (E_t A \wedge (E_t A \Rightarrow_s I^*_s B))$$

where, as before, it is understood (i) that s is the conventional signalling (or communication) system that the agents t, v and b have adopted (for the purposes of their trade communication) and (ii) that the scope formula B may exhibit a range of different forms, depending on which type of communicative act $E_t A$ is.

Then we may represent the trusting beliefs that v and b have, vis-à-vis t, in the following way

$$B_v(Says_t B/A \rightarrow B) \tag{10}$$

$$B_b(Says_t B/A \rightarrow B) \tag{11}$$

And we may also wish to add that v and b and t are mutually aware that v and b have these trusting beliefs.[11]

Acknowledgements

Much of the research reported here has been carried out within the EU project ALFEBIITE (IST-1999-10298), the EU Working Group iTRUST (IST-2001-34910) and the EU 6th Framework Integrated Project TrustCoM (`www.eu-trust-com.com`). The financial support of the EU is gratefully acknowledged, as are the anonymous DEON06 reviewers of this paper, who provided a number of incisive and useful comments.

References

1. Verdicchio, M., Colombetti, M.: A logical model of social commitment for agent communication. In Dignum, F., ed.: Advances in Agent Communication. Lecture Notes in Computer Science, Volume 2922. Springer-Verlag, Berlin, Heidelberg, New York (2004) 128–145
2. Singh, M.P.: Agent communication languages: Rethinking the principles. IEEE Computer **31**(12) (1998) 40–47

[11] For an outline account of a logic of mutual belief, see [4].

3. Jones, A.J., Parent, X.: Conventional signalling acts and conversation. In Dignum, F., ed.: Advances in Agent Communication. Lecture Notes in Computer Science, Volume 2922. Springer-Verlag, Berlin, Heidelberg, New York (2004) 1–17
4. Jones, A.J., Parent, X.: A convention-based approach to agent communication languages. forthcoming (2006)
5. Stenius, E.: Mood and language game. Synthese **17** (1967) 254–274
6. Searle, J.R.: Speech Acts. Cambridge University Press, Cambridge, England (1969)
7. Jones, A.J.: Communication and Meaning — An Essay in Applied Modal Logic. Volume 168 of Synthese Library. D. Reidel, Dordrecht, Holland (1983)
8. Jones, A.J., Sergot, M.J.: A formal characterisation of institutionalised power. Journal of the Interest Group in Pure and Applied Logic (IGPL) **4**(3) (1996) 427–443 Reprinted in [25, pages 349–367].
9. Jones, A.J.: On normative-informational positions. In Lomuscio, A., Nute, D., eds.: Deontic Logic in Computer Science, Proceedings of the 7th Int Workshop on Deontic Logic in Computer Science, DEON 2004. Springer Verlag, Berlin, Germany (2004) 182–190
10. Searle, J.R., Vanderveken, D.: Foundations of Illocutionary Logic. Cambridge University Press, Cambridge, England (1985)
11. Austin, J.L.: How to Do Things with Words. Oxford at the Clarendon Press, Oxford, England (1962)
12. Skyrms, B.: Evolution of the Social Contract. Cambridge University Press, Cambridge, UK (1996)
13. Skyrms, B.: The Stag Hunt and the Evolution of Social Structure. Cambridge University Press, Cambridge, UK (2004)
14. Cangelosi, A., Parisi, D., eds.: Simulating the evolution of language. Springer-Verlag New York, Inc., New York, NY, USA (2002)
15. Jones, A.J., Kimbrough, S.O.: A note on modelling speech acts as signalling conventions. In Kimbrough, S.O., Wu, D.J., eds.: Formal Modelling in Electronic Commerce. International Handbooks on Information Systems. Springer, Berlin, Germany (2005) 325–342
16. Jones, A.J., Kimbrough, S.O.: A convention-based approach to a formal language for business communication. draft manuscript, University of Pennsylvania, Philadelphia, PA (2006)
17. Kimbrough, S.O.: EDI, XML, and the transparency problem in electronic commerce. In Kimbrough, S.O., Wu, D.J., eds.: Formal Modelling in Electronic Commerce: Representation, Inference, and Strategic Interaction. International Handbooks on Information Systems. Springer, Berlin, Germany (2005) 201–227
18. Kimbrough, S.O., Tan, Y.H.: On lean messaging with unfolding and unwrapping for electronic commerce. International Journal of Electronic Commerce **5**(1) (2000) 83–108
19. Kimbrough, S.O., Moore, S.A.: On automated message processing in electronic commerce and work support systems: Speech act theory and expressive felicity. ACM Transactions on Information Systems **15**(4) (October 1997) 321–367
20. Kimbrough, S.O.: Reasoning about the objects of attitudes and operators: Towards a disquotation theory for representation of propositional content. In: Proceedings of ICAIL '01, International Conference on Artificial Intelligence and Law. (2001)
21. Kimbrough, S.O., Yang, Y.: On representing special languages with FLBC: Message markers and reference fixing in SeaSpeak. In Kimbrough, S.O., Wu, D.J., eds.: Formal Modelling in Electronic Commerce. International Handbooks on Information Systems. Springer, Berlin, Germany (2005) 297–324 ISBN 3-540-21431-3.

22. Kimbrough, S.O.: A note on interpretations for federated languages and the use of disquotation. In Gardner, A., ed.: Proceedings of the Tenth International Conference on Artificial Intelligence and Law (ICAIL-2005), Bologna, Italy, In cooperation with ACM SIGART and The American Association for Artificial Intelligence (2005) 10–19
23. Kimbrough, S.O.: A note on the Good Samaritan paradox and the disquotation theory of propositional content. In Horty, J., Jones, A.J., eds.: Proceedings of ΔEON'02, Sixth International Workshop on Deontic Logic in Computer Science. (May 2002) 139–148
24. Abrahams, A.S., Bacon, J.M., Eyers, D.M., Jones, A.J., Kimbrough, S.O.: Introducing the fair and logical trade project. In: Workshop on Contract Architectures and Languages (CoALa2005). (2005)
25. Valdés, E.G., et al., eds.: Normative Systems in Legal and Moral Theory – Festschrift for Carlos E. Alchourrón and Eugenio Bulygin. Duncker & Humblot, Berlin, Germany (1997)

Permissions and Uncontrollable Propositions in DSDL3: Non-monotonicity and Algorithms

Souhila Kaci[1] and Leendert van der Torre[2]

[1] Centre de Recherche en Informatique de Lens (C.R.I.L.)–C.N.R.S, France
[2] University of Luxembourg

Abstract. In this paper we are interested in non-monotonic extensions of Bengt Hansson's standard dyadic deontic logic 3, known as DSDL3. We study specificity principles for DSDL3 with both controllable and uncontrollable propositions. We introduce an algorithm for minimal specificity which not only covers obligations but also permissions, and we discuss the distinction between weak and strong permissions. Moreover, we introduce ways to combine algorithms for minimal and maximal specificity for DSDL3 with controllable and uncontrollable propositions, based on 'optimistic' and 'pessimistic' reasoning respectively.

1 Introduction

Hansson's standard dyadic deontic logic 3 [9], known as DSDL3, is an extension of standard deontic logic, SDL, also known as system KD, with dyadic obligations. It has been called a defeasible deontic logic because it does not satisfy unrestricted strengthening of the antecedent, the derivation of $O(p|q \wedge r)$ from $O(p|q)$. Spohn's axiom in his axiomatization of DSDL3 [18] informs us that strengthening of the antecedent only holds conditional to a permission, where $P(p|q) = \neg O(\neg p|q)$:

$$P(r|q) \rightarrow (O(r \rightarrow p|q) \leftrightarrow O(p|q \wedge r))$$

Monotonic and non-monotonic extensions to DSDL3 have been studied to strengthen the antecedent. The former has been studied using notions of settledness or necessity by, for example, Prakken and Sergot [16]. The latter has been directly inspired by the interpretation of DSDL3 as a theory of default conditionals, or more generally as a framework for non-monotonic logic following the work of Shoham [17] and Kraus, Lehmann and Magidor [11]. The main approach in this setting to strengthen the antecedent is based on the so-called minimal specificity principle by, amongst others, Lehmann and Magidor [13] and Boutilier [4]. These non-monotonic extensions are accompanied by efficient algorithms [15], though these algorithms have the drawback to be defined only for sets of dyadic obligations, not for more complex formulae such as, for example, permissions.

DEON2006 has a special focus on deontic notions in the theory, specification and implementation of artificial normative systems, such as electronic institutions, norm-regulated multi-agent systems, and artificial agent societies more generally. In the context of agent theory, Boutilier studies non-monotonic DSDL3 extended with the distinction between controllable and uncontrollable propositions [5]. Though this distinction

L. Goble and J.-J.C. Meyer (Eds.): DEON 2006, LNAI 4048, pp. 161–174, 2006.

originates from the areas of discrete event systems and control theory, Boutilier uses it as a simple theory of decision (or action) in qualitative decision theory. It has been further developed by, for example, Lang *et al.* [12] and Cholvy and Garion [8].

In this paper we are interested in the following questions:

1. How can we extend non-monotonic DSDL3 with permissions?
2. What is the relevance of the distinction between controllable and uncontrollable propositions in non-monotonic DSDL3?

Despite the work in non-monotonic extensions of DSDL3 for default conditionals [4, 1], desires [12], and preferences [2, 10] in artificial intelligence, somewhat surprisingly the extension of existing algorithms to permission seems to have received less attention. Apparently, whereas permission plays a central role in deontic logic, the analogous negation of default conditionals, absence of desires, and non-strict preference are of less interest in the other research areas. Though there are related extensions, such as ones dealing with equalities, the only extension of algorithms concerned with permission we are aware of has been proposed by Booth and Paris [3]. However, their algorithm is inefficient as it requires the construction of a potentially large number of pre-orders. Our algorithm constructs only the minimal specific pre-order.

When a distinction between controllable and uncontrollable propositions is introduced in DSDL3, one may revisit the use of the minimal specificity principle. For example, another option would be to use the *maximal* specificity principle, which does not assume that worlds are as normal as possible, or gravitate towards the ideal, but which assumes that worlds are as abnormal as possible, or gravitate towards the worst. We argue that whereas the 'optimistic' reasoning underlying the minimal specificity principle may make sense for controllable propositions, because, for example, any rational agent will see to it that the best state will be realized, for uncontrollable propositions a more 'pessimistic' attitude may be used as well. We also study the combination of both kinds of reasoning.

In this paper we do not discuss the advantages and disadvantages of DSDL3, nor of non-monotonic DSDL3, since they have already been discussed extensively during the last 35 years. For the same reason we do not present the usual examples again, but we focus on the logical properties of the system. Next to SDL, DSDL3 is probably the best known deontic logic, and the most successful logic developed in deontic logic and used outside this area (in particular in artificial intelligence). We also do not discuss its well known relation to preference logic, due to the fact that "the best q are p" is equivalent to "$p \wedge q$ is preferred to $\neg p \wedge q$" in several preference logics, see for example [20]. However, we believe that the preference-based reading of DSDL3 suggests that the 'optimistic' reading of non-monotonic DSDL3 may be arbitrary, because in this representation, there does not seem to be a reason why we compare the best $p \wedge q$ worlds, and not the worst ones (or, for example, the ones in the middle).

The layout of this paper is as follows. In Section 2 we repeat the definitions of non-monotonic DSDL3, and in Section 3 we present the algorithm to compute the most specific pre-order satisfying a set of obligations and permissions. In Section 4 we repeat the distinction between controllable and uncontrollable propositions, and we present the algorithms for the uncontrollable case. In Section 5 we consider the merging of the two kinds of obligations.

2 Non-monotonic Extension of DSDL3

Norm specifications consist of obligations and permissions. $O(p|q)$ is read as 'p is obligatory if q' and $P(p|q)$ is read as 'p is permitted if q'.

Definition 1 (Norm specification). *Let $\mathcal{A}$ be a finite set of propositional atoms, and $\mathcal{L}$ a propositional logic based on $\mathcal{A}$. A norm specification is a set of norms $\mathcal{C} = \mathcal{C}_O \cup \mathcal{C}_P$ where for $p_i, q_i, p'_j, q'_j \in \mathcal{L}$:*

$$\mathcal{C}_O = \{C_i = O(p_i|q_i) \mid i = 1 \dots n\}$$

$$\mathcal{C}_P = \{C'_j = P(p'_j|q'_j) \mid j = 1 \dots m\}$$

The norms are interpreted on a total pre-order on the propositional valuations (or worlds).

Definition 2 (Monotonic semantics). *Let $\mathcal{A}$ and $\mathcal{L}$ be as before, let "worlds" W be the set of propositional valuations of $\mathcal{L}$, and $\succeq$ a total pre-order on W. Let $|\phi|$ be the set of propositional models of ϕ. We write $w \succ w'$ for $w \succeq w'$ without $w' \succeq w$, and we write $\max(p, \succeq)$ for $\{w \in |p| \mid \forall w' \in |p|$ we have $w \succeq w'\}$. Satisfiability is defined as follows:*

$\langle W, \succeq\rangle \models O(p|q)$ *iff* $\max(q, \succeq) \subseteq |p|$, *which is equivalent to stating that* $\forall \omega \in \max(p \wedge q, \succeq), \forall \omega' \in \max(p \wedge \neg q, \succeq)$, *we have* $\omega \succ \omega'$.

Moreover, we define

$\langle W, \succeq\rangle \models P(p|q)$ *iff* $\langle W, \succeq\rangle \models \neg O(\neg p|q)$ *which is equivalent to stating that* $|p \wedge q| \neq \emptyset$, *and* $\forall \omega \in \max(p \wedge q, \succeq), \forall \omega' \in \max(p \wedge \neg q, \succeq)$, *we have* $\omega \succeq \omega'$.

A total pre-order $\succeq$ is a model of (satisfies) a norm specification $\mathcal{C}$ iff it satisfies each norm in the specification $\mathcal{C}$. We write $\mathcal{M}(\mathcal{C})$ for the set of models of $\mathcal{C}$.

For an infinite set of propositional atoms $\mathcal{A}$, a more sophisticated definition proposed by Lewis [14] and popularized in AI by Boutilier [4], deals with infinite descending chains. They define $\langle W, \succeq\rangle \models O(p|q)$ iff $|q| = \emptyset$, or there exists a $p \wedge q$ world w such that there does not exist a $\neg p \wedge q$ world w' with $w' \geq w$.

In the algorithm we do not use the total pre-order $\succeq$ directly, but we use an equivalent representation as an ordered partition, defined as follows. E_1 is the set of ideal worlds, and E_n is the set of worst worlds.

Definition 3 (Ordered partition). *A sequence of sets of worlds of the form $(E_1, \cdots, E_n)$ is an ordered partition of W iff $\forall i$, E_i is nonempty, $E_1 \cup \cdots \cup E_n = W$ and $\forall i, j$, $E_i \cap E_j = \emptyset$ for $i \neq j$. An ordered partition of W is associated with pre-order $\succeq$ on W iff $\forall \omega, \omega' \in W$ with $\omega \in E_i, \omega' \in E_j$ we have $i \leq j$ iff $\omega \succeq \omega'$.*

In this section we compare total pre-orders based on the so-called *minimal specificity principle* which is also known as System Z or gravitating towards the ideal.

Definition 4 (Preference semantics). *Let $\succeq$ and $\succeq'$ be two total pre-orders on a set of worlds W represented by ordered partitions $(E_1, \cdots, E_n)$ and $(E'_1, \cdots, E'_m)$ respectively. We say that $\succeq$ is at least as specific as $\succeq'$, written as $\succeq' \sqsubseteq \succeq$, iff $\forall \omega \in W$, if $\omega \in E_i$ and $\omega \in E'_j$ then $i \leq j$. $\succeq$ is less specific as $\succeq'$, written as $\succeq \sqsubset \succeq'$, iff $\succeq \sqsubseteq \succeq'$ without $\succeq' \sqsubseteq \succeq$. $\succeq$ is the least specific pre-order among a set of pre-orders $\mathcal{O}$ if there is no $\succeq'$ in $\mathcal{O}$ such that $\succeq' \sqsubset \succeq$.*

The following example illustrates minimal specificity.

Example 1. Consider the single obligation $O(p|q)$. Applying the minimal specificity principle gives the following model $\succeq = (|p \wedge q| \cup |p \wedge \neg q| \cup |\neg p \wedge \neg q|, |\neg p \wedge q|)$. The ideal worlds in this model are those which do not violate the obligation. More precisely, worlds in $|p \wedge q|$ belong to the set of ideal worlds since they fulfill the obligation, but worlds in $|p \wedge \neg q|$ and $|\neg p \wedge \neg q|$ are ideal too since they do not violate the rule even if they do not fulfill it.

Shoham [17] defines non-monotonic consequences of a logical theory as all formulas which are true in the 'preferred' models of the theory. An attractive property is case is which there is only one 'preferred' model, because in that case it can be decided whether a formula non-monotonically follows from a logical theory by calculating the unique 'preferred' model, and testing whether the formula is satisfied by the 'preferred' model. Likewise, finding all non-monotonic consequences can be found by calculating the unique 'preferred' model and characterizing all formulas satisfied by this model.

Definition 5 (Non-monotonic entailment). *A norm specification C preferentially implies $O(p|q)$ (or $P(p|q)$) if and only if for least specific models of C are also a model of $O(p|q)$ (or $P(p|q)$).*

The following example illustrates non-monotonic entailment, which can be used to reason about violations or exceptions.

Example 2 (Continued). The norm specification consisting of the obligation $O(p|q)$ preferentially implies $O(p|q \wedge r)$, but the norm specification consisting of both $O(p|q)$ and $O(\neg p|q \wedge r)$ does not preferentially imply $O(p|q \wedge r)$.

3 Algorithm for Obligations and Permissions

The algorithm to calculate the least specific pre-order of a norm specification is given in Algorithm 1.1. The basic idea of the algorithm is to construct the least specific pre-order by calculating the sets of worlds of the ordered partition, going from ideal to the worst worlds. It extends the known algorithm [15, 2] for obligations with the second line to check whether the individual permissions are consistent, the inner while loop, to deal with permissions, and the second removal clause in the end to take care of the removal of permissions. Given a norm specification, let $\mathcal{C} = \mathcal{C}_O \cup \mathcal{C}_P$ where

$$\mathcal{C}_O = \{C_i : O(p_i|q_i)|i = 1 \cdots n\} \text{ and } \mathcal{C}_P = \{C'_j : P(p'_j|q'_j)|j = 1 \cdots m\}$$

and let $\mathcal{L} = \{(L(C_i), R(C_i)) : C_i \in \mathcal{C}_O\} \cup \{(L(C'_j), R(C'_j)) : C'_j \in \mathcal{C}_P\}$, where $L(C_i) = |p_i \wedge q_i|$, $R(C_i) = |\neg p_i \wedge q_i|$, $L(C'_j) = |p'_j \wedge q'_j|$ and $R(C'_j) = |\neg p'_j \wedge q'_j|$.

Algorithm 1.1. Handling obligations and permissions

```
begin
    if any L(C'_i) = ∅ then Stop (inconsistent constraints);
    m = 0 ; W = set of all models of ℒ ;
    while W ≠ ∅ do
        - m ← m + 1, i = 1;
        - E_m = {ω : ∀(L(C_i), R(C_i)) ∈ ℒ_C, ω ∉ R(C_i)} ;
        while i = 1 do
            i=0;
            for each (L(C'_j), R(C'_j) in ℒ_C do
                if (L(C'_j) ∩ E_m = ∅ and R(C'_j) ∩ E_m ≠ ∅) then E_m = E_m\R(C'_j); i=1;
        - if E_m = ∅ then Stop (inconsistent constraints);
        - W = W − E_m ;
        - remove from ℒ_C each (L(C_i), R(C_i)) such that L(C_i) ∩ E_m ≠ ∅ ;
        - remove from ℒ_C each (L(C'_j), R(C'_j)) such that L(C'_j) ∩ E_m ≠ ∅;
    return (E_1, ···, E_m)
end
```

If we consider the case without permissions, then the algorithm calculates the next equivalence class of the partitioning E_m by taking all worlds which do not violate one of the obligations. Once an obligation is satisfied by an equivalence class, it no longer constrains the construction of the preorder, and can be removed.

With permissions, the construction is complicated since we cannot directly define the equivalence class E_m. The definition of E_m in line 6 of the algorithm is therefore an upper bound of this class. To make sure that all permissions are satisfied, thereafter some worlds may have to be removed from E_m. Moreover, once some worlds are removed, it may be the case that permissions which were already checked are now violated, so we have to reconsider them too (for which we use the variable j). Removal of permissions is analogous to the removal of obligations.

In the remainder of this section, we prove that the algorithm calculates the least specific pre-order.

Lemma 1. *The total pre-order computed by algorithm 1 belongs to the set of least specific pre-orders of $\mathcal{C}$.*

Proof. This can be checked by construction. Since the set of constraints is finite, the algorithm terminates. Since E_m cannot be the empty set, the sequence is an ordered partition. Let $\succeq = (E_1, \cdots, E_n)$ be this total pre-order. Suppose that $\succeq$ doesn't belong to the set of least specific pre-orders of $\mathcal{C}$, i.e., for some $\omega \in E_j$ we could have put ω in E_i with $i < j$. However $\omega \notin E_j$ because either:

obligations in $\mathcal{C}_O$. *$\omega \in E_j$ means that ω falsifies obligations in $\mathcal{C}_O$ which are not falsified by worlds in E_i with $i < j$. So if we put ω in E_i with $i < j$, we get a contradiction,*

permissions in $\mathcal{C}_P$. *Following the algorithm, $\omega \in E_i$ because otherwise there is some permission $P(p'|q')$ in $\mathcal{C}_P$ for which the best worlds of $p' \wedge q'$ are in E_k and the best worlds of $\neg p' \wedge q'$ are in E_l with $l < k$ which is a contradiction.*

To show the uniqueness of the least specific pre-order of $\mathcal{C}$, we follow the line of the proof given in [1]. We first define the maximum of two preference orders.

Definition 6. *Let $\succeq$ and $\succeq'$ be two preference orders represented by their well ordered partitions $(E_1, \cdots, E_n)$ and $(E'_1, \cdots, E'_m)$ respectively. We define the $\mathcal{MAX}$ operator by $\mathcal{MAX}(\succeq, \succeq') = (E''_1, \cdots, E''_{min(n,m)})$, such that $E''_1 = E_1 \cup E'_1$ and $E''_k = (E_k \cup E'_k) - (\bigcup_{i=1,\cdots,k-1} E''_i)$ for $k = 2, \cdots, min(n,m)$, and the empty sets E''_k are eliminated by renumbering the non-empty ones in sequence.*

Lemma 2 proves the uniqueness of the least specific pre-order in $\mathcal{M}(\mathcal{C})$.

Lemma 2. *If there is a minimal specific pre-order, then it is unique.*

Proof. *We first show that $\mathcal{MAX}(\succeq, \succeq') \in \mathcal{M}(\mathcal{C})$ (1). Let $\succeq = (E_1, \cdots, E_h)$, $\succeq' = (E'_1, \cdots, E'_{h'})$, $\succeq'' = (E''_1, \cdots, E''_{min(h,h')})$, and $P(p|q) \in \mathcal{C}$. $\succeq, \succeq' \in \mathcal{M}(\mathcal{C})$, i.e., $\succeq \models P(p|q)$ and $\succeq' \models P(p|q)$. In other words, $\max(p \wedge q, \succeq) \subseteq E_i$ and $\max(p \wedge \neg q, \succeq) \subseteq E_j$ such that $i \leq j$ and $\max(p \wedge q, \succeq') \subseteq E'_k$ and $\max(p \wedge \neg q, \succeq') \subseteq E'_l$ such that $k \leq l$. Following Definition 6, $\max(p \wedge q, \succeq'') \subseteq E''_{\min(i,k)}$ and $\max(p \wedge \neg q, \succeq'') \subseteq E''_{\min(j,l)}$. Since $i \leq j$ and $k \leq l$ we have $\min(i,k) \leq \min(j,l)$. We conclude $\succeq'' \models P(p|q)$. The proof for $O(p|q)$ is analogous and can be found in [1]. Consequently, $\mathcal{MAX}(\succeq, \succeq') \in \mathcal{M}(\mathcal{C})$.*

Moreover, we have that $\mathcal{MAX}(\succeq, \succeq')$ is less specific than or identical to both $\succeq$ and $\succeq'$ (2), the proof can be found also in [1].

Finally, we prove that the lemma follows from the two items by contradiction. So suppose that there are two distinct minimal specific orders $\succeq$ and $\succeq'$. Then according to item (1), $\mathcal{MAX}(\succeq, \succeq')$ is also a model of the preference specification and according to item (2), it is less specific than either $\succeq$ or $\succeq'$. Contradiction.

We can now conclude:

Theorem 1. *Algorithm 1 computes the least specific model of $\mathcal{M}(\mathcal{C})$.*

Proof. *Following Lemma 1 it computes a preference order which belongs to the set of the least specific models and following Lemma 2, this preference order is unique.*

4 Ought-to-Be and Ought-to-Do

Some approaches introduce a full fledged logic of actions in theories of rational decision, but Boutilier [5] introduces the distinction between controllable and uncontrollable propositions from discrete event systems and control theory in his qualitative decision theory. This relatively simple approach to actions has reached some popularity, see [12, 7, 19]. The reason is that this abstract representation of actions – which are typically called decision variables – lets us focus on other aspects of decision making than the usual issues concerning causality, frame axioms, etc.

In the context of deontic logic, the distinction between controllable and uncontrollable propositions can be used as a simple way to distinguish and study the relation between ought-to-be and ought-to-do obligations. Consider a dynamic deontic logic. Dynamic logic contains expressions like $[\alpha]p$, which can be read as 'after doing or executing α, p holds, and dynamic deontic logic contains expressions $O(\alpha)$, expressing an ought-to-do obligation for α, and $O(p)$, expressing an ought-to-be obligation for p. Now assume that we add propositions $done(\alpha)$ for every action statement α, together with the axiom $[\alpha]done(\alpha)$. In that case, we may say that $O(done(\alpha))$ is a kind of ought-to-do obligation. Summarizing, if we have $O(p)$ – which is short for $O(p|\top)$ for any tautology $\top$ – for uncontrollable p, then we may call it an ought-to-be obligation, and if we have $O(x)$ for controllable x, then we may call it an ought-to-do obligation.

Having made the distinction between the two kinds of obligations, we are now faced by the question whether their logic is distinct. Neither Boutilier nor the other researchers working on controllable and uncontrollable propositions seem to have introduced distinct logics or distinct non-monotonic extensions to represent the two kinds of obligations.

When we consider DSDL3 and the related minimal specificity principle, we may observe that both of them are 'optimistic', in the following sense. First, the logic of $O(p|q)$ only considers the best or ideal worlds. Second, the non-monotonic extension of the minimal specificity principle assumes that each world is as good as possible. But why not select a more 'pessimistic' approach? Note that the notion of 'optimistic' and 'pessimistic' should be read metaphorically, referring to psychological or decision-theoretic notions.

For controllable propositions, this choice seems justified to us. The agent can control the truth value of the propositions, and therefore he or she should see to it that the best world will be realized. But for uncontrollable propositions, it is less clear. the choice of the best worlds seems rather arbitrary. Moreover, in decision making, there is often a trend to reason pessimistically about the environment.

Therefore, in the remainder of this paper we study 'pessimistic' kinds of reasoning for ought-to-be obligations. The 'pessimistic' alternatives are that $O(p|q)$ no longer means that the best q worlds are p worlds, but that the worst q worlds are $\neg p$ worlds. Moreover, instead of assuming that worlds are as good as possible, we assume that worlds are as bad as possible. As one may expect, the 'pessimistic' definition and the 'pessimistic' specificity principle go well together.

From now on, we write O^+ and P^+ to refer to the usual kinds of 'optimistic' obligations and permissions of DSDL3, as studied thus far in this paper. Moreover, we introduce new 'pessimistic' obligations and permissions, which we write as O^- and P^-.

Definition 7 (Norm specification). *Let C and $\mathcal{U}$ be two disjoint finite sets of controllable resp. uncontrollable propositional atoms, and $\mathcal{L}$ a propositional logic based on $C \cup \mathcal{U}$. A norm specification is a set of norms $\mathcal{C} = \mathcal{C}_O^+ \cup \mathcal{C}_P^+ \cup \mathcal{C}_O^- \cup \mathcal{C}_P^-$, where the 'optimistic' $\mathcal{C}_O^+ \cup \mathcal{C}_P^+$ are defined using C only, and $\mathcal{C}_O^- \cup \mathcal{C}_P^-$ using U only.*

As before, the norms are interpreted on a total pre-order. The semantics are straightforward.

Definition 8 (Monotonic semantics). *Satisfiability is defined as follows:*
$\langle W, \succeq \rangle \models O^+(p|q)$ *iff* $\max(q, \succeq) \subseteq |p|$
$\langle W, \succeq \rangle \models O^-(p|q)$ *iff* $\min(q, \succeq) \subseteq |\neg p|$
Moreover, we define
$\langle W, \succeq \rangle \models P^+(p|q)$ *iff* $\langle W, \succeq \rangle \models \neg O^+(\neg p|q)$
$\langle W, \succeq \rangle \models P^-(p|q)$ *iff* $\langle W, \succeq \rangle \models \neg O^-(\neg p|q)$

The non-monotonic semantics based on maximal specificity principle are straightforward too, and the maximal specificity algorithm is simply the dual of the minimal specificity algorithm. The basic idea of the algorithm is to construct the most specific pre-order by calculating the sets of worlds of the ordered partition, going from worst to the ideal worlds. As can easily be verified, we obtained algorithm 2 by replacing left hand side and right hand side in various places. Moreover, the pre-order is constructed from worst to ideal class, so in the last line we have to reverse the order of the classes. Let $\mathcal{C} = \mathcal{C}_O \cup \mathcal{C}_P$ where

$$\mathcal{C}_O = \{C_i : O(p_i|q_i)|i = 1 \cdots n\} \text{ and } \mathcal{C}_P = \{C'_j : P(p'_j|q'_j)|j = 1 \cdots m\}$$

We put $\mathcal{L} = \{(L(C_i), R(C_i)) : C_i \in \mathcal{C}_O\} \cup \{(L(C'_j), R(C'_j)) : C'_j \in \mathcal{C}_P\}$, where $L(C_i) = |p_i \wedge q_i|$, $R(C_i) = |\neg p_i \wedge q_i|$, $L(C'_j) = |p'_j \wedge q'_j|$ and $R(C'_j) = |\neg p'_j \wedge q'_j|$.

Algorithm 1.2. Handling ought-to-be obligations and permissions

```
begin
    if any R(C'_i) = ∅ then Stop (inconsistent constraints);
    m = 0 ; W = set of all models of 𝓛 ;
    while W ≠ ∅ do
        – m ← m + 1, i = 1;
        – E_m = {ω : ∀(L(C_i), R(C_i)) ∈ 𝓛_C, ω ∉ L(C_i)} ;
        while i = 1 do
            i=0;
            for each (L(C'_j), R(C'_j) in 𝓛_C do
                if (L(C'_j) ∩ E_m ≠ ∅ and R(C'_j) ∩ E_m = ∅) then E_m = E_m \ L(C'_j); i=1;
        – if E_m = ∅ then Stop (inconsistent constraints);
        – W = W − E_m ;
        – remove from 𝓛_C each (L(C_i), R(C_i)) such that R(C_i) ∩ E_m ≠ ∅ ;
        – remove from 𝓛_C each (L(C'_j), R(C'_j)) such that R(C'_j) ∩ E_m ≠ ∅);
    return (E'_1, ···, E'_l) s.t. ∀1 ≤ h ≤ l, E'_h = E_{l−h+1}
end
```

The 'pessimistic' approach may be criticized, just as we have criticized the 'optimistic' approach, and more sophisticated approaches may be developed. However, our more general point is that one may reason in a different way with ought-to-be and ought-to-do obligations, or with controllable and uncontrollable propositions - where the above suggestion is just an instance of that general idea. Any approach that deals with these two kinds of obligations in a different way has to solve the problem we address in the following section: how can these approaches be combined?

5 Merging Ought-to-Be and Ought-to-Do

Figure 1 visualizes our approach to combine two distinct ways to reason with ought-to-be and ought-to-do in DSDL3. Our approach is based on 'optimistic' reasoning about controllables, and 'pessimistic' reasoning about uncontrollables. The former represent the agent's rationality to choose the optimal state if he has the power to do so, and the latter represents Wald criterion: the decision-maker selects that strategy which is associated with the best possible worst outcome.

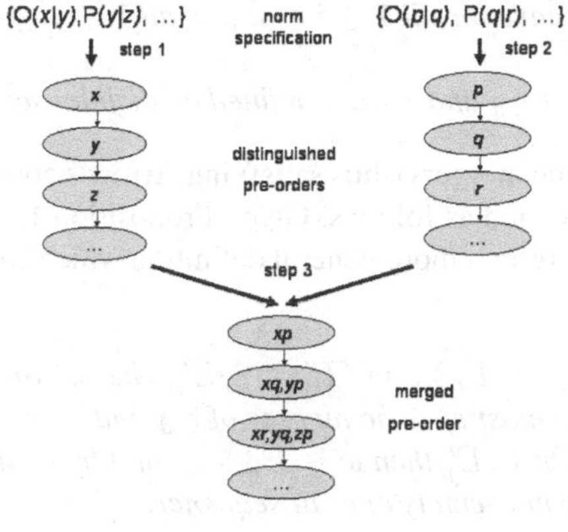

Fig. 1. Combining 'optimistic' and 'pessimistic' norms

The norms in the norm specifications are interpreted as constraints on total pre-orders on worlds. Moreover, there are non-monotonic reasoning mechanisms to calculate distinguished pre-orders from the norm specifications. There is an 'optimistic' algorithm to calculate the unique distinguished total pre-order from the 'optimistic' norm specification (step 1), and a 'pessimistic' algorithm to calculate the unique distinguished total pre-order from the 'pessimistic' norm specification (step 2). Since we need a single preference order for decision making, we need to merge the two total pre-orders (step 3). We distinguish symmetric and a-symmetric mergers.

In this section we consider the merger of the least specific pre-order satisfying the 'optimistic' norm specification, and the most specific pre-order satisfying the 'pessimistic' norm specification. From now on, let $\mathcal{L}$ be a propositional language on disjoint sets of controllable and uncontrollable propositional atoms $\mathcal{C} \cup \mathcal{U}$. A norm specification consists of an 'optimistic' and a 'pessimistic' norm specification, i.e., 'optimistic' norms on controllables, and 'pessimistic' norms on uncontrollables. In general, let $\succeq$ be the merger of a pre-order $\succeq_d$ generated by 'optimistic' reasoning about ought-to-do obligations and a pre-order $\succeq_b$ generated by 'pessimistic' reasoning about ought-to-be obligations. We assume that the following conditions hold, known in economic theory as Arrows' conditions:

Definition 9. *Let $\succeq_d$, $\succeq_b$ and $\succeq$ be three total-preorders on the same set. $\succeq$ is a merger of $\succeq_d$ and $\succeq_b$ if and only if the following three conditions hold:*

If $w_1 \succ_d w_2$ *and* $w_1 \succ_b w_2$ *then* $w_1 \succ w_2$
If $w_1 \succ_d w_2$ *and* $w_1 \succeq_b w_2$ *then* $w_1 \succeq w_2$
If $w_1 \succeq_d w_2$ *and* $w_1 \succ_b w_2$ *then* $w_1 \succeq w_2$

Given two arbitrary pre-orders, there are many possible mergers. We therefore again consider distinguished pre-orders below.

The two minimal and maximal specific pre-orders of 'optimistic' and 'pessimistic' preference specifications satisfy the property that no two sets are disjoint.

Proposition 1. *Let* $(E_1, \cdots, E_n)$ *and* $(E'_1, \cdots, E'_m)$ *be the ordered partitions of* $\succeq_d$ *and* $\succeq_b$ *respectively. We have for all* $1 \leq i \leq n$ *and all* $1 \leq j \leq m$ *that* $E_i \cap E'_j \neq \emptyset$.

Proof. Due to the fact that $\succeq_d$ *and* $\succeq_b$ *are defined on disjoint sets of variables.*

The least and most specific mergers (thus satisfying Arrow's conditions) are unique and identical, and can be obtained as follows. Given Proposition 1, thus far nonempty sets E''_k do not exist, but we prefer a more general definition which can also be used in other mergers.

Proposition 2. *Let* $(E_1, \cdots, E_n)$ *and* $(E'_1, \cdots, E'_m)$ *be the ordered partitions of* $\succeq_d$ *and* $\succeq_b$ *respectively. The most specific merger of* $\succeq_d$ *and* $\succeq_b$ *is* $\succeq = (E''_1, \cdots, E''_{n+m})$ *such that if* $\omega \in E_i$ *and* $\omega \in E'_j$ *then* $\omega \in E''_{i+j-1}$, *and by eliminating nonempty sets* E''_k *and renumbering the non-empty ones in sequence.*

The most specific merger is illustrated by the following example.

Example 3. Consider the 'optimistic' preference specification $O(p)$ and the 'pessimistic' preference specification $O(m)$, where p and m stand respectively for "I have to work on a project in order to get money" and "my boss has given me money to pay my conference fee".

We have $\succeq_d = (\{mp, \neg mp\}, \{m\neg p, \neg m\neg p\})$ and $\succeq_b = (\{mp, m\neg p\}, \{\neg mp, \neg m\neg p\})$. The most specific merger is $\{\{mp\}, \{\neg mp, m\neg p\}, \{\neg m\neg p\}\}$.

Analogously we may also consider the product rule (...then $\omega \in E''_{i*j}$), or other symmetric mergers.

The minimax merger gives priority to the preorder associated to the 'optimistic' preference specification and computed following the minimal specificity principle ($\succeq_d$) over the one associated to the 'pessimistic' preference specification and computed following the maximal specificity principle ($\succeq_b$). Indeed alternatives are first ordered w.r.t. $\succeq_d$ and only in the case of equality $\succeq_b$ is considered.

Definition 10 (Minimax merger). $w_1 \succ w_2$ *iff* $w_1 \succ_1 w_2$ *or* $w_1 \sim_1 w_2$ *and* $w_1 \succ_2 w_2$.

The minimax merger can be defined as follows.

Proposition 3. *Let* $(E_1, \cdots, E_n)$ *and* $(E'_1, \cdots, E'_m)$ *be the ordered partitions of* $\succeq_d$ *and* $\succeq_b$ *respectively. The result of merging* $\succeq_d$ *and* $\succeq_b$ *is* $\succeq = (E''_1, \cdots, E''_{n\times m})$ *such that if* $\omega \in E_i$ *and* $\omega \in E'_j$ *then* $\omega \in E''_{(i-1)*m+j}$.

Example 4. (continued) The minimax merger of the norm specification is $\{\{mp\}, \{\neg mp\}, \{m\neg p\}, \{\neg m\neg p\}\}$.

The principle of this merger is similar to minimax merger. The dictator here is the pre-order associated to the 'pessimistic' preference specification and computed following the maximal specificity principle.

Definition 11. $w_1 \succ w_2$ *iff* $w_1 \succ_2 w_2$ *or* $w_1 \sim_2 w_2$ *and* $w_1 \succ_1 w_2$.

Example 5. Consider the 'optimistic' $O(p)$ and the 'pessimistic' $O(m)$. The merger is $\{\{mp\}, \{m\neg p\}, \{\neg mp\}, \{\neg m\neg p\}\}$.

The problem of handling preferences on controllable variables and uncontrollable variables separately is that it is not possible to express *interaction* between the two kinds of variables. For example my decision on whether I will work hard to finish a paper (which is a controllable variable) depends on the uncontrollable variable "money", decided by my boss. If my boss accepts to pay the conference fees then I have to work hard to finish the paper. We therefore consider in the remainder of this paper preference formulas with controllable and uncontrollable variables.

A general approach would be to define 'optimistic' and 'pessimistic' preference specifications on any combination of controllables and uncontrollables, such as an 'optimistic' preference $O^-(p \wedge x)$ or even $O^+(p)$. However, this approach blurs the idea that 'optimistic' reasoning is restricted to controllables, and 'pessimistic' reasoning is restricted to uncontrollables. Mixed 'optimistic' and 'pessimistic' norms are defined as follows.

Definition 12 (Mixed norm specification). *A conditional 'optimistic' obligation is a formula of the form* $\{O(x_i|p_i \wedge y_i) \mid i = 1, \cdots, n, p_i \in \mathcal{L}_\mathcal{U}, x_i, y_i \in \mathcal{L}_\mathcal{C}\}$.

We merge the two pre-orders using the symmetric merger operator since there is no reason to give priority either to $\succeq_d$ or to $\succeq_b$.

6 Related Work

6.1 Permission

One may wonder whether there is a role for permissive norms in deontic logic, once we have obligations. The issue is whether there is a need to distinguish between the absence of a prohibition and an explicit permission. Several researchers have doubted whether there is such a need, which may explain why the study of permission has gotten so little attention in the deontic logic literature, despite the fact that deontic logic started by von Wright's observation that the relation between obligation and permission is analogous to the relation between necessity and possibility, and despite the fact that the earliest papers used permission instead of its dual obligation.

The distinction between weak and strong permission is well known in deontic logic, and the permission in DSDL3 is a standard example of a weak permission. Basically, the argument is that a weak permission is only the absence of an obligation, and this is precisely how permissions are defined in DSDL3. Moreover, we have as a theorem $O(p|q) \vee P(\neg p|q)$, indicating that every proposition is normed; a typical property of weak permission. However, it is a priori less clear whether these arguments still hold in non-monotonic DSDL3. We therefore consider two more detailed discussions on the distinction between weak and strong permissions.

However, there has been a convincing argument that permissions are a distinct kind of norms, besides obligations. Bulygin [6] observes that in a setting with higher and lower authorities, a higher authority needs to issue strong permissions to delimit the power of the lower authorities. So when there are two agents only and we consider DSDL3 as a compact specification language used by agent 2 (an authority) to define a pre-order for agent 1 (his ideal and sub-ideal states), then we only have to use obligations. However, as Bulygin argues, the picture is completely different when we consider three agents, for example in hierarchical normative systems, where a higher authority agent 0 limits the pre-orders agent 1 can prescribe to agent 2. Agent 0 can say now, for example, that agent 1 is not allowed to oblige p for agent 2. Bulygin's game among three agents is challenging for the development of deontic logic for normative multi-agent systems, but its implications for non-monotonic DSDL3 are not clear to us at this moment.

In Makinson and van der Torre's analysis of permission in their input/output logic framework, the distinction between weak and strong permissions has been made explicit (because the set of norms in the input/output logic framework is explicit). Maybe the norm specification used in non-monotonic DSDL3 may also be seen as such an explicitly represented set of norms.

6.2 Non-monotonic Logic

Our algorithm generalizes the algorithm of Benferhat *et al.* [2] which captures "equal preferences" , denoted $p = q$, which stands for "all best p worlds are q worlds and all best q worlds are p worlds". These equivalences can be represented in our framework by two non-strict preferences $p \geq q$ and $q \geq p$, but our non-strict preferences cannot be represented in their framework.

7 Summary

We study specificity principles for non-monotonic extensions of Bengt Hansson's standard dyadic deontic logic 3, known as DSDL3, with both controllable and uncontrollable propositions. This extension is important in artificial normative systems, maybe more than in alternative applications of DSDL3 in default reasoning, qualitative decision or preference logic. Permissions play an important role in normative multi-agent systems, for example when higher and lower authorities are distinguished in normative multi-agent systems. Moreover, the distinction between ought-to-be and ought-to-do is central in agent theory too.

We introduce an efficient algorithm for minimal specificity which not only covers obligations but also permissions. The extension with permissions is more complicated, because we cannot directly define the equivalence classes of the pre-order, but we need a second loop in the algorithm to deal with the permissions. The algorithm may be further extended with other kinds of obligations and permissions, for example with ceteris paribus obligations (p is obliged if each p world is preferred to each $\neg p$ world, ceteris paribus).

Moreover, we introduce ways to combine ought-to-be and ought-to-do obligations in DSDL3 extended with the distinction between controllable and uncontrollable propositions. We illustrate our approach for algorithms for minimal and maximal specificity for DSDL3 with controllable and uncontrollable propositions, based on 'optimistic' and 'pessimistic' reasoning respectively. Alternative ways to combine ought-to-be and ought-to-do obligations in non-monotonic DSDL3 are subject of further research.

An assumption of this paper has been that DSDL3 is established as a deontic logic, and that its non-monotonic mechanisms are needed to deal with either violations or exceptions. However, a referee has suggested to us that in a deontic logic it does not seem to make sense to assume that worlds are as good as possible and that agents will see to it that the best (or worst) worlds are realized. Moreover, he or she observes that these criticisms seem to only apply to deontic uses of DSDL3. It may thus be that we have not sufficiently separated applications of DSDL3 to deontic reasoning from applications to default reasoning and decision making. We have left this issue for further research.

References

1. S. Benferhat, D. Dubois, and H. Prade. Possibilistic and standard probabilistic semantics of conditional knowledge bases. *Logic and Computation*, 9:6:873–895, 1999.
2. S. Benferhat, D. Dubois, and H. Prade. Towards a possibilistic logic handling of preferences. *Applied Intelligence*, 14(3):303–317, 2001.
3. R. Booth and J. B. Paris. A note on the rational closure of knowledge bases with both positive and negative knowledge. *Journal of Logic, Language and Information*, 7(2), 1998.
4. C. Boutilier. Conditional logics of normality : a modal approach. *Artificial Intelligence*, 68:87–154, 1994.
5. C. Boutilier. Toward a logic for qualitative decision theory. In *Proceedings of the 4th International Conference on Principles of Knowledge Representation, (KR'94)*, pages 75–86, 1994.
6. E. Bulygin. Permissive norms and normative systems. In A. Martino and F. S. Natali, editors, *Automated Analysis of Legal Texts*, pages 211–218. Publishing Company, Amsterdam, 1986.
7. L. Cholvy and C. Garion. Deriving individual obligations from collective obligations. In *Procs of AAMAS 2003*, pages 962–963, 2003.
8. L. Cholvy and C. Garion. Desires, norms and constraints. In *Procs of AAMAS 2004*, pages 724–731, 2004.
9. Bengt Hansson. An analysis of some deontic logics. *Nos*, 3:373–398, 1969. Reprinted in Hilpinen (1971), pages 121 -147.
10. S. Kaci and L. van der Torre. Algorithms for a nonmonotonic logic of preferences. In *Procs of 8th European Conference on Symbolic and Quantitative Approaches to Reasoning with Uncertainty*, LNCS 3571, pages 281–292. Springer, 2005.
11. S. Kraus, D. Lehmann, and M. Magidor. Nonmonotonic reasoning, preferential models and cumulative logics. *Artificial Intelligence*, 44 (1):167– 207, 1990.
12. J. Lang, L. Van Der Torre, and E. Weydert. Utilitarian desires. *Autonomous Agents and Multi-Agent Systems*, 5:329–363, 2002.

13. D. Lehmann and M. Magidor. What does a conditional knowledge base entail? *Artificial Intelligence*, 55(1):1–60, 1992.
14. D. Lewis. *Counterfactuals*. Blackwell, 1973.
15. J. Pearl. System Z: A natural ordering of defaults with tractable applications to default reasoning. In R. Parikh. Eds, editor, *Proceedings of the 3rd Conference on Theoretical Aspects of Reasoning about Knowledge (TARK'90)*, pages 121–135. Morgan Kaufmann, 1990.
16. H. Prakken and M.J. Sergot. Dyadic deontic logic and contrary-to-duty obligations. In D. Nute, editor, *Defeasible Deontic Logic*, volume 263 of *Synthese Library*, pages 223–262. Kluwer, 1997.
17. Y. Shoham. Nonmonotonic logics: Meaning and utility. In *Procs of IJCAI 1987*, pages 388–393, 1987.
18. Wolfgang Spohn. An analysis of Hanssons dyadic deontic logic. *Journal of Philosophical Logic*, 4:237 - 252, 1975.
19. W. van der Hoek and M. Wooldridge. On the logic of cooperation and propositional control. *Artif. Intell.*, 164(1-2):81–119, 2005.
20. L. van der Torre and Y. Tan. Contrary-to-duty reasoning with preference-based dyadic obligations. *Annals of Mathematics and Artificial Intelligence*, 27:49–78, 1999.

Conflicting Obligations in Multi-agent Deontic Logic

Barteld Kooi and Allard Tamminga

Vakgroep Theoretische Filosofie, Faculteit der Wijsbegeerte, Rijksuniversiteit Groningen, Oude Boteringestraat 52, 9712 GL Groningen, The Netherlands
{B.P.Kooi, A.M.Tamminga}@rug.nl

Abstract. Extending John Horty's multi-agent deontic logic to moral reasoning with subjective utilities, we provide a language and semantics to study moral reasoning with sentences like 'Group $\mathcal{G}$ of agents ought see to it that ϕ in the interest of group $\mathcal{F}$'. We illustrate our deontic logic with a new formal analysis of the Prisoner's Dilemma, thereby showing that games can be studied fruitfully with our deontic logic. Finally, we prove a characterization theorem on conflicting obligations.

1 Introduction

J.L. Austin once suggested that "before we consider what actions are good or bad, right or wrong, it is proper to consider first what is meant by, and what not, (..) the expression 'doing an action' or 'doing something' " (Austin 1957, p. 178). In the same vein, G.H. von Wright deems a logic of action a necessary requirement for deontic logic.[1] For want of a sufficiently sophisticated logic of action – von Wright's proposals were pioneering, but primitive –, Austin's suggestion hardly found response among deontic logicians.[2] Over the last twenty years, however, the (modal) logic of action has made significant progress, covering philosophical inquiries into the basic notions of agency and (meta)logical investigations of various modal logics of action, known as *stit* theory.[3]

In his groundbreaking *Agency and Deontic Logic* (2001), John Horty substitutes modern *stit* theory for von Wright's logic of action, and thus makes an important contribution to deontic logic. On the basis of *stit* theory, objective utilities, and the concept of dominance from decision theory, Horty develops a new semantics for deontic formulas of the form $\odot[\Gamma\ cstit\!: A]$, interpreted informally as "Group (of agents) Γ ought to see to it that A".

In the present paper we generalize Horty's deontic logic to subjective utilities. Unlike Horty's objective utilities, in our deontic logic one and the same evaluation index – moment/history pairs in Horty's approach, possible worlds in

[1] See von Wright (1963), p. vii and von Wright (1966), p. 134.

[2] Thus, in his pioneering study on deontic logic, Fred Feldman writes: "we avoid all the problems about the nature of actions and their alternatives" (Feldman 1986, p. 13).

[3] For a textbook exposition of *stit* theory, see Belnap, Perloff and Xu (2001).

L. Goble and J.-J.C. Meyer (Eds.): DEON 2006, LNAI 4048, pp. 175–186, 2006.

ours – may have different utilities for different agents. Moreover, *pace* Horty, we do not assume that a deontic formula always must be evaluated with respect to the utilities of *all* agents, although, of course, it must be possible to do so. Hence, in evaluating a deontic formula we must know whose utilities are pertinent to the evaluation, i.e. we must know what the group of agents is whose utilities we have to consider in evaluating a deontic formula. We shall refer to this group of agents as the "interest group". Accordingly, the basic expressions of our deontic logic are of the form $\odot_{\mathcal{G}}^{\mathcal{F}}\phi$, interpreted informally as "Group $\mathcal{G}$ ought to see to it that ϕ in the interest of group $\mathcal{F}$".

One of the merits of introducing subjective utilities and interest groups in deontic logic is that the resulting multi-agent deontic logic enables us to analyse single-shot games in remarkable detail.[4] To substantiate this claim, we shall illustrate our deontic logic with an analysis of the well-known Prisoner's Dilemma. We shall show that the dilemma can be completely translated into our model theory and that our semantics rules that both agents ought to confess in their own interest, but also that both agents ought not to confess in their collective interest. Hence, agents may have conflicting obligations, depending on the interest groups in whose interest they act. We conclude our exposition of our multi-agent deontic logic with a formal characterization of the necessary and sufficient conditions for conflicts of obligations, thereby giving a formal answer to a question central to ethical theory.

In Section 2 we introduce our multi-agent deontic logic. We provide a language and semantics, and illustrate the semantics with the Prisoner's Dilemma. In Section 3 we characterize the situation in which conflicts of obligations can only occur. In Section 4 conclusions are drawn.

2 Multi-agent Deontic Logic

In multi-agent contexts, where different (groups of) agents may attach different utilities to certain states of affairs, moral reasoning may seem to escape all attempts at formalization. We introduce subjective utilities and interest groups to cope with the vagaries of multi-agent moral reasoning. Adopting the basic notions of Belnap, Perloff, Xu and Horty's *stit* logic of action, we omit, for the sake of exposition, their underlying branching-time models and restrict our attention to models in which there is one moment only. In this way, we simplify the semantics of our logic considerably and focus directly on multi-agent moral reasoning with subjective utilities and interest groups.

2.1 Language

Definition 1. *The language $\mathfrak{L}$ is built from a countable set $\mathfrak{P}$ of propositional variables $\{p_1, p_2, \ldots\}$ and a finite set A of individual agents $\{a_1, \ldots, a_n\}$. Let*

[4] Repeated games might be studied fruitfully from a deontic point of view, if we include the branching-time framework in our model theory.

$p \in \mathfrak{P}$. Let $\mathcal{F}, \mathcal{G} \subseteq \mathsf{A}$ be sets of agents. Then $\mathfrak{L}$ is given by the following rule in Backus-Naur Form (BNF)[5]*:*

$$\phi ::= p \mid \neg\phi \mid \phi \wedge \phi \mid \Box\phi \mid [\mathcal{G}]\phi \mid \odot_{\mathcal{G}}^{\mathcal{F}}\phi$$

The three modalities $\Box$, $[\mathcal{G}]$, and $\odot_{\mathcal{G}}^{\mathcal{F}}$ are interpreted as the standard necessity operator, the Chellas-von Kutschera *stit* operator, and our new deontic operator, respectively:

$\Box\phi$ It is necessary that ϕ.
$[\mathcal{G}]\phi$ Group $\mathcal{G}$ sees to it that ϕ.
$\odot_{\mathcal{G}}^{\mathcal{F}}\phi$ Group $\mathcal{G}$ ought to see to it that ϕ in the interest of group $\mathcal{F}$.

The language $\mathfrak{L}$ allows us to formalize various sentences relevant to multi-agent moral reasoning. An egoistic obligation like 'Agent a ought to see to it that ϕ in his own interest' can be formalized as $\odot_a^a\phi$. If $a \neq b$, then an altruistic obligation like 'Agent a ought to see to it that ϕ in the interest of the agent b' may be formalized as $\odot_a^b\phi$. In utilitarianism, the well-being of the community as a whole is pertinent to the evaluation of obligations. Since A denotes the group of all agents, a utilitarian obligation like 'Agent a ought to see to it that ϕ in everybody's interest' is formalized as $\odot_a^{\mathsf{A}}\phi$. The other limiting case would be a formula $\odot_a^{\emptyset}\phi$, which is a formal rendering of the sentence 'Agent a ought to see to it that ϕ in nobody's interest' (our semantics implies that such empty obligations are true if and only if ϕ is a tautology). Obviously, combinations of operators are allowed as well. Hence, a sentence like 'Agent a is able to see to it that ϕ' may be formalized as $\Diamond[a]\phi$. Hence, the Kantian principle that 'ought' implies 'can' may be formalized as $\odot_{\mathcal{G}}^{\mathcal{F}}\phi \to \Diamond[\mathcal{G}]\phi$. It is, according to our semantics, a valid formula.

From the perspective of ethical theory, it is particularly interesting to study the conditions under which obligations for (groups of) agents may conflict. We shall show that it is perfectly possible for a single agent (or a single group of agents) to have conflicting obligations, if these are obligations with respect to different interest groups. In our formal analysis of the Prisoner's Dilemma, both 'Agent a ought to confess in his own interest' and 'Agent a ought not to confess in the collective interest of himself and his partner in crime' are true in the same model. More generally, it may be asked whether it is possible to characterize the circumstances under which obligations with respect to the same interest group never conflict, that is, to characterize the circumstances under which the following formula is a contradiction:

$$\odot_{\mathcal{G}_1}^{\mathcal{F}}\phi \wedge \odot_{\mathcal{G}_2}^{\mathcal{F}}\neg\phi.$$

We conclude our paper with a precise answer to this question.

[5] Logical rigour would demand that we draw a sharp distinction between (1) the *names* of (sets of) agents and (2) the *objects* that are being named, *i.e.*, the (sets of) agents themselves. We waive this distinction and thereby avoid unnecessary complications, as our present aims can be reached without it.

2.2 Agency and Ability

Now we have set forth the language $\mathfrak{L}$ for our multi-agent deontic logic, we provide a formal semantics for it. The standard semantics for *stit* logics relies on branching-time frames, tree-like structures that represent possible future developments of the present. For expository reasons, we shall omit these branching-time frames: a fully-fledged branching-time semantics would needlessly obfuscate our present aims. Hence, we shall interpret the Chellas-von Kutschera *stit* operator $[\mathcal{G}]$ and our deontic operator $\odot_{\mathcal{G}}^{\mathcal{F}}$ in rather standard possible worlds models, leaving aside all temporal considerations.[6] This policy amounts to a logical study of actions at a single moment in time. Hence, a formula of the form $[\mathcal{G}]\phi$ is true if the group of agents $\mathcal{G}$ performs an action that constrains the set of possible worlds to worlds in which ϕ is true.

We render an action of a single agent a as the choice of an option from a's set of possible choices, where a's set of possible choices, denoted by $\mathsf{Choice}(a)$, is a partition of the set W of possible worlds. We shall refer to elements of $\mathsf{Choice}(a)$ as a's "choices". Furthermore, for every agent a it holds that all of a's choices are real options, that is, no choice of a can be obstructed by choices of the other agents. Formally, this demand can be met by requiring that the intersection of every possible combination of each agent's chosen options is nonempty.[7]

Definition 2 (Individual Agent's Choices). *Let* W *be a set of possible worlds and let* A *be a finite set of agents. Then* $\mathsf{Choice} : \mathsf{A} \mapsto \wp(\wp(\mathsf{W}))$ *is a* choice function for A, *if*

(i) for all $a \in \mathsf{A}$ *it holds that* $\mathsf{Choice}(a)$ *is a partition of* W,
(ii) $\bigcap_{a\in\mathsf{A}} s(a) \neq \emptyset$ *for every selection function* $s : \mathsf{A} \mapsto \wp(\mathsf{W})$ *with* $s(a) \in \mathsf{Choice}(a)$ *for all* $a \in \mathsf{A}$.

For example, if $\mathsf{W} = \{w_1, w_2, w_3, w_4\}$ and $\mathsf{A} = \{a, b\}$, then $\mathsf{Choice}(a) = \{\{w_1, w_2\}, \{w_3, w_4\}\}$ and $\mathsf{Choice}(b) = \{\{w_1, w_3\}, \{w_2, w_4\}\}$ is a choice function for A, since both $\mathsf{Choice}(a)$ and $\mathsf{Choice}(b)$ are partitions of W and for all four possible selection functions s with $s(a) \in \mathsf{Choice}(a)$ and $s(b) \in \mathsf{Choice}(b)$ it holds that $s(a) \cap s(b) \neq \emptyset$.[8]

As is clear from the language presented in the previous section we also wish to model actions performed by groups of agents. When actions are viewed as constraining the set of possible worlds, then the actions of a group of agents are the ways in which that group can constrain the set of possible worlds. We take it as quite natural that the individual choice function determines the set of choices assigned to a group. This leads to the following definition which extends the function Choice to groups of agents.

---

[6] For a standard treatment of branching-time models for *stit* logics, we refer to Belnap, Perloff and Xu (2001) and Horty (2001). Note that our semantics can easily be extended to Belnap, Perloff, Xu and Horty's branching-time frames.

[7] This is called "the condition of independence of agents" (Horty 2001, p. 31).

[8] $s_1(a) = \{w_1, w_2\}$, $s_1(b) = \{w_1, w_3\}$; $s_2(a) = \{w_1, w_2\}$, $s_2(b) = \{w_2, w_4\}$; $s_3(a) = \{w_3, w_4\}$, $s_3(b) = \{w_1, w_3\}$; $s_4(a) = \{w_3, w_4\}$, $s_4(b) = \{w_2, w_4\}$.

Definition 3 (Collective Choice Functions). *Let $\mathcal{G} \subseteq \mathsf{A}$ and let S be the set of selection functions $s : \mathcal{G} \mapsto \wp(\mathsf{W})$ such that $s(a) \in \mathsf{Choice}(a)$ for all $a \in \mathcal{G}$. Then*

$$\mathsf{Choice}(\mathcal{G}) = \left\{ \bigcap_{a \in \mathsf{A}} s(a) : s \in S \right\}.$$

So group $\mathcal{G}$'s set of possible choices is completely determined by the possible choices of all the agents in the group $\mathcal{G}$. For instance, in the example above $\mathsf{Choice}(a, b) = \{\{w_1\}, \{w_2\}, \{w_3\}, \{w_4\}\}$.

From the point of view of possible worlds semantics for modal logic one can think of a partition as being generated by an equivalence relation, with which we can associate an $S5$ modality. When two worlds are in the same element of the partition assigned to a group of agents, it means that these worlds are *choice-equivalent* for that group.

Definition 4. *Let $\mathcal{G} \subseteq \mathsf{A}$. The choice equivalence relation $\sim_{\mathcal{G}} \subseteq W \times W$ is defined to be:*

$$w \sim_{\mathcal{G}} w' \text{ iff } \exists K (K \in \mathsf{Choice}(\mathcal{G}) \text{ and } w, w' \in K).$$

If $w \sim_{\mathcal{G}} w'$ we say that w and w' are choice-equivalent for $\mathcal{G}$.

The choice equivalence relation associated with a group of agents $\mathcal{G}$ is the intersection of all the relations assigned to the members of $\mathcal{G}$. This relation provides the semantics of the $[\mathcal{G}]$ modality.

2.3 Utilities

In order to determine whether an action is right or better than another action, we follow Horty and take a consequentialist approach. The idea is that a normative theory distinguishes the ideal from the nonideal worlds, or generally imposes some order on the set of possible worlds. This has led Horty to develop utilitarian models in which each history is assigned some value by a utility function which encodes the order on the worlds. As before, we leave the temporal issues aside and thus assign values to possible worlds, but we also deviate from Horty in another respect.

We view the values assigned to the possible worlds as utilities in the game-theoretic or economical sense. In this view it is obvious that different agents can assign different values to the same worlds. Consequentialist normative theories, in particular utilitarianism, use these utilities to determine the value of actions. The value of a world depends both on which individuals one counts as part of the moral community and also on the relative weight one assigns to the interests of each of the members in the community. In this paper we leave it open which individuals are members of the moral community, but, in order to keep things simple, we let the interests of each individual in that community count equally. This leads to a simple calculation of the average utility when determining the value of a possible world. Hence, we extend the utility function U to groups of agents by stipulating that

$$\mathsf{U}_{\mathcal{G}}(w) = \frac{1}{|\mathcal{G}|} \sum_{a \in \mathcal{G}} \mathsf{U}_a(w)$$

where $|\mathcal{G}|$ denotes the cardinality of $\mathcal{G}$, and let $\mathsf{U}_{\emptyset}(w) = 0$.

2.4 $\mathcal{F}$-Dominance

In this section, we shall define and investigate the notion of $\mathcal{F}$-dominance, a notion that will be central to our semantics for formulas of the form $\odot_{\mathcal{G}}^{\mathcal{F}} \phi$. When a group $\mathcal{G}$ performs a collective action by choosing an option K from $\mathsf{Choice}(\mathcal{G})$, it constrains the set W of possible worlds to the set K of possible worlds. It may be, however, that the agents who are not members of $\mathcal{G}$ (and thus are members of the group $\mathsf{A} - \mathcal{G}$) perform a collective action L, thereby constraining the set K to the set of possible worlds $K \cap L$. Hence, $\mathcal{G}$ usually will not be able to fully determine the outcome of its collective actions, since the final outcome also depends on the actions of agents in $\mathsf{A} - \mathcal{G}$. Hence, the outcome of a collective action is, in this sense, uncertain. Nevertheless, we can define an $\mathcal{F}$-dominance relation $\succeq_{\mathcal{G}}^{\mathcal{F}}$ over $\mathcal{G}$'s choices. If K and K' both are in $\mathsf{Choice}(\mathcal{G})$, then, intuitively, $K \succeq_{\mathcal{G}}^{\mathcal{F}} K'$ is true if and only if K promotes the interests of group $\mathcal{F}$ at least as well as K', whatever the collective action of the agents in $\mathsf{A} - \mathcal{G}$ may be. Hence, we insert an interest group $\mathcal{F}$ in Horty's Definitions 4.1 and 4.5 (Horty 2001, p. 60 and p. 68) to define $\mathcal{F}$-dominance:

Definition 5 ($\mathcal{F}$-Dominance). *Let $\mathcal{F}, \mathcal{G} \subseteq \mathsf{A}$. Let $K, K' \in \mathsf{Choice}(\mathcal{G})$. Then $K \succeq_{\mathcal{G}}^{\mathcal{F}} K'$ (K* weakly $\mathcal{F}$-dominates K' for $\mathcal{G}$*) is defined to be:*

$$K \succeq_{\mathcal{G}}^{\mathcal{F}} K' \text{ iff for all } S \in \mathsf{Choice}(\mathsf{A} - \mathcal{G}) \text{ and for all } w, w' \in \mathsf{W} \text{ it holds that if } w \in K \cap S \text{ and } w' \in K' \cap S, \text{ then } \mathsf{U}_{\mathcal{F}}(w) \geq \mathsf{U}_{\mathcal{F}}(w')$$

As usual, $K \succ_{\mathcal{G}}^{\mathcal{F}} K'$ if and only if $K \succeq_{\mathcal{G}}^{\mathcal{F}} K'$ and $K' \not\succeq_{\mathcal{G}}^{\mathcal{F}} K$. We shall say that K strongly $\mathcal{F}$-dominates K' for $\mathcal{G}$, *if $K \succ_{\mathcal{G}}^{\mathcal{F}} K'$.*

It is easy to check that $\mathcal{F}$-Dominance has the following properties:[9]

Lemma 1 (Properties of $\mathcal{F}$-Dominance). *Let $\mathcal{F}, \mathcal{G} \subseteq \mathsf{A}$. Let $K, K', K'' \in \mathsf{Choice}(\mathcal{G})$. Then*

(i) If $K \succ_{\mathcal{G}}^{\mathcal{F}} K'$, then $K \succeq_{\mathcal{G}}^{\mathcal{F}} K'$
(ii) If $K \succeq_{\mathcal{G}}^{\mathcal{F}} K'$ and $K' \succeq_{\mathcal{G}}^{\mathcal{F}} K''$, then $K \succeq_{\mathcal{G}}^{\mathcal{F}} K''$
(iii) If $K \succeq_{\mathcal{G}}^{\mathcal{F}} K'$ and $K' \succ_{\mathcal{G}}^{\mathcal{F}} K''$, then $K \succ_{\mathcal{G}}^{\mathcal{F}} K''$
(iv) If $K \succ_{\mathcal{G}}^{\mathcal{F}} K'$ and $K' \succeq_{\mathcal{G}}^{\mathcal{F}} K''$, then $K \succ_{\mathcal{G}}^{\mathcal{F}} K''$
(v) If $K \succ_{\mathcal{G}}^{\mathcal{F}} K'$ and $K' \succ_{\mathcal{G}}^{\mathcal{F}} K''$, then $K \succ_{\mathcal{G}}^{\mathcal{F}} K''$
(vi) If $K \succ_{\mathcal{G}}^{\mathcal{F}} K'$, then $K' \not\succ_{\mathcal{G}}^{\mathcal{F}} K$
(vii) $K \not\succ_{\mathcal{G}}^{\mathcal{F}} K$.

[9] Compare Horty's Proposition 4.7 (Horty 2001, p. 69).

In our theorem on conflicting obligations (in Section 3), we shall need a lemma to infer $\mathcal{F}$-dominance relations between choices of a group $\mathcal{G}_1 \cup \mathcal{G}_2$ from $\mathcal{F}$-dominance relations between choices of a subgroup $\mathcal{G}_1$.

Lemma 2. *Let $\mathcal{F}, \mathcal{G}_1, \mathcal{G}_2 \subseteq \mathsf{A}$ such that $\mathcal{G}_1 \cap \mathcal{G}_2 = \emptyset$. Let $K, K' \in \mathsf{Choice}(\mathcal{G}_1)$ and $L \in \mathsf{Choice}(\mathcal{G}_2)$. Then*

$$\text{If } K \succeq^{\mathcal{F}}_{\mathcal{G}_1} K', \text{ then } (K \cap L) \succeq^{\mathcal{F}}_{(\mathcal{G}_1 \cup \mathcal{G}_2)} (K' \cap L).$$

Proof. Assume that $K \succeq^{\mathcal{F}}_{\mathcal{G}_1} K'$. Then for all $M \in \mathsf{Choice}(\mathsf{A} - \mathcal{G}_1)$ and for all $w, w' \in \mathsf{W}$ it holds that if $w \in K \cap M$ and $w' \in K' \cap M$, then $\mathsf{U}_{\mathcal{F}}(w) \geq \mathsf{U}_{\mathcal{F}}(w')$. Suppose that $N \in \mathsf{Choice}(\mathsf{A} - (\mathcal{G}_1 \cup \mathcal{G}_2))$ and $w \in K \cap L \cap N$ and $w' \in K' \cap L \cap N$. It holds that $L \cap N \in \mathsf{Choice}(\mathsf{A} - \mathcal{G}_1)$. Hence, $\mathsf{U}_{\mathcal{F}}(w) \geq \mathsf{U}_{\mathcal{F}}(w')$. Therefore, $(K \cap L) \succeq^{\mathcal{F}}_{(\mathcal{G}_1 \cup \mathcal{G}_2)} (K' \cap L)$. □

2.5 Semantics

The language introduced in Section 2.1 is interpreted in consequentialist models $\mathfrak{M} = \langle \mathsf{W}, \mathsf{A}, \mathsf{Choice}, \mathsf{V}, \mathsf{U} \rangle$, which consist of a set of possible worlds, a finite set of agents, a choice function for the agents, a valuation that assigns a set of possible worlds to each atomic proposition, and a utility function that assigns a value for each agent to each world.

Most of the semantics is standard. The notion of $\mathcal{F}$-dominance underlies our semantical rule for formulas of the form $\odot^{\mathcal{F}}_{\mathcal{G}} \phi$. The idea is that a group ought to see to it that ϕ iff each action the group can take that does not lead to ϕ is strongly dominated by an action that does lead to ϕ, moreover this second action is only dominated by actions that also lead to ϕ. More precisely: a group $\mathcal{G}$ ought to see to it that ϕ in the interest of a group $\mathcal{F}$ if and only if every choice K in $\mathsf{Choice}(\mathcal{G})$ that does not guarantee ϕ is strongly $\mathcal{F}$-dominated for $\mathcal{G}$ by a choice K' in $\mathsf{Choice}(\mathcal{G})$ that does guarantee ϕ and every choice K'' in $\mathsf{Choice}(\mathcal{G})$ that weakly $\mathcal{F}$-dominates K' for $\mathcal{G}$ also guarantees that ϕ.

Definition 6 (Semantics). *Let $\mathfrak{M} = \langle \mathsf{W}, \mathsf{A}, \mathsf{Choice}, \mathsf{V}, \mathsf{U} \rangle$ be a consequentionalist model and let $w \in \mathsf{W}$. Let $p \in \mathfrak{P}$ and let $\phi, \psi \in \mathfrak{L}$. Then*

(i) $\mathfrak{M}/w \models p$ *iff* $w \in \mathsf{V}(p)$

(ii) $\mathfrak{M}/w \models \neg\phi$ *iff* $\mathfrak{M}/w \not\models \phi$

(iii) $\mathfrak{M}/w \models \phi \wedge \psi$ *iff* $\mathfrak{M}/w \models \phi$ *and* $\mathfrak{M}/w \models \psi$

(iv) $\mathfrak{M}/w \models \Box\phi$ *iff for all w' in* W *it holds that* $\mathfrak{M}/w' \models \phi$

(v) $\mathfrak{M}/w \models [\mathcal{G}]\phi$ *iff for all w' in* W *with* $w \sim_{\mathcal{G}} w'$ *it holds that* $\mathfrak{M}/w' \models \phi$

(vi) $\mathfrak{M}/w \models \odot^{\mathcal{F}}_{\mathcal{G}} \phi$ *iff for all K in* $\mathsf{Choice}(\mathcal{G})$ *with* $K \not\subseteq [\![\phi]\!]$ *there is a K' in* $\mathsf{Choice}(\mathcal{G})$ *with* $K' \subseteq [\![\phi]\!]$*, such that (1)* $K' \succ^{\mathcal{F}}_{\mathcal{G}} K$ *(2) for all K'' in* $\mathsf{Choice}(\mathcal{G})$ *with* $K'' \succeq^{\mathcal{F}}_{\mathcal{G}} K'$ *it holds that* $K'' \subseteq [\![\phi]\!]$

We write $\mathfrak{M} \models \phi$, if for all w in W it holds that $\mathfrak{M}/w \models \phi$. Moreover, we write $\models \phi$, if for all $\mathfrak{M}$ it holds that $\mathfrak{M} \models \phi$.

These semantics generalize Horty's semantics. Horty's class of 'utilitarian models' is a subclass of our consequentionalist models. When all the agents in A agree on the values of all possible worlds, then $\mathsf{U}_a(w) \geq \mathsf{U}_a(w')$ if and only if $\mathsf{U}_{\mathcal{G}}(w) \geq \mathsf{U}_{\mathcal{G}}(w')$ for all $w, w' \in \mathsf{W}$, all $a \in \mathsf{A}$, and all $\mathcal{G} \subseteq \mathsf{A}$.

The semantics for formulas of the form $\odot_{\mathcal{G}}^{\mathcal{F}} \phi$ also allows one to deal with the case where a group can perform infinitely many actions where there are no optimal actions (that are not dominated by other actions). In such a case where there are infinitely many actions, the rule says that a group ought to see to it that ϕ iff from some point onwards the better actions all see to it that ϕ. If there are only finitely many actions (or the dominance relation is well founded) one could simplify the semantics by saying that a group ought to see to it that ϕ iff each optimal action the group can take ensures that ϕ.

2.6 Validities

The semantics defined in the previous section makes the $\odot_{\mathcal{G}}^{\mathcal{F}}$ operator a fairly standard deontic operator.

Lemma 3 (Validities). *Let $\phi, \psi \in \mathfrak{L}$. Then*

$D_{\mathcal{G}}^{\mathcal{F}}\odot$ $\models \odot_{\mathcal{G}}^{\mathcal{F}} \phi \to \Diamond[\mathcal{G}]\phi$
$RE_{\mathcal{G}}^{\mathcal{F}}\odot$ *If* $\models \phi \leftrightarrow \psi$, *then* $\models \odot_{\mathcal{G}}^{\mathcal{F}} \phi \leftrightarrow \odot_{\mathcal{G}}^{\mathcal{F}} \psi$
$N_{\mathcal{G}}^{\mathcal{F}}\odot$ *If* $\models \phi$, *then* $\models \odot_{\mathcal{G}}^{\mathcal{F}} \phi$
$M_{\mathcal{G}}^{\mathcal{F}}\odot$ $\models \odot_{\mathcal{G}}^{\mathcal{F}}(\phi \wedge \psi) \to (\odot_{\mathcal{G}}^{\mathcal{F}} \phi \wedge \odot_{\mathcal{G}}^{\mathcal{F}} \psi)$
$C_{\mathcal{G}}^{\mathcal{F}}\odot$ $\models (\odot_{\mathcal{G}}^{\mathcal{F}} \phi \wedge \odot_{\mathcal{G}}^{\mathcal{F}} \psi) \to \odot_{\mathcal{G}}^{\mathcal{F}}(\phi \wedge \psi)$

Proof. The proofs of $D_{\mathcal{G}}^{\mathcal{F}}\odot$, $RE_{\mathcal{G}}^{\mathcal{F}}\odot$, $N_{\mathcal{G}}^{\mathcal{F}}\odot$, $M_{\mathcal{G}}^{\mathcal{F}}\odot$ are straightforward. For a proof of $C_{\mathcal{G}}^{\mathcal{F}}\odot$, see Horty (2001), pp. 166-167. □

This set of validities does not capture the whole logic as a complete axiomatization would. We do not attempt to give such an axiomatization in this paper. One can see that validities concerning the relation between different deontic operators (i.e. for operators $\odot_{\mathcal{G}_1}^{\mathcal{F}_1}$ and $\odot_{\mathcal{G}_2}^{\mathcal{F}_2}$, where all these groups may be different) do not occur in the list. In Section 3 we study one relation between such operators.

2.7 The Prisoner's Dilemma

Let us illustrate our semantics for multi-agent deontic logic with an analysis of the Prisoner's Dilemma. We take its standard version from Osborne and Rubinstein (1994), p. 17. The Prisoner's Dilemma is a two-player strategic game, represented by the following pay-off matrix:

	Don't confess	Confess
Don't confess	$3, 3$	$0, 4$
Confess	$4, 0$	$1, 1$

In this game, the Nash-equilibrium is reached when both players confess and the outcome is $\langle 1, 1\rangle$. Many find this solution counterintuitive, as the outcome $\langle 1, 1\rangle$ differs from the Pareto-efficient outcome $\langle 3, 3\rangle$.

If we read p as "a confesses" and q as "b confesses", we can translate the payoff matrix of the Prisoner's Dilemma into our model theory as the consequentionalist model $\mathfrak{M} = \langle \mathsf{W}, \mathsf{A}, \mathsf{Choice}, \mathsf{V}, \mathsf{U}\rangle$, where $\mathsf{W} = \{w_1, w_2, w_3, w_4\}$, $\mathsf{A} = \{a, b\}$, $\mathsf{Choice}(a) = \{\{w_1, w_2\}, \{w_3, w_4\}\}$, $\mathsf{Choice}(b) = \{\{w_1, w_3\}, \{w_2, w_4\}\}$, $\mathsf{V}(p) = \{w_3, w_4\}$, $\mathsf{V}(q) = \{w_2, w_4\}$, and

$$\begin{array}{llll} \mathsf{U}_a(w_1) = 3 & \mathsf{U}_a(w_2) = 0 & \mathsf{U}_a(w_3) = 4 & \mathsf{U}_a(w_4) = 1 \\ \mathsf{U}_b(w_1) = 3 & \mathsf{U}_b(w_2) = 4 & \mathsf{U}_b(w_3) = 0 & \mathsf{U}_b(w_4) = 1. \end{array}$$

First, our semantics for multi-agent deontic logic rules that in the present model the agents a and b ought to see to it that the Nash-equilibrium is reached, if they base their obligations on their own individual interests only, *i.e.*,

$$\mathfrak{M} \models (\odot^a_a p) \wedge (\odot^b_b q).$$

Note that $\mathfrak{M} \models \odot^a_a p$ if and only if for each $K \in \mathsf{Choice}(a)$ with $K \not\subseteq [\![p]\!]$ there is a $K' \in \mathsf{Choice}(a)$ with $K' \subseteq [\![p]\!]$, such that $K' \succ^a_a K$, and for each $K'' \in \mathsf{Choice}(a)$ with $K'' \succeq^\alpha_\alpha K'$ it holds that $K'' \subseteq [\![p]\!]$. By the definition of $\mathfrak{M}$, we have $\mathsf{Choice}(a) = \{\{w_1, w_2\}, \{w_3, w_4\}\}$. It holds that $\{w_1, w_2\} \not\subseteq [\![p]\!]$ and $\{w_3, w_4\} \subseteq [\![p]\!]$. Hence, we only need to check (i) $\{w_3, w_4\} \succ^a_a \{w_1, w_2\}$ and (ii) $\{w_3, w_4\} \succeq^a_a \{w_3, w_4\}$.

Ad (i). It holds that $\{w_3, w_4\} \succ^a_a \{w_1, w_2\}$ if and only if $\{w_3, w_4\} \succeq^a_a \{w_1, w_2\}$ and $\{w_1, w_2\} \not\succeq^a_a \{w_3, w_4\}$. The first conjunct holds if and only if for all $K'' \in \mathsf{Choice}(b)$ and for all $w, w' \in \mathsf{W}$ it holds that if $w \in \{w_3, w_4\} \cap K''$ and $w' \in \{w_1, w_2\} \cap K''$, then $\mathsf{U}_a(w) \geq \mathsf{U}_a(w')$. Since $\mathsf{Choice}(b) = \{\{w_1, w_3\}, \{w_2, w_4\}\}$, this is indeed the case.

The second conjunct holds if and only if there is a $K'' \in \mathsf{Choice}(b)$ and there are $w, w' \in \mathsf{W}$ such that $w \in \{w_1, w_2\} \cap K''$ and $w' \in \{w_3, w_4\} \cap K''$ and $\mathsf{U}_a(w) < \mathsf{U}_a(w')$. Any $K'' \in \mathsf{Choice}(b)$ suffices.

Ad (ii). It holds that $\{w_3, w_4\} \succeq^a_a \{w_3, w_4\}$ if and only if for all $K'' \in \mathsf{Choice}(b)$ and for all $w, w' \in \mathsf{W}$ it holds that if $w \in \{w_3, w_4\} \cap K''$ and $w' \in \{w_3, w_4\} \cap K''$, then $\mathsf{U}_a(w) \geq \mathsf{U}_a(w')$. As for each $K'' \in \mathsf{Choice}(b)$ it holds that $\{w_3, w_4\} \cap K''$ is a singleton, the requirements hold trivially.

Second, our semantics rules that in the present model the agents a and b ought to see to it that the Pareto-efficient outcome is reached, if they base their obligations on the collective interests of the group $\{a, b\}$, *i.e.*,

$$\mathfrak{M} \models \odot^{a,b}_{a,b}(\neg p \wedge \neg q).$$

Third, it is not the case that agents a and b individually ought to see to it that the Pareto-efficient outcome is reached, if they base their obligations on the collective interest of the group $\{a, b\}$, *i.e.*,

$$\mathfrak{M} \not\models (\odot^{a,b}_a(\neg p \wedge \neg q)) \vee (\odot^{a,b}_b(\neg p \wedge \neg q)).$$

This is exactly what is intuitively required, as a is unable to see to it that $\neg q$ and b is unable to see to it that $\neg p$. As we have seen in the previous section, this must be so as 'ought' implies 'can'.

Fourth, the agent a ought to see to it that $\neg p$, if a bases his obligations on the collective interests of the group $\{a, b\}$. Likewise, the agent b ought to see to it that $\neg q$ in the interest of $\{a, b\}$. Hence,

$$\mathfrak{M} \models (\odot_a^{a,b} \neg p) \wedge (\odot_b^{a,b} \neg q).$$

Therefore, our deontic logic provides a precise description of the different senses of what agents ought to do, once they have chosen to maximize their individual utility or the utility of a group to which they belong. It does not, of course, *solve* the Prisoner's Dilemma: our logic does not prescribe agents to opt for the interests of a group to which they belong rather than to follow their individual interests, nor the other way round.

3 On Conflicts of Obligations

Let us finally present our characterization theorem stating the exact conditions under which the formula $\odot_{\mathcal{G}_1}^{\mathcal{F}} \phi \wedge \odot_{\mathcal{G}_2}^{\mathcal{F}} \neg\phi$ is a contradiction.[10] With our theorem, we give a partial answer to a question raised by von Wright: "In order to answer the question whether a norm and its negation-norm are ... mutually exclusive, we ought to give criteria for the possible co-existence of norms" (von Wright 1963, p. 140).

Theorem 1. *Let* $\mathcal{G}_1, \mathcal{G}_2 \in \mathsf{A}$. *Then*

$$\models \neg((\odot_{\mathcal{G}_1}^{\mathcal{F}} \phi) \wedge (\odot_{\mathcal{G}_2}^{\mathcal{F}} \neg\phi)) \text{ iff } \mathcal{G}_1 \subseteq \mathcal{G}_2 \text{ or } \mathcal{G}_2 \subseteq \mathcal{G}_1 \text{ or } \mathcal{G}_1 \cap \mathcal{G}_2 = \emptyset.$$

Proof. ($\Leftarrow$) [Case 1] Assume $\mathcal{G}_1 \subseteq \mathcal{G}_2$. Suppose $w \models (\odot_{\mathcal{G}_1}^{\mathcal{F}} \phi) \wedge (\odot_{\mathcal{G}_2}^{\mathcal{F}} \neg\phi)$. We first show that ($\star$) there is a $M^\star \in \mathsf{Choice}(\mathcal{G}_2)$ with $M^\star \not\subseteq [\![\neg\phi]\!]$. Let $M \in \mathsf{Choice}(\mathcal{G}_2)$. If $M \not\subseteq [\![\neg\phi]\!]$, we are done. Otherwise, suppose $M \subseteq [\![\neg\phi]\!]$. Note that $M = K \cap L$ with $K \in \mathsf{Choice}(\mathcal{G}_1)$ and $L \in \mathsf{Choice}(\mathcal{G}_2 - \mathcal{G}_1)$. Then $K \not\subseteq [\![\phi]\!]$. Hence, since $w \models \odot_{\mathcal{G}_1}^{\mathcal{F}} \phi$, there is a $K^\star \in \mathsf{Choice}(\mathcal{G}_1)$ with $K^\star \subseteq [\![\phi]\!]$. Let $M^\star = K^\star \cap L$. Then $M^\star \in \mathsf{Choice}(\mathcal{G}_2)$ and $M^\star \not\subseteq [\![\neg\phi]\!]$. Therefore, ($\star$) has been established.

Hence, since $w \models \odot_{\mathcal{G}_2}^{\mathcal{F}} \neg\phi$, there is a $M' \in \mathsf{Choice}(\mathcal{G}_2)$ with $M' \subseteq [\![\neg\phi]\!]$, such that $M' \succ_{\mathcal{G}_2}^{\mathcal{F}} M^\star$ and for all $M'' \in \mathsf{Choice}(\mathcal{G}_2)$ with $M'' \succeq_{\mathcal{G}_2}^{\mathcal{F}} M'$ it holds that $M'' \subseteq [\![\neg\phi]\!]$. Note that $M' = K' \cap L'$ with $K' \in \mathsf{Choice}(\mathcal{G}_1)$ and $L' \in \mathsf{Choice}(\mathcal{G}_2 - \mathcal{G}_1)$. Then $K' \not\subseteq [\![\phi]\!]$. Hence, since $w \models \odot_{\mathcal{G}_1}^{\mathcal{F}} \phi$, there is a $K'' \in \mathsf{Choice}(\mathcal{G}_1)$ with $K'' \subseteq [\![\phi]\!]$ and $K'' \succ_{\mathcal{G}_1}^{\mathcal{F}} K''$. By Lemma 2, it holds that $K'' \cap L' \succeq_{\mathcal{G}_2}^{\mathcal{F}} K' \cap L'$. Note that $K'' \cap L' \subseteq [\![\phi]\!]$. Finally, substituting $K'' \cap L'$ for M'', we find that $K'' \cap L' \subseteq [\![\neg\phi]\!]$. Contradiction. Therefore, $\models \neg((\odot_{\mathcal{G}_1}^{\mathcal{F}} \phi) \wedge (\odot_{\mathcal{G}_2}^{\mathcal{F}} \neg\phi))$.

[Case 2] Assume $\mathcal{G}_2 \subseteq \mathcal{G}_1$. Analogous to Case 1.

[Case 3] Assume $\mathcal{G}_1 \cap \mathcal{G}_2 = \emptyset$. Suppose $w \models (\odot_{\mathcal{G}_1}^{\mathcal{F}} \phi) \wedge (\odot_{\mathcal{G}_2}^{\mathcal{F}} \neg\phi)$. By $D_{\mathcal{G}}^{\mathcal{F}} \odot$ of Lemma 2.6, $w \models \Diamond[\mathcal{G}_1]\phi$ and $w \models \Diamond[\mathcal{G}_2]\neg\phi$. Hence, there are a $K \in \mathsf{Choice}(\mathcal{G}_1)$

[10] For a discussion of characterizations of frame conditions by modal formulas, see van Benthem (1983) and van Benthem (1984).

with $K \subseteq [\![\phi]\!]$ and a $L \in \mathsf{Choice}(\mathcal{G}_2)$ with $L \subseteq [\![\neg\phi]\!]$. By our assumption and Definition 3, $K \cap L \neq \emptyset$. Contradiction. Therefore, $\models \neg((\odot^{\mathcal{F}}_{\mathcal{G}_1}\phi) \wedge (\odot^{\mathcal{F}}_{\mathcal{G}_2}\neg\phi))$.[11]

($\Rightarrow$) Assume $\mathcal{G}_1 \not\subseteq \mathcal{G}_2$, $\mathcal{G}_2 \not\subseteq \mathcal{G}_1$, and $\mathcal{G}_1 \cap \mathcal{G}_2 \neq \emptyset$. Then there are (at least) three agents. We define $\mathfrak{M} = \langle \mathsf{W}, \mathsf{A}, \mathsf{Choice}, \mathsf{V}, \mathsf{U}\rangle$. Let $\mathsf{W} = \{w_i : 1 \leq i \leq 8\}$. Let $\mathsf{A} = \{a, b, c\}$. Let $\mathsf{Choice}(a) = \{\{w_1, w_2, w_3, w_4\}, \{w_5, w_6, w_7, w_8\}\}$, $\mathsf{Choice}(b) = \{\{w_1, w_3, w_5, w_7\}, \{w_2, w_4, w_6, w_8\}\}$, $\mathsf{Choice}(c) = \{\{w_1, w_2, w_5, w_6\}, \{w_3, w_4, w_7, w_8\}\}$. Let $\mathsf{V}(p) = \{w_2, w_4\}$. Let $\mathsf{U}_a(w_i) = 0$, if $i \in \{1, 5, 6, 8\}$ and $\mathsf{U}_a(w_i) = 1$, if $i \in \{2, 3, 4, 7\}$. This situation is represented by the following matrices:

0 w_1	1 p w_2
1 w_3	1 p w_4

0 w_5	0 w_6
1 w_7	0 w_8

where a chooses between the left and right matrix, b chooses between the left and right columns, and c chooses between the upper and lower rows. Note that for all $K \in \mathsf{Choice}(a, b)$ with $K \neq \{w_2, w_4\}$ it holds that $\{w_2, w_4\} \succ^a_{a,b} K$ and that for all $L \in \mathsf{Choice}(b, c)$ with $L \neq \{w_3, w_7\}$ it holds that $\{w_3, w_7\} \succ^a_{b,c} L$. Hence, $\mathfrak{M} \models \odot^a_{a,b}p$ and $\mathfrak{M} \models \odot^a_{b,c}\neg p$. Hence, $\mathfrak{M} \not\models \neg((\odot^a_{a,b}p) \wedge (\odot^a_{b,c}\neg p))$. Therefore, $\not\models \neg((\odot^{\mathcal{F}}_{\mathcal{G}_1}\phi) \wedge (\odot^{\mathcal{F}}_{\mathcal{G}_2}\neg\phi))$. □

Hence, only when there are at least three agents a, b, and c, such that $\{a, b\} \subseteq \mathcal{G}_1$ and $\{b, c\} \subseteq \mathcal{G}_2$, can a model be built in which it does not hold that $\odot^{F}_{\mathcal{G}_1}\phi \wedge \odot^{F}_{\mathcal{G}_2}\neg\phi$. The agent b cannot make a principled choice from $\mathsf{Choice}(b)$ to maximize the interest of group $\mathcal{F}$. If b is taken to belong to group $\mathcal{G}_1$, he has to choose $\{w_2, w_4, w_6, w_8\}$ to maximize $\mathcal{F}$'s interest. On the other hand, if b is seen as a member of group $\mathcal{G}_2$, he must rather choose $\{w_1, w_3, w_5, w_7\}$ to maximize $\mathcal{F}$'s interest. Obviously, b cannot choose both options. The agent b is wearing two hats here.

4 Conclusion

In Section 2.7 we showed that, when one takes the Prisoner's Dilemma as a moral situation, one agent can have conflicting obligations with respect to *different* interests groups (himself, and the group of all agents). In Section 3 we showed that a conflict of obligations between two groups who act in the interests of *one and the same* group can only occur in situations involving at least three agents. The agents that are members of both acting groups are in a dilemma. Do they coordinate their actions with the one group or with the other? The fact that this situation only occurs when there are at least three agents, and the fact that there are two agents involved in the Prisoner's Dilemma emphasizes the importance of studying multi-agent situations in deontic logic. Conflicts of obligations only occur in multi-agent context.

[11] Compare Horty's proof of Case 3 (Horty 2001, p. 48).

References

Austin, J.L. (1957). A Plea for Excuses. In J.L. Austin. *Philosophical Papers*, Second Edition, (pp. 175–204). Oxford, 1970: Oxford University Press.

Belnap, N., M. Perloff, & M. Xu (2001). *Facing the Future. Agents and Choices in Our Indeterminist World.* New York: Oxford University Press.

Feldman, F. (1986). *Doing the Best We Can. An Essay in Informal Deontic Logic.* Dordrecht: D. Reidel Publishing Company.

Horty, J.F. (2001). *Agency and Deontic Logic.* New York: Oxford University Press.

Osborne, M. & A. Rubinstein (1994). *A Course in Game Theory.* Cambridge MA: MIT Press.

van Benthem, J. (1983). *Modal Logic and Classical Logic.* Naples: Bibliopolis.

van Benthem, J. (1984). Correspondence Theory. In D. Gabbay & F. Guenther (eds.). *Handbook of Philosophical Logic*, Volume II (pp. 167–247). Dordrecht: D. Reidel Publishing Company.

von Wright, G.H. (1963). *Norm and Action. A Logical Enquiry.* London: Routledge & Kegan Paul.

von Wright, G.H. (1966). The Logic of Action – A Sketch. In N. Rescher (ed.). *The Logic of Decision and Action* (pp. 121–136). Pittsburgh: University of Pittsburgh Press.

Intermediate Concepts in Normative Systems

Lars Lindahl and Jan Odelstad

lars.lindahl@jur.lu.se, jod@hig.se

Abstract. In legal theory, a well-known idea is that an intermediate concept like "ownership" joins a set of legal consequences to a set of legal grounds. In our paper, we attempt to make the idea of a joining between grounds and consequences more precise by using an algebraic representation of normative systems earlier developed by the authors. In the first main part, the idea of intermediate concepts is presented and earlier discussions of the subjects are outlined. Subsequently, in the second main part, we introduce a more rigorous framework and develop the formal theory. In the third part, the formal framework is applied to examples and some remarks on a methodology of intermediate concepts are given.

1 The Problem of Intermediaries

1.1 Introduction

The role played by concept formation in philosophy and science has been varying. After some decades of rather low interest, there are signs indicating that the situation is changing. The aim of the present paper is to contribute to the study of this field. More specifically, our contribution aims at presenting a framework for analysing the role of what we call "intermediaries" as links between conceptual structures.

In [5], we presented a first working model for analysing the notion of intermediary. The present paper is different in several respects. The framework to be developed is based on the theory of Boolean algebra instead of lattice theory. The structures dealt with are not necessarily finite. The basic kind of relations considered are quasi-orderings rather than partial orderings as was the case in our previous paper, where partial orderings were introduced by a transition to equivalence classes. The framework is abstract in the sense that the main results are not tied to a specific interpretation in terms of conditions as was the case in the earlier paper.[1] Thus, the case where the domains of the orderings have conditions, or equivalence classes of conditions, as their members only plays the part of one of several models for the theory.

The first part of the paper presents the background of the idea of intermediaries. The second part introduces the formal framework. In the third part, the formal tools are used to clarify different types of intermediaries in concept formation.

[1] For our previous development of the abstract theory, see, in particular, [6] with further references. Cf. [8].

L. Goble and J.-J.C. Meyer (Eds.): DEON 2006, LNAI 4048, pp. 187–200, 2006.

1.2 Legal Concepts as Intermediaries

Facts, Deontic Positions and Intermediaries. Legal rules attach obligations, rights, deontic positions to facts, i.e., actions, events, circumstances. Deontic positions are, so we might say, legal consequences of these facts:

Facts	*Deontic positions*
Events, actions, circumstances	Obligations, claims, powers etc.

Facts and deontic positions are objects of two different sorts; we might call them Is-objects and Ought-objects. In a legal system, when Ought-objects are said to be "attached to" or to be "consequences of" Is-objects, there is sense of direction. In a legal system, inferences and arguments go from Is-objects to Ought-objects, not vice versa.

In the scheme just shown, something very essential is missing, namely the great bulk of more specific legal concepts. A few examples are: property, tort, contract, trust, possession, guardianship, matrimony, citizenship, crime, responsibility, punishment. These concepts are links between grounds on the left hand side and normative consequences on the right hand side of the scheme just given:

Facts	*Links*	*Deontic positions*
Events	Ownership	Obligations
Actions	Valid contract	Claims
Circumstances	Citizenship (etc.)	Powers (etc.)

Using this three-column scheme, we might say that ownership, valid contract, citizenship etc. are attached to certain facts, and that deontic positions, in turn, are attached to these legal positions.

To exemplify: Among the facts justifying an assertion that there is a valid contract between two parties are: that the parties have made an agreement, that they were in a sane state of mind when agreeing, that no force or deceit was used by any of them in the process, and so on. The deontic positions attached to there being a valid contract between them depend on what they have agreed on but are formulated in terms of claims and duties, legal powers etc. In the example, the facts are stated in terms of communicative acts, mental states and other descriptive notions, while the deontic positions are stated in normative or deontic terms.

Wedberg and Ross on Ownership. In the 1950's, each of the two Scandinavians Wedberg and Ross proposed the idea that a legal term such as "ownership", or "x is the owner of y at time t" is a syntactical tool serving the purpose of economy of expression of a set of legal rules. In the same year 1951, when Ross published his well-known essay "Tû-Tû" in a Danish Festschrift [10][2], Wedberg published an essay on the same theme in the Swedish journal *Theoria*. Possibly, the two authors arrived at these ideas independently of each other.[3] In any case no priority can be established.

[2] English translation [11].

[3] Cf [12] at p. 266, footnote 15, and [11] at p. 822, footnote 6.

As an example, the function of the term "ownership" is illustrated as follows by Ross [10], [11]:

$$\left.\begin{array}{l} F_1 \rightarrow \\ F_2 \rightarrow \\ F_3 \rightarrow \\ \vdots \\ F_p \rightarrow \end{array}\right\} \quad O \rightarrow \quad \left\{\begin{array}{l} C_1 \\ C_2 \\ C_3 \\ \vdots \\ C_n \end{array}\right.$$

Ross's scheme is aimed at representing a set of legal rules concerning ownership in a particular legal system (for example the rules on ownership in Danish law at a specific time). In the picture, the letters are to be interpreted as follows:

$F_1 - F_p$ **for:** x has lawfully purchased y, x has inherited y, x has acquired y by prescription, and so on.

$C_1 - C_n$ **for:** judgment for recovery shall be given in favor of x against other persons retaining y in their possession, judgment for damages shall be given in favor of x against other persons who culpably damage y, if x has raised a loan from z that it is not repaid at the proper time, z shall be given judgment for satisfaction out of y, and so on.

The letter "O" is a link between the left hand side and the right hand side. It can be read "x is the owner of y".

In Ross's scheme, the number of implications to ownership from the grounds for ownership is p (since the grounds are $F_1, \ldots, F_p$); similarly the number of implications from ownership to consequences of ownership is n (since there are n consequences). Therefore, the total number of implications in the scheme is $p + n$. On the other hand, if the rules were formulated by attaching each C_j among the consequences to each F_i among the grounds, the number of rules would be $p \cdot n$. Consequently, by the formulation in the scheme, the number of rules is reduced from $p \cdot n$ to $p + n$, a number that is much smaller.[4] In this way, economy of expression is obtained.

The similarities between Wedberg's and Ross's ideas are striking. Both use the example of ownership. Central ideas propounded by both of them are: By use of the linking term, the number $p \cdot n$ of rules is reduced to $p + n$, and, the linking term has no independent meaning (Wedberg) or has no semantical reference (Ross).

In our view, there is a great difference between speaking of an expression like "O is the property of P at t" as meaningless and speaking of it as being without independent meaning. The latter way of speaking goes well together with the view that the term has meaning but that this meaning consists precisely in its occurrence and use in inference rules linking the term to facts, on one hand, and to deontic consequences on the other.

---

[4] [12] pp. 273 f.

1.3 Intermediaries in Non-legal Contexts

Michael Dummett's Example. Dummett distinguishes between the conditions for applying a term and the consequences of its application. According to Dummett both are part of the meaning. Dummett exemplifies by the use of the term "Boche" as a pejorative term.

> The condition for applying the term to someone is that he is of German nationality; the consequences of its application are that he is barbarous and more prone to cruelty than other Europeans. We should envisage the minimal joinings in both directions as sufficiently tight as to be involved in the very meaning of the word: neither could be severed without altering its meaning. Someone who rejects the word does so because he does not want to permit a transition from the grounds for applying the term to the consequences of doing so. The addition of the term 'Boche' to a language which did not previously contain it would produce a non-conservative extension, i.e., one in which certain statements which did not contain the term were inferable from other statements not containing it which were not previously inferable. [1] at p. 454.[5]

Dummett's example illustrates how the use of a word is determined by two rules (I) and (II):[6]

(I) Rule linking a concept a to an intermediary m:

For all $x, y:$ If $a(x, y)$ then $m(x, y)$.

(II) Rule linking intermediary m to a concept b:

For all $x, y:$ If $m(x, y)$ then $b(x, y)$.

If the standpoint "meaning is use" is adopted, it can be held that the meaning of m is given by two rules (I) and (II) together. To understand the meaning of an intermediary m is to know how it is used in such a pair of rules.

Dummett's example is not concerned with a legal system and with an inference from facts to deontic positions. We note, however, that the antecedent "being of German nationality" in (I) and the consequent "being more prone ... etc" in (II) are conditions of "different kinds".

---

[5] Since the example is interesting from a philosophical point of view, we use it even though it has the disagreeable feature of being offensive to German nationals.

[6] The rules (I) and (II) can be compared to the rules of introduction and rules of elimination, respectively, in Gentzen's theory of natural deduction in [2]. If this comparison is made, (I) is regarded as an introduction rule and (II) as an elimination rule for m. An obvious difference is that while Gentzen's introduction rules and elimination rules are rules of inference, the rules (I) and (II) are formulated in "if, then" sentences of predicate logic. A reason for the difference is, of course, that Gentzen aims at providing a theory for predicate logic, and, therefore, the language of predicate logic itself is not admissible within his theory.

Dummett intends his example to illustrate a non-conservative extension. In Section 3, where applications of our formal framework is discussed, we will indicate how this idea is expressed within our framework.

Other well-known examples, outside the area of connections from descriptive to normative, are the connection from physical to mental and the connection from chemical to biological. At a very general level, in empirical science, there is the problem of the connection from observable to theoretical.[7]

In some of the cases where the connection of different kinds is problematic, the notion of supervenience is used for clarifying the nature of the connection. Existing theories of supervenience, seem to us, however, to yield at best a very partial insight into the nature of the relation in view. In particular, they do not provide much information about the specific interrelations between parts of the two different structures.

2 The Formal Framework

2.1 Introduction

As stated in Section 1.1 above, we distinguish between the abstract level of formal analysis (to be dealt with in the present section), where a general algebraic framework is developed, and the level of applications where the abstract theory is used as a tool for analysing different conceptual structures (Section 3).

At the abstract algebraic level, the notion "intermediary" will not be used. In the algebraic theory, however, a technical notion "intervenient" will be defined. In Section 3, the notion "intervenient" will be used as a tool for analysis of what, informally, is called "intermediaries". More precisely, in Section 3, we will distinguish different types of intermediaries and indicate how intermediaries can be interrelated.

The algebraic theory contains a number of definitions of technical terms. Before going into this theory, it is appropriate briefly to suggest how the algebraic theory can be used for analysing a normative system with intermediaries.

Let C be a non-empty set. We say that $\mathcal{N} = \langle B, \wedge, ', \rho \rangle$ is a *supplemented Boolean algebra freely generated by* C if $\langle B, \wedge, ' \rangle$ is a Boolean algebra freely

[7] An interesting approach to the problem of intermediate terms in mechanics was outlined in the nineteenth century by Henri Poincaré. Poincaré pointed out that a proposition like (1) "the stars obey Newton's laws" can be broken up into two others, namely (2) "gravitation obeys Newton's laws" and (3) "gravitation is the only force acting on the stars". Among these, proposition (2) is a definition and not subject to the test of experiment, while (1) is subject to such a test. "Gravitation", according to Poincaré, is an intermediary. Poincaré maintains that in science, when there is a relation between two facts A and B, an intermediary C is often introduced by the formulation of one relationship between A and C, and another between C and B. The relation between A and C, then, is often elevated to a principle, not subject to revision, while the relation between C and B is a law, subject to such revision. See [9], pp. 124 f., in the chapter "Is science artificial?"

generated by C and ρ is a binary relation on B.[8] The partial ordering determined by the Boolean algebra $\langle B, \wedge, ' \rangle$ is a subset of ρ. An application can be that $\mathcal{N}$ is a normative system expressed in terms of a set of conditions B and a relation ρ such that, for $a, b \in B$, $a\rho b$ holds if and only if a implies b in the normative system $\mathcal{N}$.

Next, let $\langle B_1, \wedge, ' \rho/B_1 \rangle$ and $\langle B_2, \wedge, ', \rho/B_2 \rangle$ be two substructures of $\langle B, \wedge, ', \rho \rangle$ where B_1 and B_2 are disjoint, except for the zero and unit constants $\bot$ and $\top$. In the application where $\mathcal{N}$ is a normative system, we can think of B_1 as a set of descriptive conditions and B_2 as a set of normative conditions. If B is a set of conditions, $\bot$ stands for the absurd condition and $\top$ for the trivial condition.

Of special interest is where B contains a subset M, disjoint from $B_1 \cup B_2$, where, for $m \in M$, there is $a \in B_1$ and $b \in B_2$ such that $a\rho m$ and $m\rho b$. In this case, given certain further requirements, m will be called an "intervenient". In the application where $\mathcal{N}$ is a normative system, we can conceive of a case where a condition m belongs neither to the set B_1 of descriptive conditions nor to the set B_2 of normative conditions but where, in $\mathcal{N}$, m is implied by a descriptive condition and implies a normative condition.

2.2 The Basic Formal Framework

Boolean Quasi-orderings, Fragments and Joinings. One formal structure that will be used in our investigation of how subsystems of different kinds are linked is that of a *Boolean quasi-ordering* (*Bqo*). Technical concepts related to *Bqo*'s, defined in previous papers are: *fragments* of *Bqo*'s, and *joinings* of elements of *Bqo*'s. For formal definitions of a *Bqo* and of these related notions, the reader is referred to [6]. A short recapitulation is as follows.

The relational structure $\langle B, \wedge, ', R \rangle$ is a *Boolean quasi-ordering (Bqo)* if $\langle B, \wedge, ' \rangle$ is a Boolean algebra and R is a binary, reflexive and transitive relation on B (i.e. R is a quasi-ordering), $\bot$ is the zero element, $\top$ is the unit element, and where R satisfies some additional requirements.[9] If $\mathcal{B} = \langle B, \wedge, ', R \rangle$ is a Boolean quasi-ordering, and $\langle B_i, \wedge, ' \rangle$ is a subalgebra of $\langle B, \wedge, ' \rangle$, and $R_i = R/B_i$, then the structure $\mathcal{B}_i = \langle B_i, \wedge, ', R_i \rangle$ is a *fragment* of $\mathcal{B}$. Let $\mathcal{B}$, $\mathcal{B}_1$, $\mathcal{B}_2$ be *Bqo*'s such that $\mathcal{B}_1$ and $\mathcal{B}_2$ are fragments of $\mathcal{B}$. A *joining* from $\mathcal{B}_1$ to $\mathcal{B}_2$ in $\mathcal{B}$ is a pair $\langle b_1, b_2 \rangle$ in $\mathcal{B}$ such that $b_1 \in B_1$, $b_2 \in B_2$, $b_1 R b_2$, not $b_1 R \bot$ and not $\top R b_2$.

Narrowness and Minimal Elements. The *narrowness-relation determined by* two quasi-orderings $\langle B_1, R_1 \rangle$ and $\langle B_2, R_2 \rangle$ is the binary relation $\trianglelefteq$ on $B_1 \times B_2$ such that $\langle a_1, a_2 \rangle \trianglelefteq \langle b_1, b_2 \rangle$ if and only if $b_1 R_1 a_1$ and $a_2 R_2 b_2$. $\langle a_1, a_2 \rangle$ is a *minimal element* in $X \subseteq B_1 \times B_2$ with respect to $\langle B_1, R_1 \rangle$ and $\langle B_2, R_2 \rangle$ if there is no $\langle x_1, x_2 \rangle \in X$ such that $\langle x_1, x_2 \rangle \triangleleft \langle a_1, a_2 \rangle$. The set of minimal elements in X

[8] For the notion of freely generated Boolean algebras, see for example [4] p.131. Instead of *freely generated* one can say *independently generated*.

[9] (1) aRb and aRc implies $aR(b \wedge c)$, (2) aRb implies $b'Ra'$, (3) $(a \wedge b)Ra$, (4) not $\top R \bot$.(Requirement (4) excludes the possibility that $R = B_1 \times B_2$, which holds for inconsistent systems.)

is denoted $\min_{R_1}^{R_2} X$. When there is no risk of ambiguity we write just $\min X$. We note that $\trianglelefteq$ is a quasi-ordering. We let $\simeq$ denote the equality part of $\trianglelefteq$ and $\lhd$ the strict part of $\trianglelefteq$. The equality part $\simeq$ is an equivalence relation and we denote the equivalence class determined by $\langle b_1, b_2\rangle \in B_1 \times B_2$ by $[b_1, b_2]\simeq$.[10]

Boolean Joining Systems (*Bjs*). Another important structure is that of a *Boolean joining-system* (*Bjs*), see [7]. A Boolean joining-system is an ordered triple $\langle \mathcal{B}_1, \mathcal{B}_2, J\rangle$ such that $\mathcal{B}_1 = \langle B_1, \wedge, ', R_1\rangle$ and $\mathcal{B}_2 = \langle B_2, \wedge, ', R_2\rangle$ are Boolean quasi-orderings and $J \subseteq B_1 \times B_2$, $J \neq \varnothing$ and three specific requirements are satisfied.[11]

If $\mathcal{B}$, $\mathcal{B}_1$ and $\mathcal{B}_2$ are Boolean quasi-orderings such that $\mathcal{B}_1$ and $\mathcal{B}_2$ are fragments of $\mathcal{B}$ and J is the set of joinings from $\mathcal{B}_1$ to $\mathcal{B}_2$ in $\mathcal{B}$, then $\langle \mathcal{B}_1, \mathcal{B}_2, J\rangle$ is a *Bjs*. Also, if $a_1, b_1 \in B_1$, $a_2, b_2 \in B_2$, and $\langle a_1, a_2\rangle \in J$, then $\langle a_1, a_2\rangle \trianglelefteq \langle b_1, b_2\rangle$ implies $\langle b_1, b_2\rangle \in J$.

Generating of Joining-Spaces. We note that if $\mathcal{B}_1$ and $\mathcal{B}_2$ are *Bqo*'s and

$$\mathcal{J} = \{J \subseteq B_1 \times B_2 | \langle \mathcal{B}_1, \mathcal{B}_2, J\rangle \text{ is a } Bjs\},$$

then $\mathcal{J}$ is a closure system.

If $\langle \mathcal{B}_1, \mathcal{B}_2, J\rangle$ is a Boolean joining-system, we call J the *joining-space* from $\mathcal{B}_1$ to $\mathcal{B}_2$ in $\langle \mathcal{B}_1, \mathcal{B}_2, J\rangle$. $\mathcal{J}$ is the family of all joining-spaces from $\mathcal{B}_1$ to $\mathcal{B}_2$. If $K \subseteq B_1 \times B_2$ let

$$[K]_{\mathcal{J}} = \cap\{ J \mid J \in \mathcal{J}, J \supseteq K\}.$$

$[K]_{\mathcal{J}}$ is the joining-space over $\mathcal{B}_1$ and $\mathcal{B}_2$ *generated by* K.[12]

If J is the joining-space from $\mathcal{B}_1$ to $\mathcal{B}_2$ generated by K but J is not generated by any proper subset of K, then we say that J is *non-redundantly generated* by K.

Connectivity. A *Bjs* $\langle \mathcal{B}_1, \mathcal{B}_2, J\rangle$ satisfies *connectivity* if whenever $\langle c_1, c_2\rangle \in J$ there is $\langle b_1, b_2\rangle \in J$ such that $\langle b_1, b_2\rangle$ is a minimal joining in $\langle \mathcal{B}_1, \mathcal{B}_2, J\rangle$ and $\langle b_1, b_2\rangle \trianglelefteq \langle c_1, c_2\rangle$.

Suppose that $\langle \mathcal{B}_1, \mathcal{B}_2, J\rangle$ is a *Bjs* that satisfies connectivity. Then

$$J = \{ \langle b_1, b_2\rangle \in B_1 \times B_2 : (\exists \langle a_1, a_2\rangle \in \min J : \langle a_1, a_2\rangle \trianglelefteq \langle b_1, b_2\rangle) \}.$$

10 The sign $\simeq$ should be written as a subscript. The reason why this is not done is typograhical.

11 The requirements are: (1) for all $b_1, c_1 \in B_1$ and $b_2, c_2 \in B_2$, $\langle b_1, b_2\rangle \in J$ and $\langle b_1, b_2\rangle \trianglelefteq \langle c_1, c_2\rangle$ implies $\langle c_1, c_2\rangle \in J$, (2) for any $C_1 \subseteq B_1$ and $b_2 \in B_2$, if $\langle c_1, b_2\rangle \in J$ for all $c_1 \in C_1$, then $\langle a_1, b_2\rangle \in J$ for all $a_1 \in lub_{R_1} C_1$, (3) for any $C_2 \subseteq B_2$ and $b_1 \in B_1$, if $\langle b_1, c_2\rangle \in J$ for all $c_2 \in C_2$, then $\langle b_1, a_2\rangle \in J$ for all $a_2 \in glb_{R_2} C_2$. (Note that the definitions of least upper bound (*lub*) and greatest lower bound (*glb*) for partial orderings are easily extended to quasi-orderings, but the *lub* or *glb* of a subset of a quasi-ordering is not necessarily unique but can consist of a set of elements.)

12 For definition and results of closure systems, see for example [3] p. 23f.

If we use the notion of an image of a set under a relation, then we can say that J is the image of $\min J$ under $\trianglelefteq$.

It is easy to see that if $\langle \mathcal{B}_1, \mathcal{B}_2, J_1 \rangle$ and $\langle \mathcal{B}_1, \mathcal{B}_2, J_2 \rangle$ are *Bjs* which satisfy connectivity and $\min J_1 = \min J_2$, then $J_1 = J_2$. Note that if we "substantially reduce" $\min J$, then the image of the new set under $\trianglelefteq$ is not J. To be more precise: Suppose that $\langle a_1, a_2 \rangle \in \min J$ and $K \subset \min J$ such that if $\langle a_1, a_2 \rangle \simeq \langle b_1, b_2 \rangle$ then $\langle b_1, b_2 \rangle \notin K$. Then it follows that the image of K under $\trianglelefteq$ is a proper subset of J.

If in a *Bjs* $\langle \mathcal{B}_1, \mathcal{B}_2, J \rangle$, $\mathcal{B}_1$ and $\mathcal{B}_2$ are complete (in a sense which is a straightforward generalization of the notion of completeness applied to Boolean algebras), then $\langle \mathcal{B}_1, \mathcal{B}_2, J \rangle$ satisfies connectivity.

Couplings and Pair Couplings. If $\langle \mathcal{B}_1, \mathcal{B}_2, J \rangle$ is a *Bjs* and the number of $\simeq$-equivalence classes defined by the elements in $\min J$ is exactly one, then the elements in $\min J$ are called *couplings*. If the number of equivalence classes defined by the elements in $\min J$ is exactly two, then sets consisting of one element from each equivalence class is called a *pair coupling*. Thus if $[b_1, b_2] \simeq$ is the only equivalence class, any $\langle a_1, a_2 \rangle \in J$ encompasses every element of $[b_1, b_2] \simeq$; similarly, if $[b_1, b_2] \simeq$ and $[c_1, c_2] \simeq$ are the only equivalence classes, any $\langle a_1, a_2 \rangle \in J$ encompasses every element of $[b_1, b_2] \simeq$ or every element of $[c_1, c_2] \simeq$.

Base of a Joining-Space and Counterparts. Note that if $\langle \mathcal{B}_1, \mathcal{B}_2\ J \rangle$ is a *Bjs* and J is generated by K, then J is also generated by $\min K$. If $\langle \mathcal{B}_1, \mathcal{B}_2, J \rangle$ is a *Bjs* and J is non-redundantly generated by K and $K \subseteq \min J$, then K is called a *base of* J in $\langle \mathcal{B}_1, \mathcal{B}_2, J \rangle$.

Suppose that $K, L \subseteq B_1 \times B_2$ and that $K \simeq$ is the set of $\simeq$-equivalence classes defined by the elements in K and $L \simeq$ is the set of $\simeq$-equivalence classes defined by the elements in L. If there is a bijection φ between $K \simeq$ and $L \simeq$ such that $\varphi(x) = y$ if and only if there is $\langle a_1, a_2 \rangle, \langle b_1, b_2 \rangle \in B_1 \times B_2$ such that $\langle b_1, b_2 \rangle \in x$ and $\langle a_1, a_2 \rangle \in y$ and $\langle a_1, a_2 \rangle \simeq \langle b_1, b_2 \rangle$, then we say that K and L are $\simeq$*-counterparts*.

If K and L are $\simeq$-counterparts, then the image of K under $\trianglelefteq$ is the same as the image of L under $\trianglelefteq$, and the sets of joinings generated by K and L are the same.

If, for a base K of J in $\langle \mathcal{B}_1, \mathcal{B}_2, J \rangle$, K and L are $\simeq$-counterparts, then we say that L *up to* $\simeq$*-equivalence* is *the* base of J in $\langle \mathcal{B}_1, \mathcal{B}_2, J \rangle$.

2.3 Intervenients

Weakest Grounds and Strongest Consequences. Suppose that $\langle B, \wedge, ', \rho \rangle$ is a supplemented Boolean algebra, $B_1, B_2 \subseteq B$ and $m \in B \backslash B_1$. Then $a_1 \in B_1$ is one of the *weakest grounds* in B_1 of m with respect to $\langle B, \wedge, ', \rho \rangle$ if $a_1 \rho m$, and it holds that if there is $b_1 \in B_1$ such that $b_1 \rho m$, then $b_1 \rho a_1$. Furthermore, $a_2 \in B_2$ is one of the *strongest consequences* of m in B_2 with respect to $\langle B, \wedge, ', \rho \rangle$ if $m \rho a_2$, and it holds that if there is $b_2 \in B_2$ such that $m \rho b_2$, then $a_2 \rho b_2$.

Definition of Intervenient. Suppose that C is a non-empty set and that $\langle B, \wedge, ' \rangle$ is the Boolean algebra freely generated by C. Suppose further that $\mathcal{N} = \langle B, \wedge, ', \rho \rangle$ where ρ is a binary relation over B, i.e. $\mathcal{N}$ is a supplemented Boolean algebra extended by the binary relation ρ (cf. above, Section 2.1). (Note that $\mathcal{N}$ is not necessarily a Boolean quasi-ordering.)

A *Bjs* $\langle \mathcal{B}_1, \mathcal{B}_2, J \rangle$ *lies within* a supplemented Boolean algebra $\langle B, \wedge, ', \rho \rangle$ if $\langle B_1, \wedge, ' \rangle$ and $\langle B_2, \wedge, ' \rangle$ are subalgebras of $\langle B, \wedge, ' \rangle$, $B_1 \cap B_2 = \{\top, \bot\}$, $\rho | B_1 = R_1$ and $\rho | B_2 = R_2$, and $\rho | (B_1 \times B_2) = J$.

Suppose that $\mathcal{N} = \langle B, \wedge, ', \rho \rangle$ is a supplemented Boolean algebra and that B_1 and B_2 are disjoint subsets of B such that $\langle B_1, \wedge, ' \rangle$ and $\langle B_2, \wedge, ' \rangle$ are subalgebras of $\langle B, \wedge, ' \rangle$. An element $m \in B \setminus (B_1 \cup B_2)$ is an *intervenient between* B_1 and B_2 in $\langle B, \wedge, ', \rho \rangle$ if there is $\langle a_1, a_2 \rangle \in \rho$ such that a_1 is a weakest ground in B_1 of m with respect to $\langle B, \wedge, ', \rho \rangle$ and a_2 is a strongest consequence in B_2 of m with respect to $\langle B, \wedge, ', \rho \rangle$. We say that the intervenient m *corresponds* to the joining $\langle a_1, a_2 \rangle$ from B_1 and B_2.

We note that in a *Bjs* $\langle \mathcal{B}_1, \mathcal{B}_2, J \rangle$ lying within $\mathcal{N}$, an intervenient m between B_1 and B_2 can be used for inferring joinings from $\mathcal{B}_1$ to $\mathcal{B}_2$. That m is an intervenient in $\langle B, \wedge, ', \rho \rangle$ between B_1 and B_2 corresponding to the joining $\langle a_1, a_2 \rangle$ in $\langle \mathcal{B}_1, \mathcal{B}_2, J \rangle$ implies that $\langle a_1, a_2 \rangle \trianglelefteq \langle b_1, b_2 \rangle$ if and only if $b_1 \rho m \rho b_2$.

The fact that, in the way shown, intervenients can be used for inferring joinings, makes it appropriate to speak of an intervenient as a "vehicle of inference".

***JoinM* and Systems of Intervenients.** Recalling the presuppositions concerning $\mathcal{N} = \langle B, \wedge, ', \rho \rangle$ above, let $M \subseteq B$ and $M \cap (B_1 \cup B_2) = \varnothing$. We say that M *produces* the set

$$K = \{ \langle b_1, b_2 \rangle \in B_1 \times B_2 \mid \exists m \in M : b_1 \rho m \rho b_2 \}.$$

The set of joinings corresponding to a set of intervenients M between B_1 and B_2 is denoted $JoinM$ where

$$JoinM = \{ \langle b_1, b_2 \rangle \in B_1 \times B_2 \mid \exists m \in M : \ m \text{ corresponds to } \langle b_1, b_2 \rangle \}$$

Note that M produces K iff K is the image of $JoinM$ under $\trianglelefteq$. We say that M *non-redundantly produces* K if M produces K but no proper subset of M produces K.

If M is a set of intervenients such that $JoinM$ is a base of J, we say that M is a base of intervenients for J. Of special interest is the case where M consists of a set of generators for the Boolean algebra $\langle B, \wedge, ' \rangle$ in $\mathcal{N}$.

Three Types of Intervenients. Suppose that m is an intervenient between B_1 and B_2 in $\langle B, \wedge, ', \rho \rangle$, corresponding to the joining $\langle a_1, a_2 \rangle$ in $\langle \mathcal{B}_1, \mathcal{B}_2, J \rangle$. Then a classification can be made according to whether $\langle a_1, a_2 \rangle$ (1) is a joining that is not a minimal joining, (2) is a minimal joining that is not a pair coupling or coupling, or (3) is a pair coupling or coupling. In case (1), we say that m corresponds to a mere joining, in case (2), that m corresponds to a mere minimal joining, and, in case (3), that m corresponds to a pair coupling or coupling.

3 Applications

3.1 The cis Models

In what follows we shall be interested in a particular model of the abstract theory of quasi-orderings, Boolean quasi-orderings, and Boolean joining-systems. This model is the model of a *condition implication structure (cis)*.[13] A *cis* model of a *Bqo* $\langle B, \wedge, ', R\rangle$ is obtained if B is a domain of *conditions*, and aRb represents that a *implies* b. Similarly, a *cis* model of a *Bjs* $\langle \mathcal{B}_1, \mathcal{B}_2, J\rangle$ is obtained if $\mathcal{B}_1, \mathcal{B}_2$ are *cis* models of *Bqo's* and, for $a_1 \in B_1$ and $a_2 \in B_2$, $a_1 J a_2$ represents that a_1 implies a_2.

In simple cases, conditions can be denoted by expressions, using the sign of the infinitive, such as "to be of German nationality", "to be a citizen of the U.S.", "to be a child of", "to be entitled to inherit", or by corresponding expressions in the ing-form, like "being of German nationality" etc. Often, however, conditions should appropriately be expressed by open sentences, like "x's promises to pay \$ y to z", "x is a citizen of state y", "x is entitled to inherit y".

If a, b are conditions, we assume that a', b' are negations of a, b respectively, that $a \wedge b$ is the conjunction of a and b, and that $a \vee b$ is the disjunction of a and b.[14]

If a *Bjs* $\langle B_1, B_2, J\rangle$ represents a normative (mini-)system, a norm in this system is represented by $a_1 J a_2$, where $a_1 \in B_1$ is descriptive, while $a_2 \in B_2$ is normative.

Dummett's "Boche" Example Once More. In our formal framework, Dummett's Boche example can be represented as follows. Let $\mathcal{N} = \langle B, \wedge, ', \rho\rangle$ be a supplemented Boolean algebra freely generated by a set C of concepts, and let $\langle \mathcal{B}_1, \mathcal{B}_2, J\rangle$ be a *Bjs* which lies within $\mathcal{N}$. The set B_1 contains conditions expressing different nationalities and B_2 conditions expressing different psychological dispositions. Let $B^{(1)}$ be B extended with the term *Boche* and $\rho^{(1)}$ an extension of ρ such that *Boche* is an intervenient in $\mathcal{N}^{(1)} = \langle B^{(1)}, \wedge, ', \rho^{(1)}\rangle$ between B_1 and B_2. $J^{(1)}$ is the extension of J as an effect of the extension of ρ to $\rho^{(1)}$. Suppose that $\langle a_1, a_2\rangle$ is a joining in $\langle \mathcal{B}_1, \mathcal{B}_2, J^{(1)}\rangle$ but not a joining in $\langle \mathcal{B}_1, \mathcal{B}_2, J\rangle$, and that the intervenient *Boche* corresponds to the joining $\langle a_1, a_2\rangle$ in $\langle \mathcal{B}_1, \mathcal{B}_2, J^{(1)}\rangle$. The question arises whether $\langle a_1, a_2\rangle$ is a mere joining or a minimal joining, perhaps a coupling or pair coupling. If Dummett's example is perceived to be such that in $\mathcal{N}^{(1)}$ *Boche* corresponds to a minimal joining, we can make an extension of the system $\mathcal{N}^{(1)}$ to a system $\mathcal{N}^{(2)} = \langle B^{(2)}, \wedge, ', \rho^{(2)}\rangle$ by adding the intervenient *Berserk* (See figure 1 below) corresponding to the joining $\langle b_1, a_2\rangle$ in $\langle \mathcal{B}_1, \mathcal{B}_2, J^{(2)}\rangle$. In $\mathcal{N}^{(2)}$ *Boche* corresponds to a mere joining, since $\langle c_1, a_2\rangle = \langle a_1 \vee b_1, a_2\rangle$ is a minimal joining in $J^{(2)}$.

---

13 The present section on condition implication structures recapitulates ideas presented in earlier papers. See, in particular, [6].

14 The procedure of forming compounds can be iterated. So, for example, $(a \wedge b) \vee c$ is a condition. A condition a is simple if it is not compound.

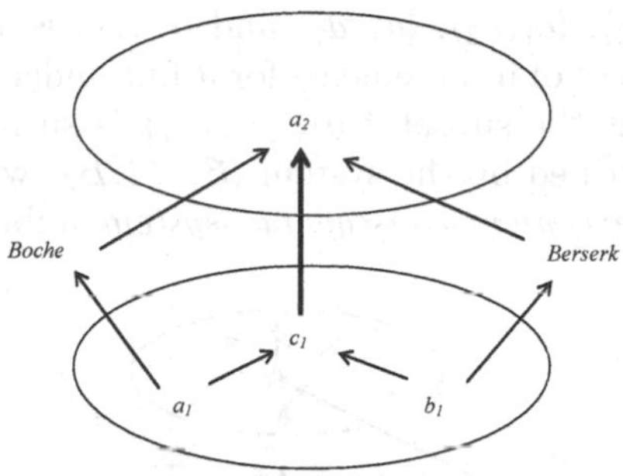

Fig. 1.

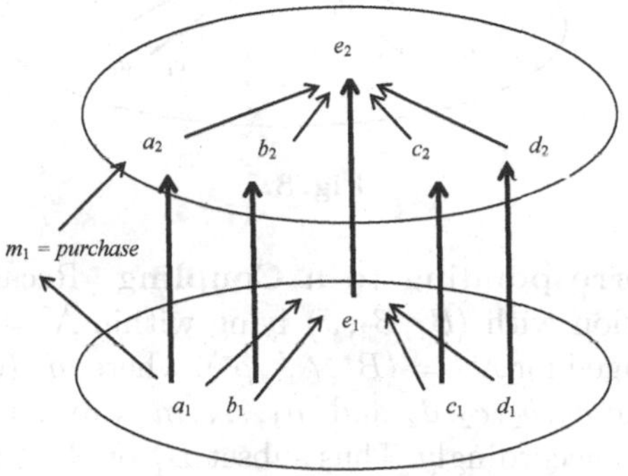

Fig. 2.

Minimal Joining and Modes for Acquiring Ownership. Next, we give a legal example concerning modes for ownership acquisition. The legal system we study is represented by the supplemented Boolean algebra $\langle B, \wedge, ', \rho \rangle$. The legal rules of ownership are expressed in terms of a set M of conditions: *purchase* m_1, *inheritance* m_2, *occupation* m_3, *specification* m_4, *ownership* m_5 (See figure 2 above). M is a subset of B. B_1 is a subset of B containing the following conditions: a_1 (making a contract etc.), b_1 (having particular kinship relationship), c_1 (appropriating something not owned), d_1 (creating a valuable thing out of worthless material), $e_1 = \langle a_1 \vee b_1 \vee c_1 \vee d_1 \rangle$. The weakest grounds in B_1 of the conditions in M with respect to $\langle B, \wedge, ', \rho \rangle$ are described by the following set G of ordered pairs: $\langle a_1, m_1 \rangle$, $\langle b_1, m_2 \rangle$, $\langle c_1, m_3 \rangle$, $\langle d_1, m_4 \rangle$, $\langle e_1, m_5 \rangle$. The strongest consequences in $B_2 \subseteq B$ of the conditions in M with respect to $\langle B, \wedge, ', \rho \rangle$ are described by the following set C of ordered pairs: $\langle m_1, a_2 \rangle$, $\langle m_2, b_2 \rangle$, $\langle m_3, c_2 \rangle$, $\langle m_4, d_2 \rangle$, $\langle m_5, e_2 \rangle$, where $e_2 = \langle a_2 \vee b_2 \vee c_2 \vee d_2 \rangle$. Note that $G \cup C \subseteq \rho$ and that M is a set of intervenients from B_1 to B_2 in $\langle B, \wedge, ', \rho \rangle$.

Let the joining-space J from $\mathcal{B}_1 = \langle B_1, \wedge, ', \rho | B_1 \rangle$ to $\mathcal{B}_2 = \langle B_2, \wedge, ', \rho | B_2 \rangle$ be characterized by G and C in the following sense: J is the the joining-space generated by $JoinM$. Then m_1 corresponds to $\langle a_1, a_2 \rangle$, m_2 corresponds to $\langle b_1, b_2 \rangle$, m_3 corresponds to $\langle c_1, c_2 \rangle$, m_4 corresponds to $\langle d_1, d_2 \rangle$ and m_5 corresponds to $\langle e_1, e_2 \rangle$.

Each of $\langle a_1, a_2\rangle$, $\langle b_1, b_2\rangle$, $\langle c_1, c_2\rangle$, $\langle d_1, d_2\rangle$ and $\langle e_1, e_2\rangle$ is a minimal joining in J. Note that M is not a base of intervenients for J but under plausible assumptions, it can be assumed that the subset $\{m_1, \ldots, m_4\}$ is such a base. Then the *Bjs* $\langle \mathcal{B}_1, \mathcal{B}_2, J\rangle$ can be described by the system $\langle \mathcal{B}_1, M, \mathcal{B}_2\rangle$ which can appropriately be called a *ground-intervenient-consequence-system*, abbreviated a *GIC-system.*

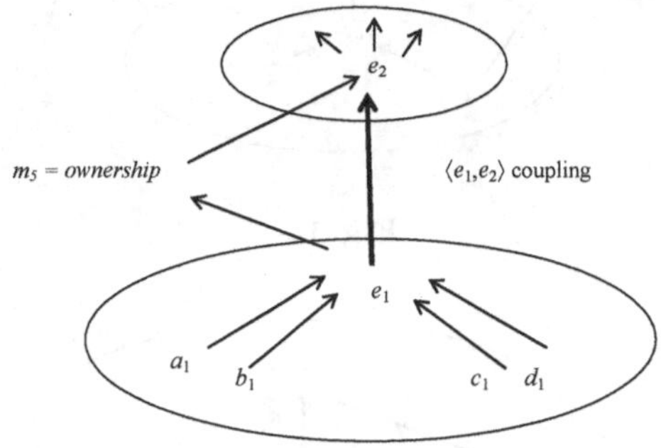

Fig. 3.

Ownership as Corresponding to a Coupling. Recalling the example of the previous subsection with $\langle \mathcal{B}_1, \mathcal{B}_2, J\rangle$ lying within $\mathcal{N} = \langle B, \wedge, ', \rho\rangle$, let $\mathcal{N} = \langle B, \wedge, ', \rho\rangle$ be exchanged for $\mathcal{N}^* = \langle B^*, \wedge, ', \rho^*\rangle$, where m_5 (ownership) is a member of B^*, but where a_2, b_2, c_2, d_2 and $m_1, \ldots, m_4$ are not members of B^* and where ρ^* is restricted accordingly. Thus subset B_1^* of B^* is as B_1 in the previous example, but a_2, b_2, c_2, d_2 are not members of subset B_2^*. In $\mathcal{N}^*$, (like in $\mathcal{N}$), e_1 is a weakest ground for m_5 and e_2 is a strongest consequence of m_5. The set M of intervenients from B_1 to B_2^*, however, has m_5 as its only member.In $\langle \mathcal{B}_1^*, \mathcal{B}_2^*, J^*\rangle$, $\langle e_1, e_2\rangle$ is the only member of min J^*. Therefore $\langle e_1, e_2\rangle$ is a *coupling* (see Section 2.2) and, in the example, m_5 (ownership) corresponds to a coupling (See figure 3 above). This system is strikingly similar to Ross's scheme, since Ross (like Wedberg) does not take into account such consequences that are specific to particular modes of acquisition such as purchase, inheritance, occupation etc.[15]

3.2 The Methodology of Intermediaries in Legal Systems

From the point of view of methodology, there is the task of formulating rational principles for constructing a system with concepts that, in a *Bjs* representation, are appropriately represented by intervenients. Three aspects to be taken into account are: (i) economy of expression, (ii) efficient inference, and (iii) adaptation to linguistic usage and commonly made distinctions.

A concept appropriately represented by an intervenient corresponding to a minimal joining, a pair coupling, or a coupling will serve the purpose of economy of expression and efficient inference. We recall the discussion concerning ownership as corresponding to a minimal joining or a coupling.

[15] Thus in the *Bjs* $\langle \mathcal{B}_1^*, \mathcal{B}_2^*, J^*\rangle$ lying within $\mathcal{N}^*$, B_2^* is generated by those simple conditions that are consequences of ownership regardless of mode of acquisition.

With regard to concepts represented by intervenients corresponding to mere joinings, considerations relating to economy of expression and efficient inference do not justify having these concepts in the system. What comes into focus is rather aspect (iii). Here, we can distinguish two situations:

One is the case where, in the appropriate representation of linguistic usage, several grounds $a_1, b_1, \ldots$ have the same strongest consequence a_2. If, in a *Bjs* representing linguistic usage, an intervenient corresponding to $\langle a_1 \vee b_1 \vee \ldots, a_2 \rangle$ does not appropriately represent linguistic usage and commonly made distinctions, this usage might sometimes be more appropriately represented by a *Bjs* with particular intervenient(s) corresponding to one or more of $\langle a_1, a_2 \rangle$, $\langle b_1, a_2 \rangle$ etc., even though these are mere joinings. Thus in Dummett's example (see above), where the intervenient *Boche* corresponds to the mere joining $\langle a_1, a_2 \rangle$.

The dual situation is where, in the representation of linguistic usage, a_1 is the weakest ground for several consequences $a_2, b_2, \ldots$ If, in the *Bjs* representation, an intervenient corresponding to $\langle a_1, a_2 \wedge b_2 \wedge \ldots \rangle$ is not an appropriate representation of usage, a better representation can sometimes be achieved with particular intervenient(s) corresponding to one or more of the mere joinings $\langle a_1, a_2 \rangle$, $\langle a_1, b_2 \rangle$ etc.

4 Conclusion

As exemplified in the foregoing, intermediate concepts (intermediaries) play an essential role in normative systems. In the paper, we have used an algebraic framework, previously developed by us, for representing normative systems. Within this framework, we have outlined a theory of intervenients including weakest grounds and strongest consequences and bases of intervenients. Also, we have taken a first step towards a typology of intervenients. This theory is intended as a means for analysing intermediate concepts, and we have sketched its application in a few cases. As a report on work in progress, we have focused on systems consisting of an algebra of grounds and an algebra of consequences and a system of intervenients between these algebras (*GIC-systems*). In further developments of the theory, we intend to extend the investigation to incorporate nets of *GIC*-systems, where the consequence-structure in one system can be the ground-structure in another, and the intervenients in one *GIC*-system can be grounds or consequences in another. Consequently, in more complex normative systems, there can be hierarchies of intervenients worth investigating.

Acknowledgement

The authors are grateful for support from Harald and Louise Ekmans Forskningsstiftelse. Jan Odelstad gratefully acknowledges financial support from Stiftelsen för kunskaps- och kompetensutveckling and the University of Gävle.[16]

[16] The paper, as well as our earlier joint papers, are the result of wholly joint work where the order of appearance of our author names has no significance.

References

1. Dummett, M. (1973). *Frege: Philosophy of Language.* London : Duckworth.
2. Gentzen, G. (1934). Untersuchungen über das logische Schließen, I. *Mathematische Zeitschrift* 39: 176-210.
3. Grätzer, G. (1979). *Universal Algebra.* 2nd ed. New York: Springer-Verlag.
4. Koppelberg, S. (1989). *Handbook of Boolean Algebras.* Vol. 1 (ed. by J.D. Monk). Amsterdam: North Holland.
5. Lindahl, L. & Odelstad, J. (1999). Intermediate Concepts as Couplings of Conceptual Structures. In H. Prakken & P. McNamara (eds.) *Norms, Logics and Informations Systems. New Studies on Deontic Logic and Computer Science.* Amsterdam: IOS Press.
6. Lindahl, L. & Odelstad, J. (2004). Normative Positions within an Algebraic Approach to Normative Systems. *Journal Of Applied Logic* 2: 63-91.
7. Odelstad, J. & Boman, M. (2004). Algebras for Agent Norm-Regulation. *Annals of Mathematics and Artificial Intelligence,* 42: 141-166.
8. Odelstad, J. & Lindahl, L. (2002). The Role of Connections as Minimal Norms in Normative Systems. In T. Bench-Capon, A. Daskalopulu and R. Winkels (eds.) *Legal Knowledge and Information Systems.* Amsterdam: IOS Press.
9. Poincaré, H., (1907). *The Value of Science.* London. (English translation of *La Valeur des sciences,* Paris 1905.)
10. Ross, A. (1951). Tû-Tû. In O.A. Borum & K. Illum (eds.) *Festskrift til Henry Ussing.* København: Juristforbundet.
11. Ross, A. (1956-57). Tû-Tû. *Harvard Law Review,* 70: 812-825. (English translation of [10].)
12. Wedberg, A. (1951). Some Problems in the Logical Analysis of Legal Science, *Theoria* 17: 246-275.

Propositional Quantifiers in Deontic Logic

Gert-Jan C. Lokhorst

Section of Philosophy,
Faculty of Technology, Policy and Management,
Delft University of Technology
g.j.c.lokhorst@tbm.tudelft.nl

Abstract. Several systems of monadic deontic logic are defined in terms of systems of alethic modal logic with a propositional constant. When the universal propositional quantifier is added to these systems, the propositional constant becomes definable in terms of the deontic operator. As a result, the meaning of this constant becomes clearer and it becomes easy to axiomatize the deontic fragments of the alethic modal systems.

1 Introduction

In 1950s Anderson [2, 3] and Kanger [9] suggested that deontic logic can be defined in terms of alethic modal logic by means of the following definition:

$$OA = L(e \supset A).$$

Here O and L are sentential operators that turn a sentence into another sentence, A is an arbitrary sentence, e is a propositional constant, and $\supset$ is a binary connective. Both Anderson and Kanger read O as "it is obligatory that," L as "it is necessary that" and $\supset$ as "materially implies." But their interpretations of the constant e differed: Kanger read it as "what morality requires" and Anderson as "the good state of affairs." Åqvist [1] later interpreted e as a "prohairetic, i.e., preference-theoretical constant" that may be read as "optimality or admissibility," Smiley [13] interpreted it "as expressing the content of an (unspecified) moral code," McNamara [11] read it as "all normative demands are met," while Van Fraassen [14] regarded it as the negation of "all hell breaks loose."

The definition $OA = L(e \supset A)$ may also be written as $OA = e \prec A$, where $\prec$ is strict implication. In 1967, Anderson [4, 5] suggested that it would be better to replace strict implication by relevant implication in this definition and accordingly redefined O by $OA = e \rightarrow A$, where $\rightarrow$ is relevant implication. The resulting system avoids the so-called "fallacies of relevance" but it is unsatisfactory because its treats O as an extensional operator in the sense that $(A \leftrightarrow B) \rightarrow (OA \leftrightarrow OB)$ is a theorem.

In 1992, Mares [12] combined both approaches and defined O as $OA = L(e \rightarrow A)$, where L is the necessity operator and $\rightarrow$ is relevant implication. He read e in the same way as Anderson did.

L. Goble and J.-J.C. Meyer (Eds.): DEON 2006, LNAI 4048, pp. 201–209, 2006.

In this paper, we will study the effects of adding a propositional quantifier to these three systems. We will show that this has two noteworthy consequences. First, the meaning of e becomes clearer. In suitably strong propositionally quantified versions of the systems, e turns out to be provably equivalent with $\forall p(Op \supset p)$ (in the classical case) or $\forall p(Op \rightarrow p)$ (in the two relevant systems). Thus, McNamara's [11] interpretation of e as "all normative demands are met" emerges as the most appropriate one. Second, it will turn out to be easy to find deontic systems in which e is defined in terms of O that have the same theorems as the alethic systems.

We start with Anderson's extensional relevant deontic logic [4, 5], then discuss intensional relevant deontic logic along the lines of Mares [12], and end with the oldest system, classical deontic logic. We conclude by indicating some other areas in which our results can be applied.

2 Relevant Logic

Definition 1 (R). *Relevant system* ***R*** *has the following axioms and rules [6, ch. V].*

R1 $A \rightarrow A$ *(Self-implication)*
R2 $(A \rightarrow B) \rightarrow ((C \rightarrow A) \rightarrow (C \rightarrow B))$ *(Prefixing)*
R3 $(A \rightarrow (A \rightarrow B)) \rightarrow (A \rightarrow B)$ *(Contraction)*
R4 $(A \rightarrow (B \rightarrow C)) \rightarrow (B \rightarrow (A \rightarrow C))$ *(Permutation)*
R5 $(A \,\&\, B) \rightarrow A$, $(A \,\&\, B) \rightarrow B$ *(&Elimination)*
R6 $((A \rightarrow B) \,\&\, (A \rightarrow C)) \rightarrow (A \rightarrow (B \,\&\, C))$ *(&Introduction)*
R7 $A \rightarrow (A \vee B)$, $B \rightarrow (A \vee B)$ *(∨Introduction)*
R8 $((A \rightarrow C) \,\&\, (B \rightarrow C)) \rightarrow ((A \vee B) \rightarrow C)$ *(∨Elimination)*
R9 $(A \,\&\, (B \vee C)) \rightarrow ((A \,\&\, B) \vee C)$ *(Distribution)*
R10 $\neg\neg A \rightarrow A$ *(Double Negation)*
R11 $(A \rightarrow \neg B) \rightarrow (B \rightarrow \neg A)$ *(Contraposition)*
→E If A *and* $A \rightarrow B$ *are theorems,* B *is a theorem (Modus Ponens)*
&I If A *and* B *are theorems,* $A \,\&\, B$ *is a theorem (Adjunction)*

Definition: $A \leftrightarrow B = (A \rightarrow B) \,\&\, (B \rightarrow A)$.

Definition 2 ($\mathbf{R}^{\forall p}$). *Propositionally quantified relevant system* $\boldsymbol{R}^{\forall p}$ *has the following axioms and axiom clause in addition to those of* ***R*** *[7, ch. VI].*

Q1 $\forall p(A \rightarrow B) \rightarrow (\forall pA \rightarrow \forall pB)$
Q2 $(\forall pA \,\&\, \forall pB) \rightarrow \forall p(A \,\&\, B)$
Q3 $\forall pA(p) \rightarrow A(B)$
Q4 $\forall p(A \rightarrow B) \rightarrow (A \rightarrow \forall pB)$ *(p not free in A)*
Q5 $\forall p(A \vee B) \rightarrow (A \vee \forall pB)$ *(p not free in A)*
Q∗ If A *is an axiom then* $\forall pA$ *is an axiom.*

Note that Q1, Q2 and Q∗ yield Generalization (Gen): if A is a theorem then $\forall pA$ is a theorem.

3 Anderson's Relevant Deontic Logic

Anderson [4, 5] defined his system of relevant deontic logic as **R** supplemented with the constant e and the following definition of O: $OA = e \to A$. He also considered an optional "axiom of avoidance" $\neg(e \to \neg e)$.

We similarly define propositionally quantified relevant deontic logic $\mathbf{R}_e^{\forall p}$ as $\mathbf{R}^{\forall p}$ supplemented with the constant e and the definition $OA = e \to A$. (We do not include the axiom of avoidance and will discuss it at the end of this section.)

Theorem 1. *$\boldsymbol{R}_e^{\forall p}$ has the following theorem: $e \leftrightarrow \forall p(Op \to p)$.*

Proof.

$\to$: 1.	$(e \to A) \to (e \to A)$	self-impl
2.	$e \to ((e \to A) \to A)$	1, permut
3.	$e \to \forall p((e \to p) \to p)$	2, Gen, Q4
4.	$e \to \forall p(Op \to p)$	3, def O
$\leftarrow$: 5.	$\forall p(Op \to p) \to \forall p((e \to p) \to p)$	def O
6.	$\forall p((e \to p) \to p) \to ((e \to e) \to e)$	Q3
7.	$(e \to e)$	self-impl
8.	$((e \to e) \to e) \to e$	7, self-impl, permut
9.	$\forall p(Op \to p) \to e$	5, 6, 8
$\leftrightarrow$: 10.	$e \leftrightarrow \forall p(Op \to p)$	4, 9, adj

Thus e says that all obligations are fulfilled (all normative demands are met). As noted above, this agrees with McNamara's reading of e [11].

Definition 3 ($\mathbf{OR}^{\forall p}$). *Propositionally quantified relevant deontic system $\boldsymbol{OR}^{\forall p}$ is $\boldsymbol{R}^{\forall p}$ supplemented with a primive operator O, a propositional constant e defined by $e = \forall p(Op \to p)$, and the following axioms in addition to those of $\boldsymbol{R}^{\forall p}$.*

OA $(A \to B) \to (OA \to OB)$
OBF $\forall pOA \to O\forall pA$
OT $O(OA \to A)$

Theorem 2. *$\boldsymbol{OR}^{\forall p}$ has the following theorem:*

OK $O(A \to B) \to (OA \to OB)$.

Proof. We first prove (Th1) $O(A \to B) \to (A \to OB)$.

1. $(A \to B) \to (A \to B)$ self-impl
2. $A \to ((A \to B) \to B)$ 1, permut
3. $A \to (O(A \to B) \to OB)$ 2, OA
4. $O(A \to B) \to (A \to OB)$ 3, permut

We also derive (Th2) $(A \to OB) \to O(A \to B)$.

1. $(OB \to B) \to ((A \to OB) \to (A \to B))$ prefixing
2. $O(OB \to B) \to O((A \to OB) \to (A \to B))$ 1, OA
3. $O((A \to OB) \to (A \to B))$ 2, OT
4. $(A \to OB) \to O(A \to B)$ 3, Th1

Next, we show that

(ROOO) *if OOA is a theorem, then OA is a theorem*

is a derivable rule.

1. OOA	premiss
2. $(OA \to A) \to (OOA \to OA)$	OA
3. $(OA \to A) \to OA$	1, 2, permut
4. $(OA \to A) \to (((OA \to A) \to OA) \to ((OA \to A) \to A))$	pref
5. $(OA \to A) \to ((OA \to A) \to A)$	3, 4, permut
6. $(OA \to A) \to A$	5, contract
7. $O(OA \to A) \to OA$	6, OA
8. OA	7, OT

We may now derive (Th3) $OOA \to OA$.

1. $OOA \to OOA$	self-impl
2. $O(OOA \to OA)$	1, Th2
3. $OO(OOA \to A)$	2, Th2, OA
4. $O(OOA \to A)$	3, rule ROOO
5. $OOA \to OA$	4, Th1

After this, theorem OK is easy:

1. $O(A \to B) \to (A \to OB)$	Th1
2. $(A \to OB) \to (OA \to OOB)$	OA
3. $(OA \to OOB) \to (OA \to OB)$	Th3
4. $O(A \to B) \to (OA \to OB)$	1–3

Theorem 3. *$\mathbf{R}_e^{\forall p}$ and $\mathbf{OR}^{\forall p}$ have the same theorems.*

Proof. First, all theorems of $\mathbf{OR}^{\forall p}$ are theorems of $\mathbf{R}_e^{\forall p}$. All cases are easy except perhaps $e \leftrightarrow \forall p(Op \to p)$, which has already been discussed. Second, all theorems of $\mathbf{R}_e^{\forall p}$ are theorems of $\mathbf{OR}^{\forall p}$. It is sufficient to prove that $OA \leftrightarrow (e \to A)$ is a theorem of $\mathbf{OR}^{\forall p}$.

$\to$:	1. $e \to (OA \to A)$	def e, Q3
	2. $OA \to (e \to A)$	1, permut
$\leftarrow$:	3. $O(e \to A) \to (Oe \to OA)$	OK
	4. $Oe \to ((e \to A) \to OA)$	3, permut
	5. $\forall pO(Op \to p)$	OT, Q$*$
	6. $O\forall p(Op \to p)$	5, OBF
	7. Oe	6, def e
	8. $(e \to A) \to OA$	4, 7
$\leftrightarrow$:	9. $OA \leftrightarrow (e \to A)$	2, 8, adj

Corollary 1. *$\mathbf{OR}^{\forall p}$ is the deontic fragment of $\mathbf{R}_e^{\forall p}$ (in the sense of Åqvist [1] and Goble [8]). $\mathbf{R}_e^{\forall p}$ is the alethic fragment of $\mathbf{OR}^{\forall p}$ (in a similar, but converse sense).*

It has been proven that $\mathbf{OR}^{\forall p}$ without propositional quantifiers and with OBF replaced by $(OA \,\&\, OB) \rightarrow O(A \,\&\, B)$ is the deontic fragment of $\mathbf{R}_e^{\forall p}$ without propositional quantifiers [8, 10]. The proof makes use of the Routley-Meyer semantics of these systems. Our proof is much shorter. Moreover, Theorem 3 shows that one cannot only reduce relevant deontic logic to alethic relevant logic, but carry out the converse reduction as well.

Finally, let us briefly discuss the axiom of avoidance $\neg(e \rightarrow \neg e)$. We define $\mathbf{R}_e^{\forall p}+$ as $\mathbf{R}_e^{\forall p}$ plus this axiom and $\mathbf{OR}^{\forall p}+$ as $\mathbf{OR}^{\forall p}$ plus $OA \rightarrow \neg O \neg A$. It can be shown that $\mathbf{R}_e^{\forall p}+$ and $\mathbf{OR}^{\forall p}+$ have the same theorems and that $\mathbf{OR}^{\forall p}+$ is the deontic fragment of $\mathbf{R}_e^{\forall p}+$, but we shall not discuss this in detail and refer to [8] for a more extended discussion.

4 Relevant Mixed Alethic-Deontic Logic

Definition 4 ($\mathbf{RKT}^{\forall p}$). *Propositionally quantified relevant alethic modal system* $\boldsymbol{RKT}^{\forall p}$ *has the following axioms and rules in addition to those of* $\boldsymbol{R}^{\forall p}$.

LK $L(A \rightarrow B) \rightarrow (LA \rightarrow LB)$
LC $(LA \,\&\, LB) \rightarrow L(A \,\&\, B)$
LBF $\forall pLA \rightarrow L\forall pA$
Nec If A *is a theorem then* LA *is a theorem*
LT $LA \rightarrow A$

Definition 5 ($\mathbf{RS4}^{\forall p}$). $\boldsymbol{RS4}^{\forall p}$ *is* $\boldsymbol{RKT}^{\forall p}$ *plus the following axiom:*

L4 $LA \rightarrow LLA$.

Definition 6 ($\mathbf{RKT}_e^{\forall p}$ and $\mathbf{RS4}_e^{\forall p}$). *Systems* $\boldsymbol{RKT}_e^{\forall p}$ *and* $\boldsymbol{RS4}_e^{\forall p}$ *have the same axioms and rules as* $\boldsymbol{RKT}^{\forall p}$ *and* $\boldsymbol{RS4}^{\forall p}$*, respectively, except that they contain a propositional constant* e *and a propositional operator* O *defined by* $OA = L(e \rightarrow A)$.

Theorem 4. $\boldsymbol{RKT}_e^{\forall p}$ *has the following theorem:* $e \leftrightarrow \forall p(Op \rightarrow p)$.

Proof.

$\rightarrow$:	1. $L(e \rightarrow A) \rightarrow (e \rightarrow A)$	LT
	2. $e \rightarrow (L(e \rightarrow A) \rightarrow A)$	1, permut
	3. $e \rightarrow \forall p(L(e \rightarrow p) \rightarrow p)$	2, Gen, Q4
	4. $e \rightarrow \forall p(Op \rightarrow p)$	3, def O
$\leftarrow$:	5. $\forall p(Op \rightarrow p) \rightarrow \forall p(L(e \rightarrow p) \rightarrow p)$	def O
	6. $\forall p(L(e \rightarrow p) \rightarrow p) \rightarrow (L(e \rightarrow e) \rightarrow e)$	Q3
	7. $L(e \rightarrow e)$	self-impl, Nec
	8. $(L(e \rightarrow e) \rightarrow e) \rightarrow e$	7, self-impl, permut
	9. $\forall p(Op \rightarrow p) \rightarrow e$	5, 6, 8
$\leftrightarrow$:	10. $e \leftrightarrow \forall p(Op \rightarrow p)$	4, 9, adj

As above, e again says that all obligations are fulfilled (all normative demands are met), which agrees with McNamara's reading of e [11]. In systems that are weaker than $\mathbf{RKT}_e^{\forall p}$ one may have a greater freedom of interpretation.

Definition 7 ($\mathbf{RMS4}^{\forall p}$). *Propositionally quantified relevant mixed alethic-deontic system* $\mathbf{RMS4}^{\forall p}$ *is* $\mathbf{RS4}^{\forall p}$ *supplemented with a primive operator* O, *a propositional constant e defined by* $e = \forall p(Op \to p)$, *and the following axioms and rules in addition to those of* $\mathbf{RS4}^{\forall p}$.

OK $O(A \to B) \to (OA \to OB)$
OBF $\forall pOA \to O\forall pA$
OT $O(OA \to A)$
LO $LA \to OA$
M4 $OA \to LOA$

Theorem 5. $\mathbf{RS4}_e^{\forall p}$ *and* $\mathbf{RMS4}^{\forall p}$ *have the same theorems.*

Proof. First, all theorems of $\mathbf{RMS4}^{\forall p}$ are theorems of $\mathbf{RS4}_e^{\forall p}$. All cases are easy except perhaps $e \leftrightarrow \forall p(Op \to p)$, which has already been discussed. Second, all theorems of $\mathbf{RS4}_e^{\forall p}$ are theorems of $\mathbf{RMS4}^{\forall p}$. It is sufficient to prove that $OA \leftrightarrow L(e \to A)$ is a theorem of $\mathbf{RMS4}^{\forall p}$.

$\to$: 1.	$e \to (OA \to A)$	def e, Q3
2.	$OA \to (e \to A)$	1, permut
3.	$L(OA \to (e \to A))$	2, Nec
4.	$LOA \to L(e \to A)$	3, LK
5.	$OA \to L(e \to A)$	4, M4
$\leftarrow$: 6.	$L(e \to A) \to O(e \to A)$	LO
7.	$O(e \to A) \to (Oe \to OA)$	OK
8.	$Oe \to (L(e \to A) \to OA)$	6, 7, permut
9.	$\forall pO(Op \to p)$	OT, Q*
10.	$O\forall p(Op \to p)$	9, OBF
11.	Oe	10, def e
12.	$L(e \to A) \to OA$	8, 11
$\leftrightarrow$: 13.	$OA \leftrightarrow L(e \to A)$	5, 12, adj

Corollary 2. $\mathbf{RMS4}^{\forall p}$ *is the alethic-deontic fragment of* $\mathbf{RS4}_e^{\forall p}$ *(in the sense of Åqvist [1] and Goble [8]).* $\mathbf{RS4}_e^{\forall p}$ *is the alethic modal fragment of* $\mathbf{RMS4}^{\forall p}$ *(in a similar, but converse sense).*

Goble [8] has proven that $\mathbf{RMS4}^{\forall p}$ without propositional quantifiers and with OBF replaced by $(OA \,\&\, OB) \to O(A \,\&\, B)$ is the alethic-deontic fragment of $\mathbf{RS4}_e^{\forall p}$ without propositional quantifiers and without LBF. He used the Routley-Meyer semantics of these systems to prove this. Our proof is much shorter. Moreover, Theorem 5 shows that one cannot only reduce mixed alethic-deontic logic to alethic modal logic, but carry out the converse reduction as well.

Finally, we remark that systems $\mathbf{RS4}_e^{\forall p}$ plus $\neg L \neg e$, on the one hand, and $\mathbf{RMS4}^{\forall p}$ plus $OA \to \neg O \neg A$, on the other, have the same theorems and that the latter system is the deontic fragment of the former, but we shall not discuss this in detail and refer to [8] for a more extended discussion.

5 Classical Mixed Alethic-Deontic Logic

Let K be an abbreviation of $A \to (B \to A)$ (the archetypical fallacy of relevance). Propositionally quantified classical modal systems $\mathbf{KT}_e^{\forall p}$ and $\mathbf{S4}_e^{\forall p}$ may be defined by $\mathbf{KT}_e^{\forall p} = \mathbf{RKT}_e^{\forall p} + K$ and $\mathbf{S4}_e^{\forall p} = \mathbf{RS4}_e^{\forall p} + K$. The proof that $e \leftrightarrow \forall p(Op \to p)$ is a theorem of these systems is identical with the proof of Theorem 4.

Propositionally quantified classical mixed alethic-deontic systems $\mathbf{MKT}^{\forall p}$ and $\mathbf{MS4}^{\forall p}$ may be defined by $\mathbf{MKT}^{\forall p} = \mathbf{RMKT}^{\forall p} + K$ and $\mathbf{MS4}^{\forall p} = \mathbf{RMS4}^{\forall p} + K$. The proof that $\mathbf{MS4}^{\forall p}$ is the alethic-deontic fragment of $\mathbf{S4}_e^{\forall p}$ is the same as the proof of Theorem 5.

It is to be noted that axiom M4 of $\mathbf{MS4}^{\forall p}$ is really required to prove this. The formula $OA \to L(\forall p(Op \to p) \to A)$ is invalid in $\mathbf{MKT}^{\forall p}$, as Fig. 1 shows. Since all theorems of $\mathbf{RMKT}^{\forall p}$ are theorems of $\mathbf{MKT}^{\forall p}$, this shows that this formula is invalid in $\mathbf{RMKT}^{\forall p}$, too.

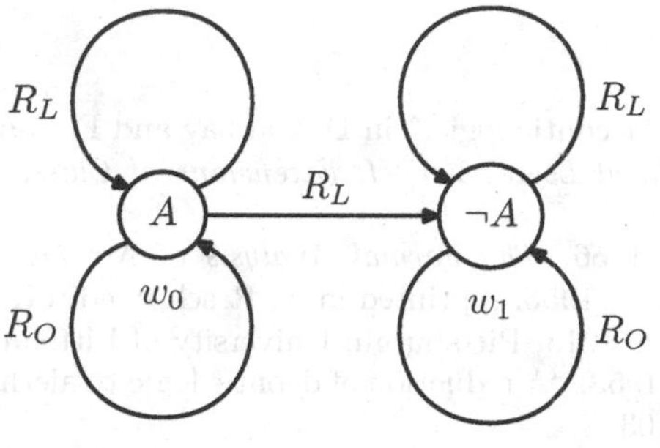

Fig. 1. An $\mathbf{MKT}^{\forall p}$-model that refutes $OA \to L(\forall p(Op \to p) \to A)$. R_L is the alethic modal accessibility relation, R_O is the deontic accessibility relation. A holds at w_0 but not at w_1. As a result, $OA \to L(\forall p(Op \to p) \to A)$ does not hold at w_0.

6 A Classical Version of Anderson's Relevant Deontic Logic

Finally, one might wonder what would happen if K were added to the extensional Andersonian systems $\mathbf{R}_e^{\forall p}$ and $\mathbf{R}_e^{\forall p}+$. The answer is simple: $A \to OA$ is a theorem of $\mathbf{R}_e^{\forall p}$ plus K and $A \leftrightarrow OA$ is a theorem of $\mathbf{R}_e^{\forall p}+$ plus K. Thus, one gets a collapse to the trivial system.

7 Further Prospects

Up to this point, we have read OA as "it is obligatory that A." But O can also be interpreted in alternative ways. As Smiley put it:

> If we define OA as $L(e \supset A)$ then to assert OA is to assert that e strictly implies A or that A is necessary *relative to* e. Since the pattern of the definition is independent of the particular interpretation that may be put on e we can say that to the extent that the standard alethic modal systems embody the idea of absolute or logical necessity, the corresponding O-systems embody the idea of *relative necessity*—necessity relative to an arbitrary proposition or body of propositions. They should therefore be appropriate for the formalisation of any modal notion that can be analysed in terms of relative necessity. Thus by applying the definition in the case where e expresses the postulates of a mathematical theory, O can be read 'it is provable that.' Where e expresses the content of a legal code, O will be read 'it is the law that'; where e is interpreted in terms of an individual's corpus of beliefs, O will be read 'so-and-so believes that.' (Smiley [13, p. 113]; he used the symbol T instead of e.)

It almost goes without saying that our formal results can be applied to these other areas as well.

References

[1] Åqvist, L., 1984, "Deontic logic," in D. Gabbay and F. Günthner, editors, *Handbook of Philosophical Logic, Vol. II: Extensions of Classical Logic*, pp. 605–714, Dordrecht: Reidel.

[2] Anderson, A. R., 1956, *The Formal Analysis of Normative Systems*, Technical Report, New Haven, 1956, reprinted in N. Rescher, editor, *The Logic of Decision and Action*, pp. 147–213, Pittsburgh: University of Pittsburgh Press, 1967.

[3] Anderson, A. R., 1958, "A reduction of deontic logic to alethic modal logic," *Mind*, vol. 67, pp. 100-103.

[4] Anderson, A. R., 1967, "Some nasty problems in the formal logic of ethics," *Noûs*, vol. 1, pp. 345–360.

[5] Anderson, A. R., 1968, "A new square of opposition: Eubouliatic logic," in *Akten des XIV. Internationalen Kongresses für Philosophie*, vol. 2, pp. 271–284, Vienna: Herder.

[6] Anderson, A. R., and N. D. Belnap, Jr., 1975, *Entailment: The Logic of Relevance and Necessity*, vol. 1, Princeton, N. J.: Princeton University Press.

[7] Anderson, A. R., N. D. Belnap, Jr., and J. M. Dunn, 1992, *Entailment: The Logic of Relevance and Necessity*, vol. 2, Princeton, N. J.: Princeton University Press.

[8] Goble, L., 2001, "The Andersonian reduction and relevant deontic logic," in B. Brown and J. Woods, editors, *New Studies in Exact Philosophy: Logic, Mathematics and Science—Proceedings of the 1999 Conference of the Society of Exact Philosophy*, pp. 213–246, Paris: Hermes Science Publications.

[9] Kanger, S., 1957, *New Foundations for Ethical Theory*, privately distributed pamphlet, reprinted in R. Hilpinen, editor, *Deontic Logic: Introductory and Systematic Readings*, pp. 36–58, Dordrecht: Reidel, 1971.

[10] Lokhorst, G. J. C., 200+, "The Andersonian reduction and Mally's deontic logic," *Notre Dame Journal of Formal Logic*, forthcoming.

[11] McNamara, P., 2006, "Deontic logic," in E. N. Zalta, editor, *The Stanford Encyclopedia of Philosophy (Spring 2006 Edition)*, URL = `http://plato.stanford.edu/archives/spr2006/entries/logic-deontic`.

[12] Mares, E. D., 1992, "Andersonian deontic logic," *Theoria*, vol. 58, 3–20.
[13] Smiley, T., 1963, "Relative necessity," *Journal of Symbolic Logic*, vol. 28, pp. 113-134.
[14] Van Fraassen, B. C., 1972, "The logic of conditional obligation," *Journal of Philosophical Logic*, vol. 1, pp. 417–438.

A Question of Trust: Assessing the Fulfillment of Commitments in Terms of Strategies

Thomas Müller

Universität Bonn, Institut für Philosophie, Lennéstr. 39, 53113 Bonn, Germany
Thomas.Mueller@uni-bonn.de

Abstract. We aim at an adequate formal description of the dynamics of commitments and trust by transferring insights about actual human practices to a formal setting. Our framework is based on Belnap's theory of agents and choices in branching time (stit theory) and his analysis of commitments in terms of strategies. The main points are that (i) commitments come in various degrees of stringency, (ii) we can define a stringency ordering on an agent's possible strategies, and that (iii) trustworthiness can be spelled out in terms of strategies: An agent is living up to a given commitment, and thus, is trustworthy with respect to that commitment, if her strategy is at least as stringent as required. Overall trustworthiness of an agent can be defined by averaging over such single case assessments.

Introduction

Human beings can enter freely into normative relations towards other human beings. Two of our most important practices in this respect are making promises and entering into contracts. Similar practices are also important when it comes to artificial normative systems; they are often taken together under the heading of *commitments*. The notion of a commitment has been analysed from a wide range of perspectives—cf. [1] for a recent contribution. In this paper, our focus is on transferring insights about details of actual human practices to a setting of artificial normative agents. We will take a lead from the formal characterisation of commitments given by Belnap et al. [2] (who acknowledge earlier, related work by Thomson [3]). Belnap specifies the content of a commitment in terms of an agent's strategy, employing a theory of agency that is based on the objectively indeterministic theory of branching time.

We will argue that looking at an agent's strategy allows us to assess in a fine-grained way the trustworthiness of an agent who is under a certain commitment. Even before fulfillment or non-fulfillment of the commitment can be assessed, the agent's strategy, which reveals itself through what the agent chooses to do, gives information on how seriously an agent takes the commitment she is subject to. The main technical contribution of this paper is the definition of a partial ordering on the set of strategies available for an agent that classifies strategies as more or less *stringent* with respect to a given commitment. That partial ordering allows us to classify the agent's behaviour as appropriate or inappropriate in

L. Goble and J.-J.C. Meyer (Eds.): DEON 2006, LNAI 4048, pp. 210–221, 2006.

relation to a given commitment, and thus, to classify the agent herself as more or less trustworthy with respect to that commitment. An estimate for the agent's overall trustworthiness can be obtained by averaging over individual assessments of trustworthiness for a number of commitments.

The main features of the model we shall describe are the following (numbers indicate the respective sections of this paper):

1. Agents and their possible choices are described in the objectively indeterministic framework of branching time. That framework allows us to model the objective uncertainty that agents have to face when making plans in a world like ours, in which there are both indeterministic external influences and indeterministic actions by other agents. The theory doesn't just assume that the agents do not *know* what the future will bring, but that, in many cases, what will happen is not determined until it happens.
2. Continuous action of agents is described via the strategy that the agent is following. A strategy allows the agent to fix defaults for future choices in an objectively indeterministic setting.
3. By looking at our (human) practices of promising vs. entering into contracts, we find that commitments come in various degrees of stringency. There are commitments not backed by sanctions, but by moral force only, and there are commitments for which there are only sanctions but no moral force. These observations open the view for an ordering among possible strategies for an agent: Given a commitment, some strategies are more stringent with respect to the commitment than others.
4. We distinguish the soft constraints that are in effect through a commitment from hard constraints used to model the effect of a possible sanction. Then, taking a lead from Belnap's analysis of promising, we arrive at a formal definition of a stringency ordering among an agent's possible strategies relative to a commitment.
5. We suggest that each commitment comes with an appropriate level of stringency. Not all commitments are alike in this respect. Thus, in assessing whether an agent deals with a commitment adequately, it seems best to (i) identify the strategy that the agent is following and (ii) compare that strategy with the required level of stringency of the given commitment. If the agent's strategy meets or surpasses the required level of stringency, we will say that the agent *is living up to her commitment.* Finally, we suggest that when it comes to assessing trustworthiness, it is the notion of living up to one's commitment that is important. Identifying an agent's strategy thus provides the basis for an appropriate description of the dynamics of commitments and trust.

1 Background: Agents and Choices in Branching Time

In this section we describe the theory of agency that forms the background of our approach. That theory is motivated and explained both formally and informally in Belnap et al.'s [2], which may be consulted for additional detail and for its many examples.[1] (In what follows, the notation of [2] has been slightly altered in a number of places.)

Branching time is a theory of objective indeterminism that was first invented by Prior [6] and that has subsequently found many applications in logic and computer science. A branching time structure is a tree-like partial ordering of moments without backward branching:

Definition 1. *A* branching time structure *is a partial ordering $\langle W, <\rangle$ such that there is no backward branching, i.e., if $x < z$ and $y < z$, then $x \leq y$ or $y \leq x$.*

The elements of W are called *moments*, and $m < m'$ is read temporally, i.e., as "m is before m'". A *history* h is a maximal linear chain (set of pairwise comparable elements) in W, and for each moment m, the set H_m is the set of histories to which m belongs. Two histories $h_1, h_2 \in H_m$ can either split at m or be undivided at m:

Definition 2. *Two histories $h_1, h_2 \in H_m$ are called* undivided at m *($h_1 \equiv_m h_2$) iff they share a moment that is properly later than m, i.e., iff there is $m' \in h_1 \cap h_2$ s.t. $m < m'$. If, on the other hand, m is maximal in $h_1 \cap h_2$, we say that h_1* splits off from h_2 at m *($h_1 \perp_m h_2$).*

The notation is sugestive: $\equiv_m$ is easily shown to be an equivalence relation, inducing a natural partition Π_m of H_m. Assuming $h \in H_m$, we write $\Pi_m\langle h\rangle$ for that unique element of Π_m that contains h.

Given a set $A = \{\alpha_1, \ldots, \alpha_n\}$ of agents, the theory of agents and choices in branching time specifies, for each agent $\alpha \in A$ and each moment m, the set of choices open to α at that moment. That set, $Choice_m^\alpha$, partitions H_m and may be more coarse-grained, but not more fine-grained, than the natural partition Π_m at m. Metaphorically, an agent has at most as much, but possibly less, control over what can happen next as nature herself. Assuming $h \in H_m$, we write $Choice_m^\alpha(h)$ for that unique member of $Choice_m^\alpha$ to which h belongs.

Our main definition is the following:

Definition 3. *A structure of* agents with choices in branching time *is a quadruple $\langle W, <, A, Choice\rangle$, where $\langle W, <\rangle$ is a branching time structure, A is a finite set of agents, and $Choice$ is a function that assigns a partition $Choice_m^\alpha$ of H_m to each pair $\langle \alpha, m\rangle$ in such a way that $Choice_m^\alpha$ is a coarse-graining of the natural partition Π_m, i.e.,*

$$\text{for all } \alpha,\, m \text{ and } h \in H_m,\ \Pi_m\langle h\rangle \subseteq Choice_m^\alpha(h).$$

We will only say a little about the formal language that is appropriate for the theory of agents and choices in branching time (cf. [2] for details). Formulae

[1] In accord with Belnap [4], we are convinced that a really successful theory of agency must not stop at the level of resolution supplied by branching time. Rather, *branching space-times* models [5] will have to be employed. As the theory of agency in branching space-times is little developed so far, in this paper we will hold on to the technically well understood theory of agency in branching time.

are built up from propositional variables by means of the truth-functional sentential connectives and modal operators for the tenses ("*was*" and "*will*") and for agency ("α *stit* : ϕ" for "agent α sees to it that ϕ"). In accord with the ideas of Prior-Thomason semantics [7], these formulae are evaluated not just at moments, but at moment-history pairs $\langle m, h\rangle$, where $m \in h$. Future-tense propositions are typically history-dependent, i.e., at moment m, the truth value of such a proposition also depends on the history of evaluation. For the purposes of this paper, the most important concept is the modality of *settled truth*, which corresponds to universal quantification over histories: ϕ is settled true at $\langle m, h\rangle$ iff ϕ is true at $\langle m, h'\rangle$ for all $h' \in H_m$. Settled truth is thus independent of the history of evaluation.

The respective inductive semantic clauses for the mentioned modal operators are:

- $m, h \models was : \phi$ iff there is $m' < m$ s.t. $m', h \models \phi$ (note that $m' \in h$ follows by backwards linearity),
- $m, h \models will : \phi$ iff there is $m' \in h$ s.t. $m < m'$ and $m', h \models \phi$, and
- $m, h \models settled : \phi$ iff for all $h' \in H_m$ we have $m, h' \models \phi$.

2 Background: Strategies

A strategy specifies defaults for an agent's future choices by fixing which choices of the agent count as in accord with the strategy. Strategies are needed to describe agency since most of what we actually do takes time. Thus, consider baking a cake, which takes about an hour. You can't at the beginning of that one hour period, m_0, act in such a way, or see to it, that no further choice of yours is required for finishing the cake. At any moment in the process, you can toss everything and leave. Nor can you make all the required future choices at the initial moment, m_0. It is a conceptual necessity that a choice can only be made once it is due—otherwise it wouldn't *be* a choice. This is not to say that an agent at m_0 is completely helpless, however. The agent can adopt a strategy, s, that prescribes default choices for future moments from m_0 on. It seems that temporally extended agency can be best described via strategies (cf. [8] for an attempt at spelling this out in a type of modal logic). Also, almost all commitments pertain to a whole array of future choices—e.g., hardly any promise can be fulfilled instantaneously.

The formal theory of strategies is laid out in detail in [2, Chap. 13]. We only give the main definitions:

Definition 4. *A* strategy for α *is a partial function* s *on moments such that for each moment* m *for which* s *is defined,* $s(m)$ *is a subset of* H_m *that respects the available choices of* α *at* m*, i.e., for every* $h \in H_m$*, if* $h \in s(m)$*, then* $Choice^{\alpha}_{m}(h) \subseteq s(m)$.

A strategy thus specifies what α should do, and in this, the strategy can give advice at most as fine-grained as the available choices for α at m allow (which, in turn, may be at most as fine-grained as the natural partition Π_m allows).

If an agent follows a strategy, some histories and moments will be admitted (the strategy will advise to stay on the history / reach the moment), and others will be excluded. Technically, we define:

Definition 5. *If s is a strategy for α, we say that*
- s admits h *iff for every m for which s is defined, if $m \in h$, then $h \in s(m)$.*
- s admits m *iff for every m' on which s is defined and for which $m' < m$ we have $m \in \bigcup s(m')$.*
- s excludes h or m *iff it does not admit it.*

The set of histories admitted by s, $Admh(s)$, and the set of admitted moments, $Admm(s)$, are defined to be

$$Admh(s) = \{h \mid s \text{ admits } h\}, \qquad Admm(s) = \{m \mid s \text{ admits } m\}.$$

The concept of an admitted moment will be used in our definition of a stringency ordering among strategies below.

By our definition, strategies are allowed to leave open choices for α—their advice does not have to be as fine-grained as possible. A strategy that gives the most detailed kind of advice possible, is called *strict*:

Definition 6. *A strategy s for α is called* strict at m *iff it is defined at m and $s(m) \in Choice^{\alpha}_{m}$. The strategy s is called* strict *iff is it strict at every moment at which it is defined.*

Strict strategies enjoy a special epistemic status: if an agent α is following a strategy that is strict at a moment m, then her actual choice at that moment reveals her strategy $s(m)$ at m in full detail. In this case, behaviour is a perfectly reliable guide to the agent's strategy. This is not so if the agent is following a non-strict strategy: at a moment m at which $s(m) \notin Choice^{\alpha}_{m}$, the actual choice that α makes (which corresponds to an element of $Choice^{\alpha}_{m}$) does not reveal $s(m)$ completely. In such a case, we would have to ask α to tell us which strategy was on her mind. However, if our task is to find out about α's strategy in order to assess her trustworthiness, asking α may appear to be circular: in order to rely on α's answer, we would have to know that we can trust her, but that is exactly what we wish to find out. Studying α's behaviour, on the other hand, does not presuppose trusting α.

At this juncture, we face a methodological decision regarding our theory of trust. It is clear that we can only ever identify an agent's strategy in full detail if that strategy is strict, and we need to know about the agent's strategy in order to assess trustworthiness (as spelled out below). May we assume that agents always follow strict strategies? It is not an easy task to weigh the pros and cons of that assumption. Assuming strict strategies allows for a smooth formal picture. Furthermore, the point can be made that action is after all our most fundamental guide to intention, or strategy [9, § 4]. On the other hand, we can never be sure that an agent is really following a strict strategy, and presupposing strict strategies from the outset may appear to impose an unnecessary restriction on our formal theory. In what follows, we will therefore try to sketch a general theory that is not confined to strict strategies.

3 Commitments: Promises vs. Contracts

Having laid out the formal background for our theory, we now turn to human practices concerning commitments. Among these, promises and contracts are the most significant examples. We suggest that taking a close look at these practices opens up the view towards an important distinction that can be made formally precise within our framework: commitments come with various degrees of stringency. That distinction will later on be used to assess trustworthiness.

Is there a difference between a promise and a contract? The terminology is certainly not established as firmly as to allow for a straight yes. However, we wish to suggest that a typical promise and a typical contract are different creatures of the normative realm, and that it is useful for our purposes to distinguish the two notions sharply. In effect we will be arguing that there is a continuum of normative relations, one pole of which can be exemplified by a certain type of promise, while the other pole corresponds to a certain type of contract.

A *normative relation* between agents α and β is a certain type of normative constraint on the behaviour of α. Minimally, the *content* of the relation specifies what counts as fulfilling or violating the norm (and what is neutral with respect to this distinction). Typically, β is also in a position to put a *sanction* on α for non-compliance (perhaps with the aid of an overarching normative metasystem such as the state and its legislation). Apart from the sanction, which is external to the normative relation, many normative relations also have a *moral dimension*: quite apart from any sanction, many norm violations are judged to be morally wrong.

Promises are normally made between people who know and trust each other at least to some extent, and in most cases, there are no enforceable external sanctions connected with promise breaking. (However, the relation between promissor and promisee is typically altered by a broken promise, which may count as a kind of sanction in some cases, and promises may give rise to *additional* legal relations that are associated with sanctions.) Promises normally concern matters of little economical importance, but of some personal, practical importance to β. They usually have a clear moral dimension, since for most promises it is judged to be morally wrong to break them.

Contracts can be made between any two persons, but certain very close relationships may make contracts appear to be inappropriate, and some legal systems do not recognise certain types of contracts between, e.g., husband and wife. Most contracts come with enforceable external sanctions backed by the law. Contracts usually concern matters of some economical importance, while the personal importance of the content of a contract may be wholly derived from its economical aspects. Not all contracts need to have a moral dimension—there seem to be contracts that it may be costly, but not morally wrong, to break.

Based on these observations, we propose to assume a continuum of normative relations characterised by the degree of stringency (pure moral bindingness) vs. external sanction involved, as follows:

1. The most solemn and most stringent case is a case of promising without any associated external sanction. In the philosophical literature, the typical example of this is a death-bed promise given without witnesses. If α promises something to dying β and there are no witnesses, there is no possibility of an external sanction. However, such promises seem to have great moral force—maybe just because of the sheer helplessness of β, who himself knows that there is no possibility of enforcing what is promised.
2. In a typical case of promising, there is no or little external sanction possible, but a relatively large degree of moral force. If α promises β to meet her at the station and fails to do so (for no good reason), then there is usually little that β can *do*, but α's behaviour will be judged morally wrong. Whether it is good to fulfill a promise is mostly a matter of morality and only to a little extent a matter of α's personal utility.
3. In a typical contract, there is a relatively large degree of external sanctioning possible, but (usually) little moral force. E.g., if α enters a contract with β to the effect that α deliver some goods at a specified time, but fails to deliver, the consequences are mostly external: α may be sued and have to pay for β's damages. This is not to say that such behaviour on α's side will be assumed to be morally neutral—there will usually be some negative assessment in this dimension too. Whether it is good to fulfill a contract is to some degree a matter of morality, but mostly a matter of α's personal utility.
4. Finally, there seem to be contracts the fulfillment or non-fulfillment of which is purely a matter of utility. A freely levelled contract between α and β gives α a choice to either comply or opt out and pay a fine. Some contracts we enter seem to be of that arbitrary, game-like kind.

When α and β enter into a normative relation, there is normally some kind of agreement as to which level of stringengy that relation should be assumed to have—and if there is not, such agreement can usually be reached by discussing the issue. (If such agreement cannot be reached, the agents may choose not to create the commitment.) Cases in which β is completely helpless (incapable of sanctioning) will tend to be more stringent morally, while free-market type agreements tend to have a low level of moral stringency.

The point of these observations about human practices is twofold. First, by showing that there are different types of normative relations in human social interaction, we wish to suggest that it may be beneficial to employ a correspondingly rich notion of normative relations in artificial normative systems, too. Certainly, distinctions of stringency of norms can be made for artificial normative systems—e.g., business transactions over the internet may be typical contracts, while giving root privileges to an agent may suggest a promissory commitment not to exploit that power in a harmful way. Secondly, the description given above already points towards a formal characterisation of stringency for strategies. We will develop that characterisation in the next section. In Section 5, we will then employ our formal framework in order to discuss the question of trustworthiness of an agent. Roughly, the idea will be that an agent is trustworthy if her strategy is appropriate with respect to the degree of strictness of the commitment she has entered.

4 On the Stringency of Strategies

In employing strategies to analyse commitments, we take a lead from Belnap's analysis of promising, which in turn is based on the theory of strategies and agents and choices in branching time. In [2, Chap. 5C], the analytical target for promising is

$$\text{at } m_0,\ \alpha \text{ promises } \beta \text{ that } p,$$

where α and β are agents, m_0 is the moment of promising, and p is the content of the promise, which is typically a future-tense proposition. The meaning of that target is spelled out in terms of a strategy that α adopts at m_0. The main idea is that from m_0 onwards, the promise-keeping strategy advises α to choose such as to make p settled true if possible, and to keep p an open possibility otherwise.

In the context of his theory of promising, Belnap distinguishes two kinds of commitments, which he calls *word-giving* vs. *promising* (taking up a distinction from [3]). Word-giving he takes to be a less stringent commitment, which is expressed by the fact that α's strategy for word-giving does not advise α to do anything until the commitment is either fulfilled or violated.[2] In the latter case, the strategy advises α to compensate β, and that is all there is to it.

We wish to suggest that Belnap's analysis points in the right direction, but that it can be improved. First of all, it appears to us that the notion of a sanction or compensation that kicks in when a commitment has been violateed, should be analysed differently. Secondly, rather than distinguishing just two types of strategies, we will be able to order strategies with respect to their stringency, thus allowing for a more fine-grained assessment of the adequacy of the strategy that an agent is following, relative to a given commitment.

4.1 Commitments and Sanctions

Commitments are normative relations between agents. Norms can be fulfilled or violated. There are two basic schemes that can be used to monitor and perhaps force compliance to norms. *Hard constraints* are such that it becomes (physically) impossible to violate the norm. E.g., in many parking garages you cannot leave without having paid—the gate just won't open. The norm to pay for parking is therefore monitored and enforced by a hard constraint. Quite another scheme is in effect in so-called *soft constraints*: Here, compliance to the norm is monitored, perhaps on a statistical basis, and non-compliance leads to sanctions if detected.

[2] Belnap et al. [2, 126f.] suggest to choose more neutral terminology. According to their (non-standard) usage, promises are "satisfied" or "infringed", and mere word-givings are "vindicated" or "impugned". We will use the standard terminology of fulfillment vs. violation, but we wish to stress that the moral implications that these words normally suggest may be absent in the case of some commitments.

E.g., in many schemes of paid on-street parking, you *can* park and retrieve your car without a valid ticket, but there will be a penalty if you are found not to have a ticket on display.

Most issues of our social lives are regulated via soft constraints. It is altogether impractical to try to enforce too many norms via hard constraints. E.g., how would you try to implement the rule not to steal via a hard constraint? This may be feasible in a few select circumstances (e.g., at vending machines), but generally, our society relies heavily on soft constraints. Indeed it seems difficult to imagine any society of human beings that would rely solely on hard constraints.

In the realm of artificial agents, hard constraints are often somewhat easier to impose than in the society of human beings—e.g., communication ports can be blocked relatively easily. However, even in a society of artificial agents, the more complex systems become, the less feasible does it become to rely on hard constraints only. Usually, hard constraints must be implemented and enforced centrally, and there is a computational and protocol overhead as well as security issues speaking against reyling on hard constraints exclusively.

Commitments are conceptually tied to soft constraints: If an agent is under a commitment, she is normally free to fulfill or violate the commitment (of course, influences outside the agent's control can have an impact on these possibilities). Once a commitment is violated, depending on the type of commitment, some sanction is appropriate. The question we now wish to address is how to model this sanction.

In human interactions, first of all, not all violations of commitments are detected. Secondly, the sanctions imposed upon (detected) norm violation are usually again enforced via soft constraints. E.g., if you fail to pay one of your bills, there will be a penalty, but it will again be up to you to pay the penalty or not. The situation can however escalate by taking a number of turns at court, and in the end, if you fail to comply persistently, you might be put in prison, which means that a truly hard contraint would be triggered. In our daily lives, we thus mostly operate with soft constraints, but legally binding norms are in the end backed by hard constraints.

In our formal model, we will employ an idealisation: We will assume that commitments are subject to soft constraints, but that upon violating a commitment, detection is certain, and a hard constraint kicks in to make sure the sanction has its desired effect. Thus, we will assume that sanctions are automatic.[3] One effect of this is that an agent must take the cost of sanctions into her utility considerations from the outset. This is how mere considerations of personal utility can lead to compliance with norms. We wish to suggest that in some cases, this will be judged good enough, whereas in other cases, it won't—it depends on the stringency of the commitment in question.

[3] This assumption is closely related to Anderson's [10] early proposal of a "reduction of deontic logic to alethic modal logic", according to which the deontic-logical "it is obligatory that p" is reduced to the alethic "necessarily, if non-p, then SANCTION". The main difference is that we do not suggest an overall reduction of anything to anything else, and that any sanction is relative to the commitment to which it is attached, whereas Anderson seems to have thought of a single, all-purpose sanction.

4.2 Ordering Strategies by Stringency

Having discussed the question of how sanctions take their effect, we can now address the question of when it is appropriate to call one strategy more stringent than another.

Stringency is always relative to a given commitment. We specify a commitment c as a quintuple $c = \langle m_0, \alpha, \beta, p, Z \rangle$, where m_0 is the moment at which the commitment is created, α is the agent who is committed, β is the addressee of the commitment, p is its content (usually, a future-tense proposition), and Z is the sanction attached to non-fulfillment.

In our approach, a commitment is monitored via soft constraints. Thus, what α *can do* is not directly influenced by the commitment she has entered—unless the commitment is violated, in which case the hard constraint enforcing the sanction Z kicks in. However, what α *does* will be more or less *appropriate* relative to the commitment c. Belnap proposes to distinguish promising, which requires a strategy that actively aims at fulfilling the commitment, from mere word-giving, which only requires a strategy that takes care of accepting the sanction—which in our case, in which sanctions are automatic, does not require any choices by the agent at all. These two strategies can be seen to indicate two poles in an ordering of strategies with respect to their stringency.

To give our (perhaps preliminary) formal definition, let a commitment $c = \langle m_0, \alpha, \beta, p, Z \rangle$ be given, and let s and s' be two strategies for α defined (partially) on the future of m_0. We define comparative stringency pointwise first: If both s and s' are defined at $m \geq m_0$, we say that s *is more stringent with respect to c at m than* s', or that s' *is less stringent with respect to c at m than* s ($s' \prec_c^m s$) iff s prescribes to choose w and s' prescribes w' (where $w \neq w'$) such that

- choice w makes p settled true, whereas w' does not, or
- choice w' makes p settled false, whereas w does not, or
- choice w (considered as a one-step strategy) admits a moment at which α can make p settled true, whereas w' does not.

Thus, a choice w at moment m is more stringent with respect to commitment c than choice w' iff overall, w more strongly favours the fulfillment of the commitment than w'.[4] We generalise this pointwise definition to full strategies:

Definition 7. *Let a commitment $c = \langle m_0, \alpha, \beta, p, Z \rangle$ be given, and let s and s' be two strategies for α defined (partially) on the future of m_0. We say that s* is more stringent with respect to c than s' *($s' \prec_c s$) iff there is some $m \geq m_0$ such that (i) $s' \prec_c^m s$ and (ii) for all m' for which $m_0 \leq m'$ and $m' < m$, the strategies coincide ($s(m') = s'(m')$).*

[4] The above definition is the natural place to take into account probability considerations, too. In this paper, however, we refrain from introducing probabilities. Cf. [11] for an attempt at introducing probabilities into the branching space-times framework, which also applies to the (simpler) theory of branching time employed here.

5 Strategies, Stringency, and Trust

How trustworthy is an agent? A normal and reasonable response to this question is that agents are trustworthy if they fulfill their commitments with high probability, which can be monitored statistically in terms of actual fulfillment. This is good as far as it goes, but we suggest that the theory of agency in branching time plus strategies sketched above, together with the stringency ordering on strategies defined in the previous section, allows for a more fine-grained assessment of trustworthiness. The key ideas are the following:

- Above we have pointed out that there are different types of commitments distinguished by their stringency. Thus we may suppose that each commitment comes with an appropriate level of stringency. As we have noticed, some (but only some) promises rely strictly on moral binding and are not backed by any means of sanctioning. On the other hand, some (but only some) contracts seem to be able to do without any moral binding altogether.—We give no theory of stringency here at all; we simply suppose that formally, each commitment is created with a specific level of stringency, specified in terms of a class of strategies that are *appropriate* for the agent who commits herself.
- On the set of strategies available for the agent at the moment of commitment, we have defined a partial ordering that tells us when one strategy is more stringent than another.
- The main idea for assessing trustworthiness of an agent with respect to a single commitment is to identify the strategy that the agent is actually following by looking at what she chooses to do. As we pointed out above, a strict strategy completely reveals itself through the agent's choices, and if the agent is following a non-strict strategy, the strict strategy read off from her actions is still our best guess as to what her strategy actually is. The initial segment of the agent's strategy thus identified may then be compared to the members of the class of appropriate strategies. If the agent's strategy is itself appropriate, or at least as stringent as one of the appropriate strategies, we say that the agent is *living up to her commitment.* Living up to one's commitment, of course, counts in favour of trust. On the other hand, if the agent's strategy is less stringent than appropriate, this counts against trusting the agent.
- Trustworthiness is not an episodic issue, but more of a character trait. Thus, individual assessments of trustworthiness should be averaged over to obtain an estimate of an agent's overall trustworthiness.

These considerations may be illustrated by daily life examples. First, we do judge agents by their strategies, not just by actual performance. We accept excuses—e.g., if α has promised you to meet you at the station, but couldn't make it because she had an accident on the way, that will not count against trusting her: her strategy was good enough (it is conceptually true that nobody has a no-accident strategy, since accidents are exactly that which cannot be planned). On the other hand, we may be disappointed by an agent's strategy even if the commitment was fulfilled: if α had forgotten about the promise and meets you at the station just by chance, this will count against trusting her. Secondly, which normative relations we are prepared to enter with a given person depends on the level of trust, and we make fine distinctions. Suppose that α has often broken her promises to you in the past. Then you may be unwilling to accept another promise by α, while you may still be ready to enter a legally binding contract with her, since a contract is backed by enforceable sanctions. When we judge whether to enter a normative relation with somebody, the important issue seems to be whether we have good reason to suppose that that person will live up to her commitment, i.e., really adopt a strategy that is appropriate to the commitment at issue.

Our formal description has shown that in the realm of artificial agents, a similarly rich notion of commitment and trust can be implemented. Future work in the form of actual case studies will be required to strengthen the altogether plausible hypothesis that our framework indeed provides an adequate basis for describing the dynamics of commitments and trust in artificial normative systems.

Acknowledgements

I would like to thank the referees for DEON2006 and Michael Perloff for helpful suggestions. Support by the Alexander von Humboldt-Stiftung is gratefully acknowledged.

References

1. Boella, G., van der Torre, L.: A game theoretic approach to contracts in multiagent systems. IEEE Transactions on Systems, Man, and Cybernetics C (2006) to appear.
2. Belnap, N., Perloff, M., Xu, M.: Facing the Future. Agents and Choices in Our Indeterminist World. Oxford: Oxford University Press (2001)
3. Thomson, J.J.: The Realm of Rights. Cambridge, MA: Harvard University Press (1990)
4. Belnap, N.: Branching histories approach to indeterminism and free will. Preprint; URL = http://philsci-archive.pitt.edu/documents/disk0/00/00/08/90 (2002)
5. Belnap, N.: Branching space-time. Synthese **92** (1992) 385–434
6. Prior, A.N.: Past, present and future. Oxford: Oxford University Press (1967)
7. Thomason, R.H.: Indeterminist time and truth value gaps. Theoria (Lund) **36** (1970) 264–281
8. Müller, T.: On the formal structure of continuous action. In Schmidt, R., Pratt-Hartmann, I., Reynolds, M., Wansing, H., eds.: Advances in Modal Logic. Volume 5. London: King's College Publications (2005) 191–209
9. Anscombe, G.E.M.: Intention. 2nd edn. Cambridge, MA: Harvard University Press (1963)
10. Anderson, A.R.: A reduction of deontic logic to alethic modal logic. Mind **67** (1958) 100–103
11. Müller, T.: Probabability theory and causation. a branching space-times analysis. British Journal for the Philosophy of Science **56** (2005) 487–520

The Deontic Component of Action Language $n\mathcal{C}+$

Marek Sergot and Robert Craven

Department of Computing, Imperial College London
{mjs, rac101}@doc.ic.ac.uk

Abstract. The action language $\mathcal{C}+$ of Giunchiglia, Lee, Lifschitz, McCain, and Turner is a formalism for specifying and reasoning about the effects of actions and the persistence ('inertia') of facts over time. An 'action description' in $\mathcal{C}+$ defines a labelled transition system of a certain kind. $n\mathcal{C}+$ (formerly known as $(\mathcal{C}+)^{++}$) is an extended form of $\mathcal{C}+$ designed for representing normative and institutional aspects of (human or computer) societies. The deontic component of $n\mathcal{C}+$ provides a means of specifying the permitted (acceptable, legal) states of a transition system and its permitted (acceptable, legal) transitions. We present this component of $n\mathcal{C}+$, motivating its details with reference to some small illustrative examples.

1 Introduction

The action language $\mathcal{C}+$ [1] is a formalism for specifying and reasoning about the effects of actions and the persistence ('inertia') of facts over time, building on a general purpose non-monotonic representation formalism called 'causal theories'. An 'action description' in $\mathcal{C}+$ is a set of $\mathcal{C}+$ rules which define a labelled transition system of a certain kind. Implementations supporting a wide range of querying and planning tasks are available, notably in the form of the 'Causal Calculator' CCALC [2]. $\mathcal{C}+$ and CCALC have been applied successfully to a number of benchmark examples in the knowledge representation literature (see e.g. [3] and the CCALC website [2]). We have used it in our own work to construct executable specifications of agent societies (see e.g. [4, 5]).

$n\mathcal{C}+$ [6, 7] is an extended form of $\mathcal{C}+$ designed for representing normative and institutional aspects of (human or computer) societies. There are two main extensions. The first is a means of expressing 'counts as' relations between actions, also referred to as 'conventional generation' of actions. This feature will not be discussed in this paper. The second extension is a way of specifying the permitted (acceptable, legal) states of a transition system and its permitted (acceptable, legal) transitions. The aim of the paper is to present this component of $n\mathcal{C}+$ and some simple illustrative examples. $n\mathcal{C}+$ was called $(\mathcal{C}+)^{++}$ in earlier presentations.

$n\mathcal{C}+$ is intended for modelling system behaviour from an external 'bird's eye' perspective, that is to say, from the system designer's point of view. It may then be verified whether properties hold or not of the system specified (a process

L. Goble and J.-J.C. Meyer (Eds.): DEON 2006, LNAI 4048, pp. 222–237, 2006.

analogous to that described in [8, 9], which concentrates on epistemic properties and communicative acts). $n\mathcal{C}+$ is not intended for representing norms from an individual agent's perspective. We have a separate development, *agent-centric* $n\mathcal{C}+$, for specifying system norms as directives that constrain an individual agent's behaviour, in a form that can be used by a (computer) agent in its internal decision-making procedures. That development will be presented elsewhere.

We have three existing implementations of the $n\mathcal{C}+$ language. The first employs the 'Causal Calculator' CCALC. As explained later in the paper, the required modifications to CCALC are minor and very easily implemented. The second implementation provides an 'event calculus' style of computation with $\mathcal{C}+$ and $n\mathcal{C}+$ action descriptions. Given an action description and a 'narrative'—a record of what events have occurred—this implementation allows all past states, including what was permitted and obligatory at each past state, to be queried and computed. The third implementation connects $\mathcal{C}+$ and $n\mathcal{C}+$ to model checking software. System properties expressed in temporal logics such as CTL can then be verified by means of standard model checking techniques (specifically the model checker NuSMV) on transition systems defined using the $n\mathcal{C}+$ language. A small example is presented in [7]. We do not discuss the implementations further for lack of space, except to explain how the CCALC method works.

Related work. Some readers may see a resemblance between $n\mathcal{C}+$ and John-Jules Meyer's Dynamic Deontic Logic [10], and other well known works based on 'modal action logics' generally (e.g. [11, 12]). There are three fundamental differences. (1) $\mathcal{C}+$ and $n\mathcal{C}+$ are not variants of dynamic logic or modal action logic. They are languages for defining specific instances of labelled transitions systems. Other languages—we refer to them as 'query languages'—can then be interpreted on these structures. Dynamic logic is one candidate, the query language in CCALC is another, but there are many other possibilities: each $\mathcal{C}+$ or $n\mathcal{C}+$ action description defines a Kripke-structure, on which a variety of (modal) query languages, including a wide range of deontic and temporal operators, can be evaluated. We do not have space to discuss any of these possibilities in detail. (2) The representation of action is quite different from that in dynamic logic and modal action logic. (3) There are important differences of detail, in particular concerning the interactions between permitted states and permitted transitions between states.

The semantical devices employed in $n\mathcal{C}+$—classification of states and transitions into green/red (good/bad, ideal/sub-ideal), violation constants, explicit names for norms, and orderings of states according to how well they comply with these norms—are all frequently encountered in the deontic logic literature. The novelty here lies, first, in the details of how they are incorporated into labelled transition systems, and second, in the way the $n\mathcal{C}+$ language is used to define these structures.

Finally, $\mathcal{C}+$ is a (recent) member of a family of formalisms called 'causal action languages' in the AI literature. Several groups have suggested encoding normative concepts in such formalisms. We have done so ourselves in other work (see e.g. [13, 4, 5]) where we have used both $\mathcal{C}+$ and the 'event calculus' for this purpose. Leon van der Torre [14] has made a suggestion along similar lines, though using a different causal action language and a different approach. See also the discussion in [12]. One feature that distinguishes $\mathcal{C}+$ from other AI action languages is that it has an explicit semantics in terms of transition systems. It thereby proves a bridge between AI formalisms and standard methods in other areas of computer science and logic. It is this feature that $n\mathcal{C}+$ seeks to exploit.

2 The Language $\mathcal{C}+$

We begin with a concise, and necessarily rather dense, summary of the $\mathcal{C}+$ language. Some features (notably 'statically determined fluents' and 'exogenous actions') are omitted for simplicity. There are also some minor syntactic and terminological differences from the version presented in [1]. See [6] for details.

A *multi-valued propositional signature* σ is a set of symbols called *constants*. For each constant c in σ there is a non-empty set $dom(c)$ of values called the *domain* of c. For simplicity, in this paper we will assume that each $dom(c)$ is finite and has at least two elements. An *atom* of a signature σ is an expression of the form $c = v$ where c is a constant in σ and $v \in dom(c)$. A *formula* φ of signature σ is any propositional compound of atoms of σ. The expressions $\top$ and $\bot$ are 0-ary connectives, with the usual interpretation.

A *Boolean* constant is one whose domain is the set of truth values $\{\mathsf{t}, \mathsf{f}\}$. If p is a Boolean constant, p is shorthand for the atom $p = \mathsf{t}$ and $\neg p$ for the atom $p = \mathsf{f}$. Notice that, as defined here, $\neg p$ is an *atom* when p is a Boolean constant.

In $\mathcal{C}+$, the signature σ is partitioned into a set σ^{f} of *fluent constants* (also known as 'state variables' in other areas of Computer Science) and a set σ^{a} of *action constants*. A *fluent formula* is a formula whose constants all belong to σ^{f}; an *action formula* is a formula containing at least one action constant and no fluent constants.

An *interpretation* of a multi-valued signature σ is a function that maps every constant c in σ to some value v in $dom(c)$; an interpretation I *satisfies* an atom $c = v$, written $I \models c = v$, if $I(c) = v$. The satisfaction relation $\models$ is extended from atoms to formulas in accordance with the standard truth tables for the propositional connectives. We write $\mathrm{I}(\sigma)$ for the set of interpretations of σ.

Transition systems. Every $\mathcal{C}+$ action description D of signature $(\sigma^{\mathrm{f}}, \sigma^{\mathrm{a}})$ defines a labelled transition system $\langle S, \mathbf{A}, R \rangle$ where

- S is a (non-empty) set of *states*, each of which is an interpretation of the fluent constants σ^{f} of D; $S \subseteq \mathrm{I}(\sigma^{\mathrm{f}})$;
- $\mathbf{A}$ is a set of *transition labels*, also called *events*; $\mathbf{A}$ is the set of interpretations of the action constants σ^{a}, $\mathbf{A} = \mathrm{I}(\sigma^{\mathrm{a}})$;
- R is a set of transitions, $R \subseteq S \times \mathbf{A} \times S$.

A *path* of length m of the labelled transition system $\langle S, \mathbf{A}, R \rangle$ is a sequence $s_0\, \varepsilon_0\, s_1 \cdots s_{m-1}\, \varepsilon_{m-1}\, s_m$ $(m \geq 0)$ such that $(s_{i-1}, \varepsilon_{i-1}, s_i) \in R$ for $i \in 1..m$.

It is convenient in what follows to represent a state by the set of fluent atoms that it satisfies, i.e., $s = \{f = v \mid s \models f = v\}$. A state is then a (complete, and consistent) set of fluent atoms. We sometimes say a formula φ 'holds in' state s or 'is true in' state s as alternative ways of saying that s satisfies φ.

Action constants in $\mathcal{C}+$ are used to name actions, attributes of actions, or properties of a transition as a whole. Since a transition label/event ε is an interpretation of the action constants σ^{a}, it is meaningful to say that ε satisfies an action formula α ($\varepsilon \models \alpha$). When $\varepsilon \models \alpha$ we say that the transition (s, ε, s') is a transition of type α. Moreover, since a transition label is an interpretation of the action constants σ^{a}, it can also be represented by the set of atoms that it satisfies.

An action description D in $\mathcal{C}+$ is a set of *causal laws*, which are expressions of the following three forms. A *static law* is an expression:

$$F \text{ if } G \tag{1}$$

where F and G are fluent formulas. Static laws express constraints on states. A state s satisfies a static law (1) if $s \models (G \rightarrow F)$. A *fluent dynamic law* is an expression:

$$F \text{ if } G \text{ after } \psi \tag{2}$$

where F and G are fluent formulas and ψ is any formula of signature $\sigma^{\mathrm{f}} \cup \sigma^{\mathrm{a}}$. Informally, (2) states that fluent formula F is satisfied by the resulting state s' of any transition (s, ε, s') with $s \cup \varepsilon \models \psi$, as long as fluent formula G is also satisfied by s'. Some examples follow. An *action dynamic law* is an expression:

$$\alpha \text{ if } \psi \tag{3}$$

where α is an action formula and ψ is any formula of signature $\sigma^{\mathrm{f}} \cup \sigma^{\mathrm{a}}$. Action dynamic laws are used to express, among other things, that any transition of type α must also be of type α' (written α' if α), or that any transition from a state satisfying fluent formula G must be of type β (written β if G).

The $\mathcal{C}+$ language provides various abbreviations for common forms of causal laws. We will employ the following in this paper.

α causes F if G expresses that fluent formula F is satisfied by any state following the occurrence of a transition of type α from a state satisfying fluent formula G. It is shorthand for the dynamic law F if $\top$ after $G \wedge \alpha$. α causes F is shorthand for F if $\top$ after α.

nonexecutable α if G expresses that there is no transition of type α from a state satisfying fluent formula G. It is shorthand for the fluent dynamic law $\bot$ if $\top$ after $G \wedge \alpha$, or α causes $\bot$ if G.

inertial f states that values of the fluent constant f persist by default (by 'inertia') from one state to the next. It is shorthand for the collection of fluent dynamic laws $f = v$ if $f = v$ after $f = v$ for every $v \in dom(f)$.

Of most interest are *definite* action descriptions, which are action descriptions in which the head of every law (static, fluent dynamic, or action dynamic) is either an atom or the symbol $\bot$, and in which no atom is the head of infinitely many laws of D. We will restrict attention to definite action descriptions in this paper.

Causal theories. The language $\mathcal{C}+$ is presented in [1] as a higher-level notation for defining particular classes of theories in a general-purpose non-monotonic formalism called 'causal theories'. For present purposes the important points are these: for every (definite) action description D and non-negative integer m there is a natural translation from D to a causal theory Γ_m^D which encodes the paths of length m in the transition system defined by D; moreoever, for every definite causal theory Γ_m^D there is a formula $comp(\Gamma_m^D)$ of (classical) propositional logic whose (classical) models are in 1-1 correspondence with the paths of length m in the transition system defined by D. Thus, one method of computation for $\mathcal{C}+$ action descriptions is to construct the formula $comp(\Gamma_m^D)$ from the action description D and then employ a (standard, classical) satisfaction solver to determine the models of $comp(\Gamma_m^D)$. This is the method employed in the 'Causal Calculator' CCALC.

A causal theory of signature σ is a set of expressions ('causal rules') of the form

$$F \Leftarrow G$$

where F and G are formulas of signature σ. F is the head of the rule and G is the body. A rule $F \Leftarrow G$ is to be read as saying that there is a cause for F when G is true (which is not the same as saying that G is the cause of F).

Let Γ be a causal theory and let X be an interpretation of its signature. The *reduct* Γ^X is the set of all rules of Γ whose bodies are satified by the interpretation X: $\Gamma^X =_{\text{def}} \{F \mid F \Leftarrow G \text{ is a rule in } \Gamma \text{ and } X \models G\}$. X is a *model* of Γ iff X is the unique model (in the sense of multi-valued signatures) of Γ^X.

Given a definite action description D in $\mathcal{C}+$, and any non-negative integer m, translation to the corresponding causal theory Γ_m^D proceeds as follows. The signature of Γ_m^D is obtained by time-stamping every fluent constant of D with non-negative integers between 0 and m and every action constant with integers between 0 and $m-1$: the (new) atom $f[i]=v$ represents that fluent $f=v$ holds at integer time i, or more precisely, that $f=v$ is satisfied by the state s_i of a path $s_0\,\varepsilon_0 \cdots \varepsilon_{m-1}\,s_m$ of the transition system defined by D; the atom $a[i]=v$ represents that action atom $a=v$ is satisfied by the transition ε_i of such a path. The domain of each timestamped constant $c[i]$ is the domain of c. In what follows, $\psi[i]$ is shorthand for the formula obtained by replacing every atom $c=v$ in ψ by the timestamped atom $c[i]=v$.

Now, for every static law F if G in D and every $i \in 0\,..\,m$, include in Γ_m^D a causal rule of the form

$$F[i] \Leftarrow G[i]$$

For every fluent dynamic law F if G after ψ in D and every $i \in 0\,..\,m{-}1$, include a causal rule of the form

$$F[i{+}1] \Leftarrow G[i{+}1] \wedge \psi[i]$$

And for every action dynamic law α if ψ in D and every $i \in 0\,..\,m{-}1$, include a causal rule of the form

$$\alpha[i] \Leftarrow \psi[i]$$

We also require the following 'exogeneity laws'. For every fluent constant f and every $v \in dom(f)$, include a causal rule:

$$f[0] = v \Leftarrow f[0] = v$$

And for every action constant a, every $v \in dom(a)$, and every $i \in 0..m-1$, include a causal rule:

$$a[i] = v \Leftarrow a[i] = v$$

It is straightforward to check [1] that the models of causal theory Γ_m^D, and hence the (classical) models of the propositional logic formula $comp(\Gamma_m^D)$, correspond 1-1 to the paths of length m of the transition system defined by the $\mathcal{C}+$ action description D. In particular, models of $comp(\Gamma_1^D)$ encode the transitions defined by D and models of $comp(\Gamma_0^D)$ the states defined by D.

3 $n\mathcal{C}+$: Coloured Transition Systems

An action description of $n\mathcal{C}+$ defines a *coloured transition system*, which is a structure of the form $\langle S, \mathbf{A}, R, S_g, R_g \rangle$ where $\langle S, \mathbf{A}, R \rangle$ is a labelled transition system of the kind defined by $\mathcal{C}+$ action descriptions, and where the two new components are

- $S_g \subseteq S$, the set of 'permitted' ('acceptable', 'ideal', 'legal') states—we call S_g the 'green' states of the system;
- $R_g \subseteq R$, the set of 'permitted' ('acceptable', 'ideal', 'legal') transitions—we call R_g the 'green' transitions of the system.

We refer to the complements $S_{red} = S - S_g$ and $R_{red} = R - R_g$ as the 'red states' and 'red transitions', respectively. Semantical devices which partition states (and here, transitions) into two categories are familiar in the field of deontic logic. For example, Carmo and Jones [15] employ a structure which has both ideal/sub-ideal states and ideal/sub-ideal transitions (unlabelled). van der Meyden's 'Dynamic logic of permission' [16] employs a structure in which transitions, but not states, are classified as 'permitted/non-permitted'. van der Meyden's version was constructed as a response to problems of Meyer's 'Dynamic deontic logic' [10] which classifies transitions as 'permitted/non-permitted' by reference only to the state resulting from a transition. 'Deontic interpreted systems' [8] classify states as 'green'/'red', where these states have further internal structure to model the local states of agents in a multi-agent context. In all of these examples (and others) the task has been to find axiomatisations of such structures in one form of deontic logic or another. Here we are concerned with a different task, that of devising a language for *defining* coloured transition systems of the form described above.

A coloured transition system $\langle S, \mathbf{A}, R, S_g, R_g \rangle$ must further satisfy the following constraint, for all states s and s' in S and all transitions (s, ε, s') in R:

$$\text{if } (s, \varepsilon, s') \in R_g \text{ and } s \in S_g \text{ then } s' \in S_g \quad (4)$$

We refer to this as the *green-green-green* constraint, or *ggg* for short. (It is difficult to find a suitable mnemonic.) The *ggg* constraint (4) expresses a kind of *well-formedness* principle: a green (permitted, acceptable, legal) transition in a green (permitted, acceptable, legal) state always leads to a green (acceptable, legal, permitted) state. What is the rationale? Since we are here classifying both states and transitions into green/red, it is natural to ask whether there are any relationships between the classification of states and the classification of transitions between them. As observed previously by Carmo and Jones [15] any such relationships are necessarily quite weak. In particular, and *contra* the assumptions underpinning John-Jules Meyer's construction of Dynamic Deontic Logic [10], a red (unacceptable, non-permitted) transition can result in a green (acceptable, permitted) state. Indeed such cases are frequent: suppose that there are two different transitions, (s, ε_1, s') and (s, ε_2, s'), between a green or red state s and a green state s'. It is entirely reasonable that the transition (s, ε_1, s') is classified as green whereas (s, ε_2, s') is classified as red. (s, ε_1, s') might represent an action by one agent, for example, and (s, ε_2, s') an action by another. This situation cannot arise if the transition system has a tree-like structure in which there is at most one transition between any pair of states, but we do not want to restrict attention to transition systems of this form. Similarly, it is easy to encounter cases in which a green (acceptable, permitted) transition can lead sensibly to a red (unacceptable, non-permitted) state: not all green (acceptable, permitted) transitions from a red state must be such that they restore the system to a green state. Some illustrations will arise in the examples later. The only plausible relationship between the classification of states and the classification of transitions, as also noted by Carmo and Jones [15], is what we called the *ggg* constraint above, if we regard it (as we do) as a required property of any well-formed system specification. Since the *ggg* constraint is so useful for the applications we have in mind, we choose to adopt it as a feature of every coloured transition system.

Note that the *ggg* constraint (4) may be written equivalently as:

$$\text{if } (s, \varepsilon, s') \in R \text{ and } s \in S_{\text{g}} \text{ and } s' \in S_{\text{red}} \text{ then } (s, \varepsilon, s') \in R_{\text{red}} \tag{5}$$

Any transition from a green (acceptable, permitted) state to a red (unacceptable, non-permitted) state must itself be red, in a well-formed system specification.

One can consider a range of other properties that we might require of a coloured transition system: that the transition relation must be serial, for example, or that there must be at least one green state, or that from every green state there must be at least one green transition, or that from every green state reachable from some specified initial state(s) there must be at least one green transition, and so on. The investigation of these, and other, properties is worthwhile but not something we undertake here. We place no restrictions on coloured transition systems, beyond the *ggg* constraint.

The language $n\mathcal{C}+$. To avoid having to specify separately which states and transitions are green and which are red, an $n\mathcal{C}+$ action description specifies those that are red and leaves the remainder to be classified as green by default. This is for convenience, and also to ensure that all states and transitions are classified completely and consistently. (One might ask why the defaults are not chosen to operate the other way round. It is very much more awkward to specify concisely what is green and allow the remainder to be red by default.)

Accordingly, the language $n\mathcal{C}+$ extends $\mathcal{C}+$ with two new forms of rules. A *state permission law* is an expression of the form

$$n\colon \textsf{not-permitted}\ F\ \textsf{if}\ G \tag{6}$$

where n is an (optional) identifier for the rule and F and G are fluent formulas. not-permitted F is a shorthand for not-permitted F if $\top$. An *action permission law* is an expression of the form

$$n\colon \textsf{not-permitted}\ \alpha\ \textsf{if}\ \psi \tag{7}$$

where n is an (optional) identifier for the rule, α is an action formula and ψ is any formula of signature $\sigma^{\text{f}} \cup \sigma^{\text{a}}$. not-permitted α is shorthand for not-permitted α if $\top$. We also allow oblig F as an abbreviation for not-permitted $\neg F$ and oblig α as an abbreviation for not-permitted $\neg\alpha$.[1]

Informally, in the transition system defined by an action description D, a state s is red whenever $s \models F \land G$ for a state permission law not-permitted F if G. All other states are green by default. A transition (s, ε, s') is red whenever $s \cup \varepsilon \models \psi$ and $\varepsilon \models \alpha$ for any action permission law not-permitted α if ψ. All other transitions are green, *subject to* the *ggg* constraint which may impose further conditions on the colouring of a given transition.

Let D be an action description of $n\mathcal{C}+$. D_{basic} refers to the subset of laws of D that are also laws of $\mathcal{C}+$. The coloured transition system defined by D has the states S and transitions R that are defined by its $\mathcal{C}+$ component, D_{basic}, and green states S_{g} and green transitions R_{g} given by $S_{\text{g}} =_{\text{def}} S - S_{\text{red}}$, $R_{\text{g}} =_{\text{def}} R - R_{\text{red}}$ where

$$\begin{aligned} S_{\text{red}} &=_{\text{def}} \{s \mid s \models F \land G \text{ for some law of the form (6) in } D\} \\ R_{\text{red}} &=_{\text{def}} \{(s, \varepsilon, s') \mid s \cup \varepsilon \models \psi,\ \varepsilon \models \alpha \text{ for some law of the form (7) in } D\} \\ &\quad\ \cup\ \{(s, \varepsilon, s') \mid s \in S_{\text{g}} \text{ and } s' \in S_{\text{red}}\} \end{aligned}$$

The second component of the R_{red} definition ensures that the *ggg* constraint is satisfied. (The state permission laws not-permitted F if G and not-permitted $(F \land G)$ are thus equivalent in $n\mathcal{C}+$; we allow both forms for convenience.) It can be shown easily [6] that the coloured transition system defined in this way is unique and satisfies the *ggg* constraint. The definition of course does not guarantee that the coloured transition system satisfies any of the other possible properties that we mentioned earlier. If they are felt to be desirable in some particular

[1] This does not raise the issue of 'action negation' as encountered in modal action logics. (See e.g. [12].) In $\mathcal{C}+$ and $n\mathcal{C}+$, α is not the name of an action but a formula expressing a property of transitions.

application, they must be checked separately as part of the specification process. (These checks are easily implemented.)

The overall effect is thus:

- a state is green unless coloured red by some static permission law;
- a transition is red if it is coloured red by some action permission law, or by the *ggg* constraint; otherwise it is green.

That the colouring of transitions is dependent on the colouring of states should *not* be interpreted as a commitment to any philosophical position about the priority of the ought-to-be and the ought-to-do, and the derivability of one from the other. It is merely a consequence of, first, adopting the *ggg* constraint as an expression of the well-formedness of a system specification, and second, of choosing to specify explicitly what is red and letting green be determined by default.

Causal theories. Any (definite) action description of $n\mathcal{C}+$ can be translated to the language of (definite) causal theories, as follows. Let D be an action description and m a non-negative integer. The translation of the $\mathcal{C}+$ component D_{basic} of D proceeds as usual. For the permission laws, introduce two new fluent and action constants, status and trans respectively, both with possible values green and red. They will be used to represent the colour of a state and the colour of a transition, respectively.

For every state permission law n: not-permitted F if G and time index $i \in 0 \mathinner{..} m$, include in Γ_m^D a causal rule of the form

$$\mathsf{status}[i] = \mathsf{red} \Leftarrow F[i] \wedge G[i] \tag{8}$$

and for every $i \in 0 \mathinner{..} m$, a causal rule of the form

$$\mathsf{status}[i] = \mathsf{green} \Leftarrow \mathsf{status}[i] = \mathsf{green} \tag{9}$$

to specify the default colour of a state. A state permission rule of the form n: oblig F if G produces causal rules of the form $\mathsf{status}[i] = \mathsf{red} \Leftarrow \neg F[i] \wedge G[i]$.

For every action permission law n: not-permitted α if ψ and time index $i \in 0 \mathinner{..} m{-}1$, include in Γ_m^D a causal rule of the form

$$\mathsf{trans}[i] = \mathsf{red} \Leftarrow \alpha[i] \wedge \psi[i] \tag{10}$$

and for every $i \in 0 \mathinner{..} m{-}1$, a causal rule of the form

$$\mathsf{trans}[i] = \mathsf{green} \Leftarrow \mathsf{trans}[i] = \mathsf{green} \tag{11}$$

to specify the default colour of a transition. An action permission law of the form n: oblig α if ψ produces causal rules of the form $\mathsf{trans}[i] = \mathsf{red} \Leftarrow \neg\alpha[i] \wedge \psi[i]$.

Finally, to capture the *ggg* constraint, include for every $i \in 0 \mathinner{..} m{-}1$ a causal rule of the form

$$\mathsf{trans}[i] = \mathsf{red} \;\Leftarrow\; \mathsf{status}[i] = \mathsf{green} \wedge \mathsf{status}[i{+}1] = \mathsf{red} \tag{12}$$

It is straightforward to show [6] that models of the causal theory Γ_m^D correspond to all paths of length m through the coloured transition system defined by D, where the fluent constant status and the action constant trans encode the colours of the states and transitions, respectively.

The translation of $n\mathcal{C}+$ into causal theories effectively treats status = red and trans = red as 'violation constants'. Notice that, although action descriptions in $n\mathcal{C}+$ can be translated to causal theories, they cannot be translated to action descriptions of $\mathcal{C}+$: there is no form of causal law in $\mathcal{C}+$ which translates to the *ggg* constraint (12). However, implementation in CCALC requires only that the causal laws (8)–(12) are included in the translation to causal theories, which is a very simple modification.

4 Examples

The examples in this section are deliberately chosen to be as simple as possible, so that in each case we can show the transition system defined in its entirety. Other examples may be found in [6, 7]. The first example illustrates the use of $n\mathcal{C}+$ in a typical (but very simple) system specification. The second is to motivate the more complicated account to come in Section 5.

Example (File system). I is some piece of (confidential) information. I, or material from which I can be derived, is stored in a file. Let x range over some set of agent names. Boolean fluent constants Kx represent that agent x has access to information I, that x 'knows' I. Boolean fluent constants Fx represent that x has read access to the file containing I. If x has read access to the file (Fx) then x knows I (Kx). Fx is inertial: both Fx and $\neg Fx$ persist by default. $\neg Kx$ persists by default but once Kx holds, it holds for ever.

Suppose, for simplicity, that there are two agents, a and b. Suppose moreover that the file is the only source of information I, in the sense that if Kx holds for any x then either Fa or Fb. This does not change the essence of the example but it reduces the number of states and simplifies the diagrams.

There are two types of acts: Boolean action constants $read(x)$ represent that x is given read access to the file containing I. Boolean action constant a *tells* b represents that a communicates to b the information I (whether or not b knows it already), and b *tells* a that b communicates it to a. In this simple example there are no actions by which read access to the file is removed once it is granted.

We can represent the above as a definite action description as follows, for x ranging over a and b.

inertial Fx	$read(x)$ causes Fx
$\neg Kx$ if $\neg Kx$ after $\neg Kx$	a *tells* b causes Kb
Kx if $\top$ after Kx	b *tells* a causes Ka
	nonexecutable a *tells* b if $\neg Ka$
Kx if Fx	nonexecutable b *tells* a if $\neg Kb$
$\bot$ if $Kx \wedge \neg Fa \wedge \neg Fb$	nonexecutable $read(x)$ if Fx

Now suppose that a is permitted to know I, and b is not. We add the following law to the action description. (Ka is permitted by default.)

$p(b)$: not-permitted Kb

The transition system defined by this action description is shown below. The labels $read(a)$, $read(b)$, a tells b, b tells a stand for the transition labels $\{read(a),\ \neg read(b),\ \neg a$ tells $b,\ \neg b$ tells $a\}$, $\{\neg read(a),\ read(b),\ \neg a$ tells $b,\ \neg b$ tells $a\}$ and so on, respectively. The label $read(a), read(b)$ is shorthand for the transition label $\{read(a), read(b), \neg a$ tells $b, \neg b$ tells $a\}$. Reflexive arcs, corresponding to the 'null event' or to transitions of type a tells b and b tells a from state $\{Fa, Ka, \neg Fb, Kb\}$ to itself, are omitted from the diagram to reduce clutter. Also omitted from the diagram are transitions of type $read(a) \wedge a$ tells b, a tells $b \wedge b$ tells a, etc. Again, this is just to reduce clutter.

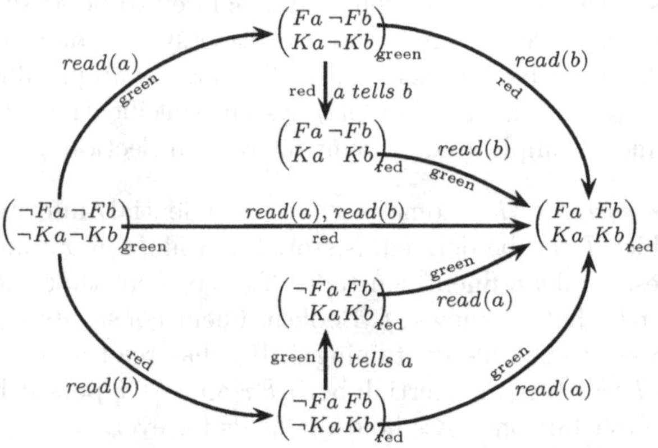

Notice that transitions of type $read(b)$ are red because of the ggg constraint, except that $read(b)$ transitions come out to be green in states where Kb already holds. If the latter is felt to be undesirable, one could add another action permission law not-permitted $read(b)$, or a state permission law not-permitted Fb. We will discuss some of these options in more detail later.

In a computerised system, b's access to information I would be controlled by the file access system. Naturally the file access system cannot determine whether b knows I: in practice, a specification of the computer system would simply say that $read(b)$ actions are nonexecutable, or simply that Fb is false. The latter can be expressed by adding the following static law to the action description:

$\perp$ if Fb

This eliminates all states in which Fb holds from the transition system. The transition system defined by this extended action description is the following:

$$\left(\begin{array}{c} \neg Fa\ \neg Fb \\ \neg Ka\ \neg Kb \end{array}\right)_{\text{green}} \xrightarrow[\text{green}]{read(a)} \left(\begin{array}{c} Fa\ \neg Fb \\ Ka\ \neg Kb \end{array}\right)_{\text{green}} \xrightarrow[\text{red}]{a \text{ tells } b} \left(\begin{array}{c} Fa\ \neg Fb \\ Ka\ \ Kb \end{array}\right)_{\text{red}}$$

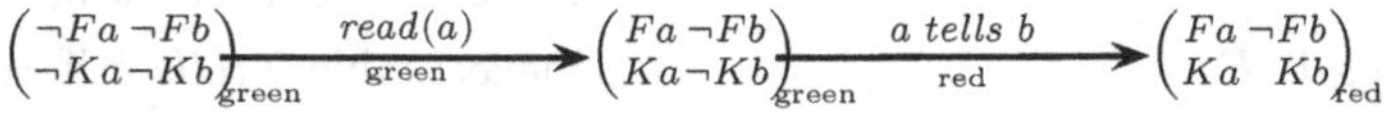

As usual, reflexive arcs are omitted from the diagram for clarity. Here, the action *read*(a) is under the control of the file access system, and *a tells b* is an action that can be performed by agent a. This difference is not explicit in the semantics of $\mathcal{C}+$ nor of $n\mathcal{C}+$. The *agent-centric* version of $n\mathcal{C}+$, alluded to in the introduction, allows such distinctions to be made.

Example (Secrets). Suppose we have agents a, b, and c, Boolean fluent constants Ka, Kb, and Kc as in the previous example, and Boolean action constants *a tells b* and *a tells c*. We ignore the file system and read access to it from now on since they play no role in this example, and we leave out other possible actions such as *b tells a*, *c tells a*, etc, to simplify the diagrams. The persistence of Ka, Kb, and Kc, and the effects of *a tells b* and *a tells c* actions are represented using $\mathcal{C}+$ laws as shown earlier.

Suppose now that b is not permitted to know I. The coloured transition system contains the following fragment:

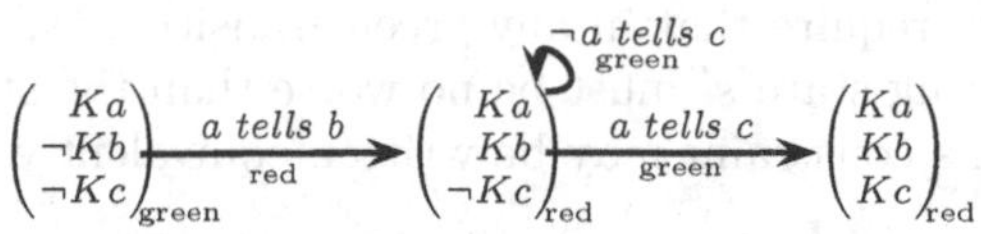

The states $\{Ka, Kb, \neg Kc\}$ and $\{Ka, Kb, Kc\}$ are red because Kb is not permitted. The transition labelled *a tells b* is red because of the *ggg* constraint. The transition labelled *a tells c* is not forced to be red by the *ggg* constraint and so becomes green by default. The (reflexive) transition labelled $\neg$*a tells c* is also green for the same reason.

But suppose now that we change the example, by adding that c is also not permitted to know I. The fragment of the transition system shown above remains unchanged (because the state $\{Ka, Kb, Kc\}$ was already red). The transition labelled *a tells c* is green even though *a tells c* results in Kc, and Kc is not permitted. We have here an instance of a general phenomenon: once a state is red, all transitions from it (including actions by all other agents) become green by default unless explicitly coloured red by action permission laws.

One possibility is to leave some transitions uncoloured, or what comes to the same thing, remove the default colouring of transitions and allow an $n\mathcal{C}+$ action description to define multiple transition systems differing in the colours assigned to some transitions. This is easy to encode, and easy to implement, but it is too weak for what we want: we would then never be able to conclude that there are (necessarily) green transitions from any red state.

Our diagnosis is that the classification of states into red/green is too crude. Why should we think that *a tells c* transitions should be inferred red even after *a tells b* has occurred and Kb holds? Because *a tells c* would lead to violation of another norm which says that Kc is not permitted. So we will introduce names for (instances of) norms and then classify states according to how well, or how badly, they comply with these norms.

5 $n\mathcal{C}+$: Graded Transition Systems

A *graded transition system* is a structure of the form: $\langle S, \mathbf{A}, R, R_{\mathrm{g}}, \prec \rangle$ where $\langle S, \mathbf{A}, R \rangle$ is a labelled transition system of the kind defined by $\mathcal{C}+$ action descriptions, and where

- $R_{\mathrm{g}} \subseteq R$ is the set of 'green' transitions;
- $\prec$ is a (strict, partial) ordering on S: $s \prec s'$ represents that state s' is worse than state s.

We refer to $R_{\mathrm{red}} = R - R_{\mathrm{g}}$ as the 'red transitions' as usual.

Notice that we have chosen to grade/rank states but not transitions: transitions are still either green or red. There may be good reasons to rank transitions as well, but we will not do so here.

As in the case of coloured transition systems, we further impose a well-formedness constraint, analogous to the *ggg* constraint. The natural generalization of *ggg* is to require that in any green transition (s, ε, s') from state s to state s', the resulting state s' must be no worse than the state s: $(s, \varepsilon, s') \in R_{\mathrm{g}}$ implies $s \not\prec s'$. This constraint may be written equivalently as:

$$\text{if } (s, \varepsilon, s') \in R \text{ and } s \prec s' \text{ then } (s, \varepsilon, s') \in R_{\mathrm{red}} \tag{13}$$

We refer to (13) as the *BRW* constraint (short for 'better-red-worse'), again apologising for the ugliness of the label.

A coloured transition system is thus a special case of a graded transition system in which $s \prec s'$ iff $s \in S_{\mathrm{g}}$ and $s' \in S_{\mathrm{red}}$. In that case the *BRW* constraint (13) takes the form: if $(s, \varepsilon, s') \in R$ and $s \in S_{\mathrm{g}}$ and $s' \in S_{\mathrm{red}}$ then $(s, \varepsilon, s') \in R_{\mathrm{red}}$, which is equivalent to the formulation of the *ggg* constraint given earlier (4).

Note that according to the *BRW* constraint, a transition (s, ε, s') is not forced to be red if $s \not\prec s'$. In particular, when s and s' are not comparable in $\prec$, the *BRW* constraint does not apply. This is deliberate. One could look for stronger well-formedness constraints than *BRW*. The requirement that $(s, \varepsilon, s') \in R_{\mathrm{g}}$ implies $s' \prec s$ is clearly much too strong, but there are several candidates stronger than *BRW* for which a plausible case can be made. We are inclined, however, not to adopt any of these stronger constraints as a fixed feature of graded transition systems.

Violation orderings. Of particular interest is the special case of graded transition systems where the ordering on states is determined by how well, or how badly, each state complies with a set of explicitly named norms.

A normative code $\mathcal{N}$ is a finite set of pairs $\langle n, o(F/G) \rangle$ where n is an identifier for a norm, and F and G are fluent formulas. Note that we do not require that norm labels are unique in $\mathcal{N}$.

The *violation set* $V_{\mathcal{N}}(s)$ of a state s in S is the set of norm identifiers in $\mathcal{N}$ that are violated in s.

$$V_{\mathcal{N}}(s) =_{\mathrm{def}} \{ n \mid \langle n, o(F/G) \rangle \in \mathcal{N} \text{ and } s \models G \land \neg F \} \tag{14}$$

Now we can define an ordering on states by comparing their violation sets. A state s is better than a state s' if the violation set of s is a proper subset of the violation set of s':

$$s \prec_{\mathcal{N}} s' \text{ iff } V_{\mathcal{N}}(s) \subset V_{\mathcal{N}}(s') \tag{15}$$

It would be easy to add weights or priorities on norms, and adjust the definition of $\prec_{\mathcal{N}}$ to take these weights into account. The details are straightforward and we omit them.

Let D be an action description of $n\mathcal{C}+$. The graded transition system defined by D is $\langle S, \mathbf{A}, R, R_{\text{g}}, \prec_{\mathcal{N}} \rangle$ where the states S, transition labels/events $\mathbf{A}$, and transitions R are exactly as in the coloured transition system described in Section 3; $R_{\text{g}} = R - R_{\text{red}}$ where the red transitions R_{red} are determined by the action permission laws and the *BRW* constraint; and where the ordering $\prec_{\mathcal{N}}$ on states is the ordering defined in (15) by the normative code $\mathcal{N}$ consisting of elements $\langle n, o(\neg F/G) \rangle$ where n: not-permitted F if G is a law in D and elements $\langle n, o(F/G) \rangle$ where n: oblig F if G is a law in D.

Encoding in causal theories. Let the Boolean fluent constant $\mathsf{viol}(n)$ represent that norm n in $\mathcal{N}$ is violated. For every state permission law n: not-permitted F if G and time index $i \in 0\,..\,m$, include in the causal theory Γ_m^D the causal rules:

$$\neg\mathsf{viol}(n)[i] \Leftarrow \neg\mathsf{viol}(n)[i] \tag{16}$$

$$\mathsf{viol}(n)[i] \Leftarrow F[i] \wedge G[i] \tag{17}$$

A state permission rule of the form n: oblig F if G produces causal rules of the form $\mathsf{viol}(n)[i] \Leftarrow \neg F[i] \wedge G[i]$.

(In place of the Boolean violation constants $\mathsf{viol}(n)[i]$ we could have used fluent constants $\mathsf{status}(n)[i]$ with possible values green and red.)

In order to encode the *BRW* constraint, it is not necessary to compute and compare violation sets for each state. Instead, we can encode the *BRW* constraint as follows. Include in Γ_m^D, for every $i \in 0\,..\,m{-}1$ and every norm identifier n in $\mathcal{N}$, the causal rules:

$$\mathsf{trans}[i] = \mathsf{green} \;\Leftarrow\; \mathsf{trans}[i] = \mathsf{green} \tag{18}$$

$$\mathsf{trans}[i] = \mathsf{red} \;\Leftarrow\; \mathsf{viol}(n)[i{+}1] \wedge \neg\mathsf{viol}(n)[i] \wedge \neg\mathsf{q}[i] \tag{19}$$

$$\neg\mathsf{q}[i] \;\Leftarrow\; \neg\mathsf{q}[i] \tag{20}$$

$$\mathsf{q}[i] \;\Leftarrow\; \mathsf{viol}(n)[i] \wedge \neg\mathsf{viol}(n)[i{+}1] \tag{21}$$

Causal rules (18) and (19) generalise the causal rules (11) and (12) used to encode the *ggg* constraint in causal theories. They make use of auxiliary constants $\mathsf{q}[i]$ defined in (20)–(21). One can easily check that in the case where the action description contains a single state permission law, causal rules (18)–(21) collapse to a form equivalent to the causal rules (11) and (12) encoding the *ggg* constraint. In the case where the action description contains no state permission law, these two sets of causal laws are trivially equivalent, since both are empty.

Example (Secrets, contd). Suppose we formulate the example of Section 4 using two state permission laws as follows:

$$p(b)\colon \textsf{not-permitted } Kb \tag{22}$$
$$p(c)\colon \textsf{not-permitted } Kc \tag{23}$$

The graded transition system defined is as follows, where the annotations on the states show the respective violation sets.

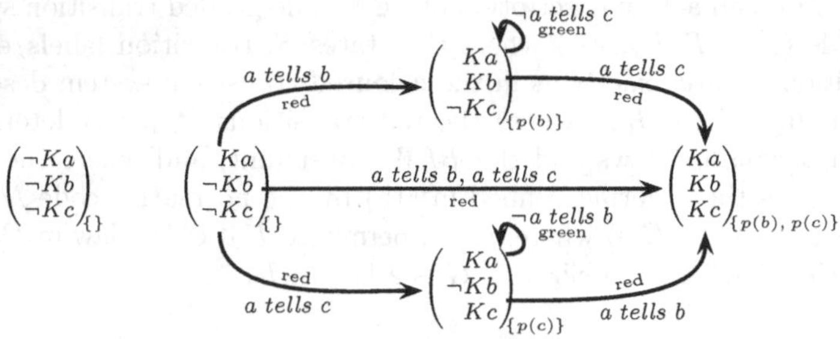

'Null events' and other reflexive arcs are omitted from the diagram for clarity.

Contrast this with a different version of the example. Suppose that instead of permission laws (22) and (23) with two explicit norms labelled $p(b)$ and $p(c)$ we specify just one explicit norm with a single label p, either in the form

$$p\colon \textsf{not-permitted } Kb \tag{24}$$
$$p\colon \textsf{not-permitted } Kc \tag{25}$$

or equivalently as a single state permission law: p: not-permitted $(Kb \vee Kc)$.

In this version of the action description, the three states $\{Ka, Kb, \neg Kc\}$, $\{Ka, \neg Kb, Kc\}$, and $\{Ka, \neg Kb, \neg Kc\}$ all have the same violation set, $\{p\}$. Since they are now not strictly worse than each other, the *BRW* constraint does not colour transitions between them red: they are all green by default. This transition system corresponds to the example in Section 4, but now with states annotated by violation sets $\{\}$ and $\{p\}$ rather than colours 'green' and 'red'.

6 Conclusion

$n\mathcal{C}+$ adds a simple deontic component to $\mathcal{C}+$, intended to support system specifications where there is a need to distinguish between acceptable/permitted and unacceptable/non-permitted system states and behaviours.

It was our intention to continue the discussion of the examples to show how $n\mathcal{C}+$ copes with (temporal) 'contrary-to-duty' structures. For instance, a natural extension of the examples would take the form '*a* must not tell *b* and *a* must not tell *c*; but if *a* tells *b* it must tell *c*, and if *a* tells *c* it must tell *b*', as in Belzer's Reykjavik scenario [17]. Another interesting variant is '*b* is not permitted to know *I* and *c* is not permitted to know *I*; but if *a* tells *b* it must tell *c*, and if *a* tells *c* it must tell *b*'. We leave that discussion for a separate paper.

Besides investigating variants of the *BRW* well-formedness constraint and other desirable properties of coloured/graded transition systems, we are developing a refined version of $n\mathcal{C}+$ to make explicit the distinction between actions and transitions, and an agent-centric $n\mathcal{C}+$ for specifying system norms as directives that constrain an individual agent's behaviour.

References

1. Giunchiglia, E., Lee, J., Lifschitz, V., McCain, N., Turner, H.: Nonmonotonic causal theories. Artificial Intelligence **153** (2004) 49–104
2. CCALC: `http://www.cs.utexas.edu/users/tag/cc.`
3. Akman, V., Erdoğan, S.T., Lee, J., Lifschitz, V., Turner, H.: Representing the Zoo World and the Traffic World in the language of the Causal Calculator. Artificial Intelligence **153** (2004) 105–140
4. Artikis, A., Sergot, M.J., Pitt, J.: Specifying electronic societies with the Causal Calculator. In Giunchiglia, F., Odell, J., Weiss, G., eds.: Agent-Oriented Software Engineering III. LNCS 2585, Springer (2003) 1–15
5. Artikis, A., Sergot, M.J., Pitt, J.: An executable specification of an argumentation protocol. In: Proc. 9th International Conference on Artificial Intelligence and Law (ICAIL'03), Edinburgh, ACM Press (2003) 1–11
6. Sergot, M.: $(\mathcal{C}+)^{++}$: An action language for modelling norms and institutions. Technical Report 2004/8, Dept. of Computing, Imperial College London (2004)
7. Sergot, M.J.: Modelling unreliable and untrustworthy agent behaviour. In Dunin-Keplicz, B., Jankowski, A., Skowron, A., Szczuka, M., eds.: Monitoring, Security, and Rescue Techniques in Multiagent Systems. Advances in Soft Computing. Springer-Verlag (2005) 161–178
8. Lomuscio, A., Sergot, M.J.: Deontic interpreted systems. Studia Logica **75**(1) (2003) 63–92
9. Lomuscio, A., Sergot, M.J.: A formalisation of violation, error recovery, and enforcement in the bit transmission problem. J. of Applied Logic **2** (2004) 93–116
10. Meyer, J.J.C.: A different approach to deontic logic: Deontic logic viewed as a variant of dynamic logic. Notre Dame J. of Formal Logic **29**(1) (1988) 109–136
11. Maibaum, T.: Temporal Reasoning over Deontic Specifications. In Meyer, J.J.C., Wieringa, R.J., eds.: Deontic Logic in Computer Science: Normative System Specification. John Wiley & Sons, Chichester, England (1993) 141–202
12. Broersen, J.: Modal Action Logics for Reasoning about Reactive Systems. PhD thesis, Vrije Universiteit Amsterdam (2003)
13. Artikis, A., Pitt, J., Sergot, M.J.: Animated specification of computational societies. In Castelfranchi, C., Johnson, W.L., eds.: Proc. 1st International Joint Conference on Autonomous Agents and Multi-Agent Systems (AAMAS'02), Bologna, ACM Press (2002) 1053–1062
14. van der Torre, L.: Causal deontic logic. In: Proceedings of the Fifth Workshop on Deontic Logic in Computer Science (Deon2000). (2000) 351–367
15. Carmo, J., Jones, A.J.I.: Deontic database constraints, violation and recovery. Studia Logica **57**(1) (1996) 139–165
16. Meyden, R.: The dynamic logic of permission. Journal of Logic and Computation **6**(3) (1996) 465–479
17. Belzer, M.: Legal reasoning in 3-D. In: Proc. 1st International Conf. on Artificial Intelligence and Law, Boston, ACM Press (1987) 155–163

A Complete and Decidable Axiomatisation for Deontic Interpreted Systems*

Alessio Lomuscio and Bożena Woźna

Department of Computer Science, University College London
Gower Street, London WC1E 6BT, United Kingdom
{A.Lomuscio, B.Wozna}@cs.ucl.ac.uk

Abstract. We solve the problem left open in [5] by providing a complete axiomatisation of deontic interpreted systems on a language that includes full CTL as well as the K_i, O_i and $\widehat{\mathrm{K}}_i^j$ modalities. Additionally we show that the logic employed enjoys the finite model property, hence decidability is guaranteed. To achieve these results we follow and extend the technique used by Halpern and Emerson in [2].

1 Introduction

Concepts based on deontic notions are increasingly being used in specification and verification of large multi-agent systems. Because of their open and self-interested nature it is unrealistic to assume that a team of engineers in a single organisation may maintain designe, engineer and control a whole multi-agent system. This makes it difficult, even a priori, to verify either off-line or at runtime that each individual agent complies with a set of specifications. It seems more feasible, instead, to permit the agents to perform incorrect/unwanted/undesirable actions, only to flag all unwanted behaviours and reason about the properties that these may bring about in the system.

In other words, by adding a suitable set of deontic notions we can aim to verify not only what properties the system enjoys when each individual agent is performing following the intended specifications (as it is traditionally done in Software Engineering), but also what consequences result from the violation of some of these specifications by some agents. This shift to a more liberal, finer grained approach requires the introduction of suitable formal machinery both in terms of specification languages and verification tools.

Deontic interpreted systems [5] have recently been introduced for this objective. In their basic form they provide a computationally grounded semantics [9] to interpret a logic capturing epistemic, temporal and correctness notions. By using this formalism it is possible to give a semantical description of key scenarios [6] and use the logic to check whether or not particular properties hold on these specifications. Specifically, deontic interpreted systems can be used to interpret a language that includes CTL modalities AU, EU, EX [2], epistemic modalities K_i

* The authors acknowledge support from the EPSRC (grant GR/S49353) and the Nuffield Foundation (grant NAL/690/G).

L. Goble and J.-J.C. Meyer (Eds.): DEON 2006, LNAI 4048, pp. 238–254, 2006.

[3], modalities representing correct functioning behaviour O_i, and modalities $\widehat{K}_i^j$ representing knowledge under the assumption of correct behaviour. Automatic model checking tools for deontic interpreted systems have been developed [4, 7] supporting the automatic verification of state spaces of the region of 10^{40} and beyond [8, 10, 7].

While the above results concern specification patters, verification tools, and concrete scenarios, important theoretical issues have so far been left open. In particular, the axiomatisation of deontic interpreted systems originally provided in [5] was limited to a language that did not include temporal operators. Furthermore, the bi-indexed modality $\widehat{K}_i^j$, whose importance in practical verification is now well recognised, was not included in the language.

The difficulty of the problem is linked to two issues. First, the modality $\widehat{K}_i^j$ is defined in terms of the intersection between two relations with different properties: modalities like these are known to be hard to treat. Second, any axiomatisation for deontic interpreted systems would have to include a logic for branching-time, but the standard procedure for axiomatising CTL involves a non-standard filtration procedure [2].

The contribution of the present work is to solve the problem left open in [5], i.e., to provide a complete axiomatisation of deontic interpreted systems on a language that includes full CTL as well as the K_i, O_i and $\widehat{K}_i^j$ modalities. Additionally we show that the logic employed enjoys the finite model property, hence it is decidable. To show these results we extend the technique originally presented by Halpern and Emerson in [2] to the richer language above.

The rest of the paper is organised as follows. In Section 2 we present syntax and semantics of the logic. Section 3 is devoted to the construction of the underlying machinery to prove the main results of the paper. Sections 4 and 5 present a decidability theorem and a completeness proof for the logic.

2 Deontic Interpreted Systems

Deontic interpreted systems [5] constitute a semantics to interpret epistemic, correctness and temporal operators in a computational setting. They extend the framework of interpreted systems [3], popularised by Halpern and colleagues in the 90s to reason about knowledge, to modalities expressing correctness and knowledge under assumptions of correctness. Technically, deontic interpreted systems provide an interpretation to the operators O_i ($O_i\phi$ representing "whenever agent i is working correctly ϕ is the case") and $\widehat{K}_i^j$ ($\widehat{K}_i^j\phi$ representing "agent i knows that ϕ under the assumption that agent j is working correctly") as well as the standard epistemic operators K_i and branching time operators of CTL already supported by interpreted systems. Semantically this is achieved simply by assuming that the local states of the agents are composed by two disjoint sets of allowed (or "green") and disallowed (or "red") local states. Loosely speaking an agent "is working correctly" whenever it is following its protocol (defined in interpreted systems as a function from local states to sets of actions) in its choice of actions. Given that the focus of this paper is to axiomatise the result trace based semantics resulting from this we refer to [5] and related papers for more details.

It also should be noted that the O_i (respectively $\widehat{K}$) operators tackled in this paper concern correctness of execution (respectively knowledge under the assumption of correctness). Although they intuitively are related to some deontic concepts, they do not refer to the standard meaning of "obligation". Consequently the widely-discussed paradoxes of deontic logic do not apply to the concepts presented here.

Let $\mathbb{N} = \{0, 1, 2, \ldots\}$, $\mathbb{N}_+ = \{1, 2, \ldots\}$, $\mathcal{PV}$ be a set of propositional variables, and $\mathcal{AG} = \{1, \ldots, n\}$ a set of agents, for $n \in \mathbb{N}_+$.

Definition 1 (Syntax). *Let $p \in \mathcal{PV}$ and $i \in \mathcal{AG}$. The language $\mathcal{L}$ is defined by the following grammar:*

$$\varphi := p \mid \neg\varphi \mid \varphi \vee \varphi \mid \mathrm{EX}\varphi \mid \mathrm{E}(\varphi \mathrm{U} \varphi) \mid \mathrm{A}(\varphi \mathrm{U} \varphi) \mid \mathrm{K}_i\varphi \mid \mathrm{O}_i\varphi \mid \widehat{\mathrm{K}}_i^j\varphi$$

The language above extends CTL [1] with a standard epistemic operator K_i [3], and two further modalities: O_i and $\widehat{\mathrm{K}}_i^j$ [5]. The formula $\mathrm{EX}\alpha$ is read as "there exists a computation path such that at the next step of the path α holds", $\mathrm{E}(\alpha \mathrm{U} \beta)$ is read as "there exists a computation path such that β eventually occurs and α continuously holds until then", $\mathrm{K}_i\alpha$ is read as "agent i knows that α", $\mathrm{O}_i\alpha$ is read as "whenever agent i is functioning correctly α holds", and $\widehat{\mathrm{K}}_i^j\alpha$ is read as "agent i knows that α under the assumption that the agent j is functioning correctly".

The remaining operators can be introduced via abbreviations as usual, i.e., $\alpha \wedge \beta \stackrel{def}{=} \neg(\neg\alpha \vee \neg\beta)$, $\alpha \Rightarrow \beta \stackrel{def}{=} \neg\alpha \vee \beta$, $\alpha \Leftrightarrow \beta \stackrel{def}{=} (\alpha \Rightarrow \beta) \wedge (\beta \Rightarrow \alpha)$, $\mathrm{AX}\alpha \stackrel{def}{=} \neg\mathrm{EX}\neg\alpha$, $\mathrm{EF}\alpha \stackrel{def}{=} \mathrm{E}(\top \mathrm{U} \alpha)$, $\mathrm{AF}\alpha \stackrel{def}{=} \mathrm{A}(\top \mathrm{U} \alpha)$, $\mathrm{EG}\alpha \stackrel{def}{=} \neg\mathrm{AF}\neg\alpha$, $\mathrm{AG}\alpha \stackrel{def}{=} \neg\mathrm{EF}\neg\alpha$, $\mathrm{A}(\alpha \mathrm{W} \beta) \stackrel{def}{=} \neg\mathrm{E}(\neg\alpha \mathrm{U} \neg\beta)$, $\mathrm{E}(\alpha \mathrm{W} \beta) \stackrel{def}{=} \neg\mathrm{A}(\neg\alpha \mathrm{U} \neg\beta)$, $\overline{\mathrm{K}}_i\alpha \stackrel{def}{=} \neg\mathrm{K}_i(\neg\alpha)$, $\overline{\mathrm{O}}_i\alpha \stackrel{def}{=} \neg\mathrm{O}_i(\neg\alpha)$.

Since most of the proofs of the paper are by induction on the length of the formula, below we give a definition of length that will be used throughout the paper.

Definition 2 (Length). *Let $\varphi \in \mathcal{L}$. The* length *of φ (denoted by $|\varphi|$) is defined inductively as follows:*

- *If $\varphi \in \mathcal{PV}$, then $|\varphi| = 1$,*
- *If φ is of the form $\neg\alpha$, $\mathrm{K}_i\alpha$, $\mathrm{O}_i\alpha$, or $\widehat{\mathrm{K}}_i^j\alpha$, then $|\varphi| = |\alpha| + 1$,*
- *If φ is of the form $\mathrm{EX}\alpha$, then $|\varphi| = |\alpha| + 2$,*
- *If φ is of the form $\alpha \vee \beta$ then $|\varphi| = |\alpha| + |\beta| + 1$,*
- *If φ is of the form $\mathrm{A}(\alpha \mathrm{U} \beta)$ or $\mathrm{E}(\alpha \mathrm{U} \beta)$, then $|\varphi| = |\alpha| + |\beta| + 2$.*

Following [5] we interpret $\mathcal{L}$ on *deontic interpreted systems*. Whenever reasoning about models and other semantic structures (such as Hintikka's structures below) we assume that each agent $i \in \mathcal{AG}$ (respectively the environment e) is associated with a set of local states L_i (respectively L_e). These are partitioned into allowed (or green) $\mathcal{G}_i$ (respectively $\mathcal{G}_e$) and disallowed (red) $\mathcal{R}_i$ (respectively $\mathcal{R}_e$) states. The selection and execution of actions on global states generates by means of a transition function runs, or computational paths, that are represented below by means of the temporal relation T. Given our current interest is presently concerned with axiomatisations we will focus at the level of models as defined below. For more details on what below we refer to [3, 5].

Definition 3 (Deontic Interpreted Systems). *A* deontic interpreted system *(or a* model*) is a tuple* $M = (S, T, (\mathrm{R}_i^K)_{i\in\mathcal{AG}}, (\mathrm{R}_i^O)_{i\in\mathcal{AG}}, (\mathrm{R}_i^j)_{i,j\in\mathcal{AG}}, \mathcal{V})$ *where* $S \subseteq \prod_{i=1}^n L_i \times L_e$ *is a set of global states with* $L_1 \supseteq \mathcal{G}_1, \ldots, L_n \supseteq \mathcal{G}_n, L_e \supseteq \mathcal{G}_e$*;* $T \subseteq S \times S$ *is a serial relation on* S*;* $\mathrm{R}_i^K \subseteq S \times S$ *is a relation for each agent* $i \in \mathcal{AG}$ *defined by:* $(s, s') \in \mathrm{R}_i^K$ *iff* $l_i(s) = l_i(s')$*, where* $l_i : S \longrightarrow L_i$ *is a function returning the local state of agent* i *from a global state;* $\mathrm{R}_i^O \subseteq S \times S$ *is a relation for each agent* $i \in \mathcal{AG}$ *defined by:* $(s, s') \in R_i^O$ *iff* $l_i(s') \in \mathcal{G}_i$*;* $\mathrm{R}_i^j \subseteq S \times S$ *is a relation for each agent* $i \in \mathcal{AG}$ *defined by:* $(s, s') \in R_i^j$ *iff* $(s, s') \in R_i^K \cap R_j^O$*;* $\mathcal{V} : S \longrightarrow 2^{\mathcal{PV}}$ *is a valuation function, which assigns to each state a set of proposition variables that are assumed to be true at that state.*

We call $F = (S, T, (\mathrm{R}_i^K)_{i\in\mathcal{AG}}, (\mathrm{R}_i^O)_{i\in\mathcal{AG}}, (\mathrm{R}_i^j)_{i,j\in\mathcal{AG}})$ a frame.

A *path* in M is an infinite sequence $\pi = (s_0, s_1, \ldots)$ of states such that $(s_i, s_{i+1}) \in T$ for each $i \in \mathbb{N}$. For a path $\pi = (s_0, s_1, \ldots)$, we take $\pi(k) = s_k$. By $\Pi(s)$ we denote the set of all the paths starting at $s \in S$.

Definition 4 (Satisfaction). *Let* M *be a model,* s *a state, and* $\alpha, \beta \in \mathcal{L}$*. The satisfaction relation* $\models$*, indicating truth of a formula in model* M *at state* s*, is defined inductively as follows:*

$(M, s) \models p$ *iff* $p \in \mathcal{V}(s)$, $(M, s) \models \alpha \wedge \beta$ *iff* $(M, s) \models \alpha$ *and* $(M, s) \models \beta$,
$(M, s) \models \neg\alpha$ *iff* $(M, s) \not\models \alpha$, $(M, s) \models \mathrm{EX}\alpha$ *iff* $(\exists \pi \in \Pi(s))(M, \pi(1)) \models \alpha$,
$(M, s) \models \mathrm{E}(\alpha \mathrm{U} \beta)$ *iff* $(\exists \pi \in \Pi(s))(\exists m \geq 0)[(M, \pi(m)) \models \beta$ *and* $(\forall j < m)(M, \pi(j)) \models \alpha]$,
$(M, s) \models \mathrm{A}(\alpha \mathrm{U} \beta)$ *iff* $(\forall \pi \in \Pi(s))(\exists m \geq 0)[(M, \pi(m)) \models \beta$ *and* $(\forall j < m)(M, \pi(j)) \models \alpha]$,
$(M, s) \models \mathrm{K}_i\alpha$ *iff* $(\forall s' \in S)$ $(s\mathrm{R}_i^K s'$ *implies* $(M, s') \models \alpha)$,
$(M, s) \models \mathrm{O}_i\alpha$ *iff* $(\forall s' \in S)$ $(sR_i^O s'$ *implies* $(M, s') \models \alpha)$,
$(M, s) \models \widehat{\mathrm{K}}_i^j\alpha$ *iff* $(\forall s' \in S)$ $(s\mathrm{R}_i^j s'$ *implies* $(M, s') \models \alpha)$.

We conclude this section with a definition of validity/satisfiability problems.

Definition 5 (Validity and Satisfiability). *Let* M *be a model and* $\varphi \in \mathcal{L}$*.* **(a)** φ *is* valid *in* M *(written* $M \models \varphi$*), if* $M, s \models \varphi$ *for all states* $s \in S$*.* **(b)** φ *is* satisfiable *in* M*, if* $M, s \models \varphi$ *for some state* $s \in S$*.* **(c)** φ *is* valid *(written* $\models \varphi$*), if* φ *is valid in all the models* M*.* **(d)** φ *is* satisfiable *if it is satisfiable in some model* M*. In this case* M *is said to be a model for* φ*.*

In the next section we prove that $\mathcal{L}$ has the *finite model property* (FMP), that is, we show that any satisfiable $\mathcal{L}$ formula is also satisfiable on a finite model. This result allows us to provide a decidability algorithm for $\mathcal{L}$ (see Section 4), which we use later on to prove that the language has a complete axiomatic system.

3 Finite Model Property (FMP)

The standard procedure for showing the FMP in modal logic is to construct a filtration of an arbitrary model of a satisfiable formula and show that this filtrated model is itself a model for the formula. As it is well-known, while this procedure produces the intended result for a number of logics, it fails in others, for instance in the case of CTL. More refined techniques for showing the FMP exist; notably the construction given in [2] via *Hintikka structures* guarantees the result. Indeed, given that the logic in study here is an extension of CTL, here we follow the procedure given in [2] and show it can be extended to extensions of CTL.

We start by defining two auxiliary structures: a *Hintikka structure* for a given $\mathcal{L}$ formula, and the *quotient structure* for a given model. As in the previous section and in rest of the paper we assume to be dealing with a set of agents defined on local states, and protocols.

Definition 6 (Hintikka structure). *A* Hintikka structure *for φ is a tuple $H = (S, T, (\mathrm{R}_i^K)_{i\in\mathcal{AG}}, (\mathrm{R}_i^O)_{i\in\mathcal{AG}}, (\mathrm{R}_i^j)_{i,j\in\mathcal{AG}}, \mathbb{L})$ where $F = (S, T, (\mathrm{R}_i^K)_{i\in\mathcal{AG}}, (\mathrm{R}_i^O)_{i\in\mathcal{AG}}, (\mathrm{R}_i^j)_{i,j\in\mathcal{AG}})$ is a frame and $\mathbb{L} : S \to 2^{\mathcal{L}}$ is a labelling function assigning a set of formulas to each state such that $\varphi \in \mathbb{L}(s)$ for some $s \in S$ and the following conditions are satisfied:*

H.1. if $\neg\alpha \in \mathbb{L}(s)$, then $\alpha \notin \mathbb{L}(s)$
H.2. if $\neg\neg\alpha \in \mathbb{L}(s)$, then $\alpha \in \mathbb{L}(s)$
H.3. if $(\alpha \vee \beta) \in \mathbb{L}(s)$, then $\alpha \in \mathbb{L}(s)$ or $\beta \in \mathbb{L}(s)$
H.4. if $\neg(\alpha \vee \beta) \in \mathbb{L}(s)$, then $\neg\alpha \in \mathbb{L}(s)$ and $\neg\beta \in \mathbb{L}(s)$
H.5. if $\mathrm{E}(\alpha\mathrm{U}\beta) \in \mathbb{L}(s)$, then $\beta \in \mathbb{L}(s)$ or $\alpha \wedge \mathrm{EXE}(\alpha\mathrm{U}\beta) \in \mathbb{L}(s)$
H.6. if $\neg\mathrm{E}(\alpha\mathrm{U}\beta) \in \mathbb{L}(s)$, then $\neg\beta \wedge \neg\alpha \in \mathbb{L}(s)$ or $\neg\beta \wedge \neg\mathrm{EXE}(\alpha\mathrm{U}\beta) \in \mathbb{L}(s)$
H.7. if $\mathrm{A}(\alpha\mathrm{U}\beta) \in \mathbb{L}(s)$, then $\beta \in \mathbb{L}(s)$ or $\alpha \wedge \neg\mathrm{EX}(\neg\mathrm{A}(\alpha\mathrm{U}\beta)) \in \mathbb{L}(s)$
H.8. if $\neg\mathrm{A}(\alpha\mathrm{U}\beta) \in \mathbb{L}(s)$, then $\neg\beta \wedge \neg\alpha \in \mathbb{L}(s)$ or $\neg\beta \wedge \mathrm{EX}(\neg\mathrm{A}(\alpha\mathrm{U}\beta)) \in \mathbb{L}(s)$
H.9. if $\mathrm{EX}\alpha \in \mathbb{L}(s)$, then $(\exists t \in S)((s,t) \in T$ and $\alpha \in \mathbb{L}(t))$
H.10. if $\neg\mathrm{EX}\alpha \in \mathbb{L}(s)$, then $(\forall t \in S)((s,t) \in T$ implies $\neg\alpha \in \mathbb{L}(t))$
H.11. if $\mathrm{E}(\alpha\mathrm{U}\beta) \in \mathbb{L}(s)$, then $(\exists \pi \in \Pi(s))(\exists n \geq 0)(\beta \in \mathbb{L}(\pi(n))$ and $(\forall j < n)\alpha \in \mathbb{L}(\pi(j)))$
H.12. if $\mathrm{A}(\alpha\mathrm{U}\beta) \in \mathbb{L}(s)$, then $(\forall \pi \in \Pi(s))(\exists n \geq 0)(\beta \in \mathbb{L}(\pi(n))$ and $(\forall j < n)\alpha \in \mathbb{L}(\pi(j)))$
H.13. if $\mathrm{K}_i\alpha \in \mathbb{L}(s)$, then $(\forall t \in S)(s\mathrm{R}_i^K t$ implies $\alpha \in \mathbb{L}(t))$
H.14. if $\neg\mathrm{K}_i\alpha \in \mathbb{L}(s)$, then $(\exists t \in S)(s\mathrm{R}_i^K t$ and $\neg\alpha \in \mathbb{L}(t))$
H.15. if $\mathrm{O}_i\alpha \in \mathbb{L}(s)$, then $(\forall t \in S)(s\mathrm{R}_i^O t$ implies $\alpha \in \mathbb{L}(t))$
H.16. if $\neg\mathrm{O}_i\alpha \in \mathbb{L}(s)$, then $(\exists t \in S)(s\mathrm{R}_i^O t$ and $\neg\alpha \in \mathbb{L}(t))$
H.17. if $\widehat{\mathrm{K}}_i^j\alpha \in \mathbb{L}(s)$, then $(\forall t \in S)(s\mathrm{R}_i^j t$ implies $\alpha \in \mathbb{L}(t))$
H.18. if $\neg\widehat{\mathrm{K}}_i^j\alpha \in \mathbb{L}(s)$, then $(\exists t \in S)(s\mathrm{R}_i^j t$ and $\neg\alpha \in \mathbb{L}(t))$
H.19. if $s\mathrm{R}_i^K t$ and $s\mathrm{R}_i^K u$ and $\mathrm{K}_i\alpha \in \mathbb{L}(t)$, then $\alpha \in \mathbb{L}(u)$
H.20. if $\mathrm{O}_i\alpha \in \mathbb{L}(s)$ and $(s\mathrm{R}_i^O t)$, then $\mathrm{O}_i\alpha \in \mathbb{L}(t)$
H.21. if $s\mathrm{R}_i^O t$ and $s\mathrm{R}_j^O u$ and $\mathrm{O}_i\alpha \in \mathbb{L}(u)$, then $\alpha \in \mathbb{L}(t)$
H.22. if $\widehat{\mathrm{K}}_i^j\alpha \in \mathbb{L}(s)$ and $(sR_i^j t)$, then $\widehat{\mathrm{K}}_i^j\alpha \in \mathbb{L}(t)$
H.23. if $s\mathrm{R}_i^j t$ and $s\mathrm{R}_i^j u$ and $\widehat{\mathrm{K}}_i^j\alpha \in \mathbb{L}(t)$, then $\alpha \in \mathbb{L}(u)$
H.24. if $\mathrm{K}_i\alpha \in \mathbb{L}(s)$, then $\widehat{\mathrm{K}}_i^j\alpha \in \mathbb{L}(s)$
H.25. if $\mathrm{O}_j\alpha \in \mathbb{L}(s)$, then $\widehat{\mathrm{K}}_i^j\alpha \in \mathbb{L}(s)$

Lemma 1 (Hintikka's Lemma for $\mathcal{L}$). *A formula $\varphi \in \mathcal{L}$ is satisfiable (i.e., φ has a model) if and only if there is a Hintikka structure for φ.*

Proof. It is easy to check that any model $M = (S, T, (\mathrm{R}_i^K)_{i\in\mathcal{AG}}, (\mathrm{R}_i^O)_{i\in\mathcal{AG}}, (\mathrm{R}_i^j)_{i,j\in\mathcal{AG}}, \mathcal{V})$ for φ is a Hintikka structure for φ, when we extend $\mathcal{V}$ to cover all formulae which are true in a state, i.e., in M we replace $\mathcal{V}$ by $\mathbb{L}$ that is defined as: $\alpha \in \mathbb{L}(s)$ if $(M, s) \models \alpha$, for all $s \in S$.

Conversely, any Hintikka structure $H = (S, T, (\mathrm{R}_i^K)_{i\in\mathcal{AG}}, (\mathrm{R}_i^O)_{i\in\mathcal{AG}}, (\mathrm{R}_i^j)_{i,j\in\mathcal{AG}}, \mathbb{L})$ for φ can be extended to form a model for φ. Namely, it is enough to restrict $\mathbb{L}$ to propositional variables only, and require that for every propositional variable p appearing in φ and for all $s \in S$ either $p \in \mathbb{L}(s)$ or $\neg p \in \mathbb{L}(s)$. □

Observe that the Hintikka structure differs from a the deontic interpreted system in that the assignment $\mathbb{L}$ is not restricted to propositional variables, nor it is required to contain p or $\neg p$ for any $p \in \mathcal{PV}$. In line with the construction in [2], we call *H*1-*H*8 *propositional consistency rules*, *H*9, *H*10, *H*13-*H*25 *local consistency rules*, and *H*11 and *H*12 *eventuality properties*.

We now proceed to define a quotient structure for a given model. The quotient construction depends on an equivalence relation of states on a given model. To define this we use the *Fischer-Ladner closure* of a formula $\varphi \in \mathcal{L}$ (denoted by $FL(\varphi)$) as $FL(\varphi) = CL(\varphi) \cup \{\neg\alpha \mid \alpha \in CL(\varphi)\}$, where $CL(\varphi)$ is the smallest set of formulas that contains φ and satisfy the following conditions:

(a). if $\neg\alpha \in CL(\varphi)$, then $\alpha \in CL(\varphi)$,
(b). if $\alpha \vee \beta \in CL(\varphi)$, then $\alpha, \beta \in CL(\varphi)$,
(c). if $\mathrm{E}(\alpha \mathrm{U} \beta) \in CL(\varphi)$, then $\alpha, \beta, \mathrm{EXE}(\alpha\mathrm{U}\beta) \in CL(\varphi)$,
(d). if $\mathrm{A}(\alpha \mathrm{U} \beta) \in CL(\varphi)$, then $\alpha, \beta, \mathrm{AXA}(\alpha\mathrm{U}\beta) \in CL(\varphi)$,
(e). if $\mathrm{EX}\alpha \in CL(\varphi)$, then $\alpha \in CL(\varphi)$,
(f). if $\mathrm{K}_i\alpha \in CL(\varphi)$, then $\alpha \in CL(\varphi)$,
(g). if $\mathrm{O}_i\alpha \in CL(\varphi)$, then $\alpha \in CL(\varphi)$,
(h). if $\widehat{\mathrm{K}}_i^j\alpha \in CL(\varphi)$, then $\alpha \in CL(\varphi)$.

Observe that for a given formula $\varphi \in \mathcal{L}$, $FL(\varphi)$ forms a finite set of formulae, as the following lemma shows (the size of a finite set A — denoted by $Card(A)$ — is defined as the number of elements of A).

Lemma 2. *Given a formula $\varphi \in \mathcal{L}$, $Card(FL(\varphi)) \leq 2(|\varphi|)$.*
*Proof.*Straightforward by induction on the length of φ. □

Definition 7 (Fischer-Ladner's equivalence relation). *Let $\varphi \in \mathcal{L}$ and $M = (S, T, (\mathrm{R}_i^K)_{i\in\mathcal{AG}}, (\mathrm{R}_i^O)_{i\in\mathcal{AG}}, (\mathrm{R}_i^j)_{i,j\in\mathcal{AG}}, \mathcal{V})$ be a model for φ. The relation $\leftrightarrow_{FL(\varphi)}$ on a set of states S is defined as follows:*

$$s \leftrightarrow_{FL(\varphi)} s' \text{ if } (\forall \alpha \in FL(\varphi))((M, s) \models \alpha \text{ iff } (M, s') \models \alpha)$$

By $[s]$ we denote the set $\{w \in S \mid w \leftrightarrow_{FL(\varphi)} s\}$.

Observe that $\leftrightarrow_{FL(\varphi)}$ is indeed an equivalence relation, so using it we can define the quotient structure for a given model for $\mathcal{L}$.

Definition 8 (Quotient structure). *Let $\varphi \in \mathcal{L}$, $M = (S, T, (\mathrm{R}_i^K)_{i\in\mathcal{AG}}, (\mathrm{R}_i^O)_{i\in\mathcal{AG}}, (\mathrm{R}_i^j)_{i,j\in\mathcal{AG}}, \mathcal{V})$ be a model for φ, and $\leftrightarrow_{FL(\varphi)}$ a Fischer-Ladner's equivalence relation. The* quotient structure of M by $\leftrightarrow_{FL(\varphi)}$ *is the tuple $M_{\leftrightarrow_{FL(\varphi)}} = (S', T', (\mathrm{R}_i'^K)_{i\in\mathcal{AG}}, (\mathrm{R}_i'^O)_{i\in\mathcal{AG}}, (\mathrm{R}_i'^j)_{i,j\in\mathcal{AG}}, \mathbb{L}')$ such that $S' = \{[s] \mid s \in S\}$, $T' = \{([s],[s']) \in S' \times S' \mid (\exists w \in [s])(\exists w' \in [s'])$ such that $(w,w') \in T\}$, $\mathrm{R}_i'^K = \{([s],[s']) \in S' \times S' \mid (\exists w \in [s])(\exists w' \in [s'])$ such that $(w,w') \in \mathrm{R}_i^K\}$, $\mathrm{R}_i'^O = \{([s],[s']) \in S' \times S' \mid (\exists w \in [s])(\exists w' \in [s'])$ such that $(w,w') \in \mathrm{R}_i^O\}$, $\mathrm{R}_i'^j = \{([s],[s']) \in S' \times S' \mid (\exists w \in [s])(\exists w' \in [s'])$ such that $(w,w') \in \mathrm{R}_i^j\}$, and $\mathbb{L}' : S' \to 2^{FL(\varphi)}$ is defined by: $\mathbb{L}'([s]) = \{\alpha \in FL(\varphi) \mid (M,s) \models \alpha\}$.*

Note that the set S' is finite as it is the result of collapsing states satisfying formulas that belong to the finite set $FL(\varphi)$. In fact we have $Card(S') \leq 2^{Card(FL(\varphi))}$. Note also that since $\mathcal{L}$ is an extension of CTL, the resulting quotient structure may not be a model. In particular, the following lemma holds.

Lemma 3. *The quotient construction does not preserve satisfiability of formulas of the form $\mathrm{A}(\alpha \mathrm{U} \beta)$, where $\alpha, \beta \in \mathcal{L}$. In particular, there is a model M for $\mathrm{A}(\top \mathrm{U} p)$ with $p \in \mathcal{PV}$ such that $M_{\leftrightarrow_{FL(\varphi)}}$ is not a model for $\mathrm{A}(\top \mathrm{U} p)$.*

Proof. [Sketch] Consider the following model $M = (S, T, R_1^K, R_1^O, R_1^1, \mathcal{V})$ for $\mathrm{A}(\top \mathrm{U} p)$, where $S = \{s_0, s_1, \ldots\}$, $T = \{(s_0, s_0)\} \cup \{(s_i, s_{i-1}) \mid i > 0\}$, $R_1^K = R_1^O = R_1^1 = S \times S$, $p \in \mathcal{V}(s_0)$ and $p \notin \mathcal{V}(s_i)$ for all $i > 0$. It is easy to observe that in the quotient structure of M, i.e., in $M_{\leftrightarrow_{FL(\mathrm{A}(\top \mathrm{U} p))}}$, two distinct states s_i and s_j, for all $i, j > 0$, will be identified. As a result of that, a cycle along which p is always false will appear in $M_{\leftrightarrow_{FL(\mathrm{A}(\top \mathrm{U} p))}}$. This implies that $\mathrm{A}(\top \mathrm{U} p)$ does not hold along the cycle. □

Although the quotient structure of a given model M by $\leftrightarrow_{FL(\varphi)}$ may not be a model, it satisfies another important property, which allows us to view it as a *pseudo-model*; it can be unwound into a proper model. This observation can be used to show that the language $\mathcal{L}$ has the FMP property. To make this idea precise, we introduce the following auxiliary definitions.

An *interior* (respectively *frontier*) node of a *directed acyclic graph* (DAG)[1] is one which has (respectively does not have) a successor. The *root* of a DAG is the node (if it exists) from which all other nodes are reachable. A *fragment* $M' = (S', T', (\mathrm{R}_i'^K)_{i\in\mathcal{AG}}, (\mathrm{R}_i'^O)_{i\in\mathcal{AG}}, (\mathrm{R}_i'^j)_{i,j\in\mathcal{AG}}, \mathbb{L}')$ of a Hintikka structure is a structure such that (S', T') generates a finite DAG whose interior nodes satisfy $H1$-$H10$ and $H13$-$H25$, and the frontier nodes satisfy $H1$-$H8$ and $H24$-$H25$. Given $M = (S, T, (\mathrm{R}_i^K)_{i\in\mathcal{AG}}, (\mathrm{R}_i^O)_{i\in\mathcal{AG}}, (\mathrm{R}_i^j)_{i,j\in\mathcal{AG}}, \mathbb{L})$ and $M' = (S', T', (\mathrm{R}_i'^K)_{i\in\mathcal{AG}}, (\mathrm{R}_i'^O)_{i\in\mathcal{AG}}, (\mathrm{R}_i'^j)_{i,j\in\mathcal{AG}}, \mathbb{L}')$, we say that M is *contained* in M', and write $M \subseteq M'$, if $S \subseteq S'$, $T = T' \cap (S \times S)$, $\mathrm{R}_i^K = \mathrm{R}_i'^K \cap (S \times S)$, $\mathrm{R}_i^O = \mathrm{R}_i'^O \cap (S \times S)$, $\mathrm{R}_i^j = \mathrm{R}_i'^j \cap (S \times S)$, $\mathbb{L} = \mathbb{L}' | S$.

Definition 9 (Pseudo-model). *Let $\varphi \in \mathcal{L}$. A* pseudo-model *$M = (S, T, (\mathrm{R}_i^K)_{i\in\mathcal{AG}}, (\mathrm{R}_i^O)_{i\in\mathcal{AG}}, (\mathrm{R}_i^j)_{i,j\in\mathcal{AG}}, \mathbb{L})$ for φ is defined in the same manner as a*

[1] Recall that a directed acyclic graph is a directed graph such that for any node v, there is no nonempty directed path starting and ending on v.

Hintikka structure for φ in Definition 6, except that condition $H12$ is replaced by the following condition $H'12$: $(\forall s \in S)$ if $\mathrm{A}(\alpha \mathrm{U} \beta) \in \mathbb{L}(s)$, then there is a fragment $(S', T', (\mathrm{R}_i'^K)_{i \in \mathcal{AG}}, (\mathrm{R}_i'^O)_{i \in \mathcal{AG}}, (\mathrm{R}_i'^j)_{i,j \in \mathcal{AG}}, \mathbb{L}') \subseteq M$ such that: (a) (S', T') generates a finite DAG with root s; (b) for all frontier nodes $t \in S'$, $\beta \in \mathbb{L}'(t)$; (c) for all interior nodes $u \in S'$, $\alpha \in \mathbb{L}'(u)$.

We have the following.

Lemma 4. *Let $\varphi \in \mathcal{L}$, $FL(\varphi)$ be the Fischer-Ladner closure of φ, $M = (S, T, (\mathrm{R}_i^K)_{i \in \mathcal{AG}}, (\mathrm{R}_i^O)_{i \in \mathcal{AG}}, (\mathrm{R}_i^j)_{i,j \in \mathcal{AG}}, \mathcal{V})$ a model for φ, and $M_{\leftrightarrow_{FL(\varphi)}} = (S', T', (\mathrm{R}_i'^K)_{i \in \mathcal{AG}}, (\mathrm{R}_i'^O)_{i \in \mathcal{AG}}, (\mathrm{R}_i'^j)_{i,j \in \mathcal{AG}}, \mathbb{L})$ the quotient structure of M by $\leftrightarrow_{FL(\varphi)}$. Then, $M_{\leftrightarrow_{FL(\varphi)}}$ is a pseudo-model for φ.*

Proof. The proof for the CTL part of $\mathcal{L}$ follows immediately from Lemma 3.8 in [2] while the cases for the other modalities can be proven as follows. Consider φ to be of the following forms:

H.13 . $\varphi = \mathrm{K}_i\alpha$. Let $(M, s) \models \mathrm{K}_i\alpha$, and $\mathrm{K}_i\alpha \in \mathbb{L}([s])$. By the definition of $\models$, we have that $(M, t) \models \alpha$ for all $t \in S$ such that $s\mathrm{R}_i^K t$. Thus, by the definitions of $\leftrightarrow_{FL(\varphi)}$ and $\mathbb{L}$, we have that $\alpha \in \mathbb{L}([t])$ for all $t \in S$ such that $s\mathrm{R}_i^K t$. Therefore, by the definition of $\mathrm{R}_i'^K$ we conclude that $(\forall [t] \in S')$ if $[s]\mathrm{R}_i'^K[t]$ then $\alpha \in \mathbb{L}([t])$. So, condition $H13$ is fulfilled.

H.14 . $\varphi = \neg\mathrm{K}_i\alpha$. Let $(M, s) \models \neg\mathrm{K}_i\alpha$, and $\neg\mathrm{K}_i\alpha \in \mathbb{L}([s])$. By the definition of $\models$, we have that $\exists t \in S$ such that $s\mathrm{R}_i^K t$ and $(M, t) \models \neg\alpha$. Thus, by the definitions of $\leftrightarrow_{FL(\varphi)}$ and $\mathbb{L}$, we have that $\neg\alpha \in \mathbb{L}([t])$. Therefore, by the definition of $\mathrm{R}_i'^K$ we conclude that $(\exists [t] \in S')$ such that $[s]\mathrm{R}_i'^K[t]$ and $\neg\alpha \in \mathbb{L}([t])$. So, condition $H14$ is fulfilled.

H.15 . $\varphi = \mathrm{O}_i\alpha$. Let $(M, s) \models \mathrm{O}_i\alpha$, and $\mathrm{O}_i\alpha \in \mathbb{L}([s])$. By the definition of $\models$, we have that $(M, t) \models \alpha$ for all $t \in S$ such that $s\mathrm{R}_i^O t$. Thus, by the definitions of $\leftrightarrow_{FL(\varphi)}$ and $\mathbb{L}$, we have that $\alpha \in \mathbb{L}([t])$ for all $t \in S$ such that $s\mathrm{R}_i^O t$. Therefore, by the definition of $\mathrm{R}_i'^O$ we conclude that $(\forall [t] \in S')$ if $[s]\mathrm{R}_i'^O[t]$ then $\alpha \in \mathbb{L}([t])$. So, condition $H15$ is fulfilled.

H.16 . $\varphi = \neg\mathrm{O}_i\alpha$. Let $(M, s) \models \neg\mathrm{O}_i\alpha$, and $\neg\mathrm{O}_i\alpha \in \mathbb{L}([s])$. By the definition of $\models$, we have that $\exists t \in S$ such that $s\mathrm{R}_i^O t$ and $(M, t) \models \neg\alpha$. Thus, by the definitions of $\leftrightarrow_{FL(\varphi)}$ and $\mathbb{L}$, we have that $\neg\alpha \in \mathbb{L}([t])$. Therefore, by the definition of $\mathrm{R}_i'^K$ we conclude that $(\exists [t] \in S')$ such that $[s]\mathrm{R}_i'^O[t]$ and $\neg\alpha \in \mathbb{L}([t])$. So, condition $H16$ is fulfilled.

H.17 . $\varphi = \widehat{\mathrm{K}}_i^j\alpha$. Let $(M, s) \models \widehat{\mathrm{K}}_i^j\alpha$, and $\widehat{\mathrm{K}}_i^j\alpha \in \mathbb{L}([s])$. By the definition of $\models$, we have that $(M, t) \models \alpha$ for all $t \in S$ such that $s\mathrm{R}_i^j t$. Thus, by the definitions of $\leftrightarrow_{FL(\varphi)}$ and $\mathbb{L}$, we have that $\alpha \in \mathbb{L}([t])$ for all $t \in S$ such that $s\mathrm{R}_i^j t$. Therefore, by the definition of $\mathrm{R}_i'^j$ we conclude that $(\forall [t] \in S')$ if $[s]\mathrm{R}_i'^j[t]$ then $\alpha \in \mathbb{L}([t])$. So, condition $H17$ is fulfilled.

H.18 . $\varphi = \neg\widehat{\mathrm{K}}_i^j\alpha$. Let $(M, s) \models \neg\widehat{\mathrm{K}}_i^j\alpha$, and $\neg\widehat{\mathrm{K}}_i^j\alpha \in \mathbb{L}([s])$. By the definition of $\models$, we have that $(\exists t \in S)$ such that $s\mathrm{R}_i^j t$ and $(M, t) \models \neg\alpha$. Thus, by the definitions of $\leftrightarrow_{FL(\varphi)}$ and $\mathbb{L}$, we have that $\neg\alpha \in \mathbb{L}([t])$. Therefore, by the definition of $\mathrm{R}_i'^j$ we conclude that $\exists [t] \in S'$ such that $[s]\mathrm{R}_i'^j[t]$ and $\neg\alpha \in \mathbb{L}([t])$. So, condition $H18$ is fulfilled.

H.19 . Let $[s]\mathrm{R}_i'^K[t]$ and $[s]\mathrm{R}_i'^K[u]$, and $\mathrm{K}_i\alpha \in \mathbb{L}([u])$. Since $[s]\mathrm{R}_i'^K[u]$ and $\mathrm{R}_i'^K$ is symmetric, we have that $[u]\mathrm{R}_i'^K[s]$. Further, since $\mathrm{R}_i'^K$ is transitive and $[u]\mathrm{R}_i'^K[s]$ and $[s]\mathrm{R}_i'^K[t]$, we have that $[u]\mathrm{R}_i'^K[t]$. Since $\mathrm{K}_i\alpha \in \mathbb{L}([u])$, by case H.13 of the proof, we have that $(\forall[w] \in S')([u]\mathrm{R}_i'^K[w]\ \textit{implies}\ \alpha \in \mathbb{L}([w]))$. Therefore, we have that $\alpha \in \mathbb{L}([t])$. So, condition $H19$ is fulfilled.

H.20 . Let $[s]\mathrm{R}_i'^O[t]$ and $\mathrm{O}_i\alpha \in \mathbb{L}([s])$. By case H.15 of the proof, we have that

$$(\forall[w] \in S')([s]\mathrm{R}_i'^O[w]\ \textit{implies}\ \alpha \in \mathbb{L}([w])) \tag{1}$$

So, in particular, we have that $\alpha \in \mathbb{L}([t])$. Consider any $[t'] \in S'$ such that $[t]\mathrm{R}_i'^O[t']$. Since $\mathrm{R}_i'^O$ is transitive, we have that $[s]\mathrm{R}_i'^O[t']$. Since (1) holds, we have that $\alpha \in \mathbb{L}([t'])$ for each $[t']$ such that $[t]\mathrm{R}_i'^O[t']$. This implies that $\mathrm{O}_i\alpha \in \mathbb{L}([t])$. So, condition $H20$ is fulfilled.

H.21 . Let $[s]\mathrm{R}_i'^O[t]$ and $[s]\mathrm{R}_j'^O[u]$, and $\mathrm{O}_i\alpha \in \mathbb{L}([u])$. By case H.15 of the proof, we have that

$$(\forall[w] \in S')([u]\mathrm{R}_i'^O[w]\ \textit{implies}\ \alpha \in \mathbb{L}([w])) \tag{2}$$

Since $\mathrm{R}_i'^O$ is $i - j$euclidean, we have that $[u]\mathrm{R}_i'^O[t]$. Therefore, since (2) holds, we have that $\alpha \in \mathbb{L}([t])$. So, condition $H21$ is fulfilled.

H.22 . Let $[s]\mathrm{R}_i'^j[t]$ and $\widehat{\mathrm{K}}_i^j\alpha \in \mathbb{L}([s])$. By case H.17 of the proof, we have that

$$(\forall[w] \in S')([s]\mathrm{R}_i'^j[w]\ \textit{implies}\ \alpha \in \mathbb{L}([w])) \tag{3}$$

So, in particular, we have that $\alpha \in \mathbb{L}([t])$. Consider any $[t'] \in S'$ such that $[t]\mathrm{R}_i'^j[t']$. Since $\mathrm{R}_i'^j$ is transitive, we have that $[u]\mathrm{R}_i'^j[t']$. Since (3) holds, we have that $\alpha \in \mathbb{L}([t'])$ for each $[t']$ such that $[t]\mathrm{R}_i'^j[t']$. This implies that $\widehat{\mathrm{K}}_i^j\alpha \in \mathbb{L}([t])$. So, condition $H22$ is fulfilled.

H.23 . Let $[s]\mathrm{R}_i'^j[t]$ and $[s]\mathrm{R}_i'^j[u]$, and $\widehat{\mathrm{K}}_i^j\alpha \in \mathbb{L}([u])$. Since $\mathrm{R}_i'^j$ is euclidean, we have that $[u]\mathrm{R}_i'^j[t]$. Since $\widehat{\mathrm{K}}_i^j\alpha \in \mathbb{L}([u])$ holds, by case H.17 of the proof, we have that $(\forall[w] \in S')([u]\mathrm{R}_i'^j[w]\ \textit{implies}\ \alpha \in \mathbb{L}([w]))$. Therefore, we have that $\alpha \in \mathbb{L}([t])$. So, condition $H23$ is fulfilled.

H.24 . $\varphi = \mathrm{K}_i\alpha$. Let $(M, s) \models \mathrm{K}_i\alpha$, and $\mathrm{K}_i\alpha \in \mathbb{L}([s])$. By the definition of $\models$, we have that $(M, t) \models \alpha$ for all $t \in S$ such that $s\mathrm{R}_i^K t$. Consider the following two sets $K(s, i) = \{t \mid (s\mathrm{R}_i^K t) \text{ and } (M, t) \models \alpha\}$ and $O(s, i, j) = \{t \in K(s, i) \mid (s\mathrm{R}_j^O t)\}$, where $i, j \in \{1 \ldots, n\}$. By the definition of $K(s, i)$ and $O(s, j)$, we have that $O(s, i, j) = \{t \mid (s\mathrm{R}_i^j t) \text{ and } M, t \models \alpha\}$. Therefore, by the definition of $\models$ we have that $(M, s) \models \widehat{\mathrm{K}}_i^j\alpha$. Thus, by the definitions of $\leftrightarrow_{FL(\varphi)}$ and $\mathbb{L}$, we have that $\widehat{\mathrm{K}}_i^j\alpha \in \mathbb{L}([s])$. So, condition $H24$ is fulfilled.

H.25 . $\varphi = \mathrm{O}_j\alpha$. Let $(M, s) \models \mathrm{O}_j\alpha$, and $\mathrm{O}_j\alpha \in \mathbb{L}([s])$. By the definition of $\models$, we have that $(M, t) \models \alpha$ for all $t \in S$ such that $s\mathrm{R}_j^O t$. Consider the following two sets $O(s, j) = \{t \mid (s\mathrm{R}_j^O t) \text{ and } (M, t) \models \alpha\}$ and $K(s, i, j) = \{t \in O(s, j) \mid (s\mathrm{R}_i^K t)\}$, where $i, j \in \{1 \ldots, n\}$. By the definition of $K(s, i, j)$ and $O(s, i)$, we have that $K(s, i, j) = \{t \mid (s\mathrm{R}_i^j t) \text{ and } M, t \models \alpha\}$. Therefore, by the definition of $\models$ we have that $(M, s) \models \widehat{\mathrm{K}}_i^j\alpha$. Thus, by the definitions of $\leftrightarrow_{FL(\varphi)}$ and $\mathbb{L}$, we have that $\widehat{\mathrm{K}}_i^j\alpha \in \mathbb{L}([s])$. So, condition $H25$ is fulfilled. □

We can now prove the main claim of the section, i.e., the fact that $\mathcal{L}$ has the finite model property.

Theorem 1 (FMP for $\mathcal{L}$). *Let $\varphi \in \mathcal{L}$. Then the following are equivalent: (1) φ is satisfiable; (2) There is a finite pseudo-model for φ; (3) There is a Hintikka structure for φ.*

Proof. [sketch] (3) $\Rightarrow$ (1) follows from Lemma 1. (1) $\Rightarrow$ (2) follows from Lemma 4. To prove (2) $\Rightarrow$ (3) it is enough to construct a Hintikka structure for φ by "unwinding" the pseudo-model for φ. This can be done in the same way as is described in [2] for the proof of Theorem 4.1. □

4 Decidability for $\mathcal{L}$

Let φ be a $\mathcal{L}$ formula, and $FL(\varphi)$ the Fischer-Ladner closure of φ. We define $\Delta \subseteq FL(\varphi)$ to be *maximal* if for every formula $\alpha \in FL(\varphi)$, either $\alpha \in \Delta$ or $\neg\alpha \in \Delta$.

Theorem 2. *There is an algorithm for deciding whether any $\mathcal{L}$ formula is satisfiable.*

Proof. Given a formula $\varphi \in \mathcal{L}$, we will construct a finite pseudo-model for φ of size less or equal $2^{2\cdot|\varphi|}$. We proceed as follows.

1. Build a structure $M' = (S', T', (\mathrm{R}_i'^K)_{i\in\mathcal{AG}}, (\mathrm{R}_i'^O)_{i\in\mathcal{AG}}, (\mathrm{R}_i'^j)_{i,j\in\mathcal{AG}}, \mathbb{L}')$ in the following way:
   - $S' = \{\Delta \mid \Delta \subseteq FL(\varphi)$ and Δ is maximal and satisfies all the propositional consistency rules$\}$;
   - $T' \subseteq S' \times S'$ is a relation such that $(\Delta_1, \Delta_2) \in T'$ iff $\neg \mathrm{EX}\alpha \in \Delta_1$ implies that $\neg\alpha \in \Delta_2$;
   - for each agent $i \in \mathcal{AG}$, $\mathrm{R}_i'^K \subseteq S' \times S'$ is a relation such that $(\Delta_1, \Delta_2) \in \mathrm{R}_i'^K$ iff $\{\alpha \mid \mathrm{K}_i\alpha \in \Delta_1\} \subseteq \Delta_2$;
   - for each agent $i \in \mathcal{AG}$, $\mathrm{R}_i'^O \subseteq S' \times S'$ is a relation such that $(\Delta_1, \Delta_2) \in \mathrm{R}_i'^O$ iff $\{\alpha \mid \mathrm{O}_i\alpha \in \Delta_1\} \subseteq \Delta_2$;
   - for each agent $i, j \in \mathcal{AG}$, $\mathrm{R}_i'^j \subseteq S' \times S'$ is a relation such that $(\Delta_1, \Delta_2) \in \mathrm{R}_i'^j$ iff $\{\alpha \mid \widehat{\mathrm{K}}_i^j\alpha \in \Delta_1\} \subseteq \Delta_2$;
   - $\mathbb{L}' : S \to 2^{FL(\varphi)}$ is a function defined by $\mathbb{L}'(\Delta) = \Delta$.

   It is easy to observe that M', as constructed above, satisfies all of the propositional consistency properties; properties $H10$, $H13$, $H15$, and $H17$ (because of the definition of T, R_i^K, R_i^O, and R_i^j respectively). Note also that since $Card(FL(\varphi)) \leq 2 \cdot |\varphi|$ (see Lemma 2), S' has at most $2^{2\cdot|\varphi|}$ elements.
2. Test the above structure M' for fulfilment of the properties $H9$, $H11$, $H'12$, $H14$, $H16$ $H18$ – $H25$ by repeatedly applying the following deletion rules until no more states in M' can be deleted.
   (a) Delete any state which has no T-successors.
   (b) Delete any state $\Delta_1 \in S'$ such that $\mathrm{E}(\alpha \mathrm{U} \beta) \in \Delta_1$ (respectively $\mathrm{A}(\alpha \mathrm{U} \beta) \in \Delta_1$) and there does not exist a fragment $M'' \subseteq M'$ such that: (i) (S'', T'')

generates a finite DAG with root Δ_1; (ii) for all frontier nodes $\Delta_2 \in S''$, $\beta \in \Delta_2$; (iii) for all interior nodes $\Delta_3 \in S''$, $\alpha \in \Delta_3$.
(c) Delete any state $\Delta_1 \in S'$ such that $\neg \mathrm{K}_i\alpha \in \Delta_1$, and Δ_1 does not have any $\mathrm{R}_i'^K$ successor $\Delta_2 \in S'$ with $\neg\alpha \in \Delta_2$.
(d) Delete any state $\Delta_1 \in S'$ such that $\neg \mathrm{O}_i\alpha \in \Delta_1$, and Δ_1 does not have any $\mathrm{R}_i'^O$ successor $\Delta_2 \in S'$ with $\neg\alpha \in \Delta_2$.
(e) Delete any state $\Delta_1 \in S'$ such that $\neg \widehat{\mathrm{K}}_i^j\alpha \in \Delta_1$, and Δ_1 does not have any $\mathrm{R}_i'^j$ successor $\Delta_2 \in S'$ with $\neg\alpha \in \Delta_2$.
(f) Delete any state $\Delta_1 \in S'$ such that $\Delta_1 \mathrm{R}_i'^K \Delta_2$ and $\Delta_1 \mathrm{R}_i'^K \Delta_3$ and $\alpha \in \Delta_2$ and $\mathrm{K}_i\neg\alpha \in \Delta_3$
(g) Delete any state $\Delta_1 \in S'$ such that $\Delta_1 \mathrm{R}_i'^O \Delta_2$ and $\mathrm{O}_i\alpha \in \Delta_1$ and $\neg O_i\alpha \in \Delta_2$.
(h) Delete any state $\Delta_1 \in S'$ such that $\Delta_1 R_i^O \Delta_2$ and $\Delta_1 R_j^O \Delta_3$ and $\mathrm{O}_i\neg\alpha \in \Delta_3$ and $\alpha \in \Delta_2$.
(i) Delete any state $\Delta_1 \in S'$ such that $\Delta_1 \mathrm{R}_i^j \Delta_2$ and $\widehat{\mathrm{K}}_i^j\alpha \in \Delta_1$ and $\neg\widehat{\mathrm{K}}_i^j\alpha \in \Delta_2$.
(j) Delete any state $\Delta_1 \in S'$ such that $\Delta_1 R_i'^j \Delta_2$ and $\Delta_1 R_i'^j \Delta_3$ and $\alpha \in \Delta_2$ and $\widehat{\mathrm{K}}_i^j\neg\alpha \in \Delta_3$.
(k) Delete any state $\Delta \in S'$ such that $\mathrm{K}_i\alpha \in \Delta$ and $\neg\widehat{\mathrm{K}}_i^j\alpha \in \Delta$.
(l) Delete any state $\Delta \in S'$ such that $\mathrm{O}_j\alpha \in \Delta$ and $\neg\widehat{\mathrm{K}}_i^j\alpha \in \Delta$.

We call the above two points a *decidability algorithm for $\mathcal{L}$.*

Claim (1). The decidability algorithm for $\mathcal{L}$ terminates.
Proof. The termination is obvious given that the initial set S' is finite.

Claim (2). Let $M = (S, T, (\mathrm{R}_i^K)_{i\in\mathcal{AG}}, (\mathrm{R}_i^O)_{i\in\mathcal{AG}}, (\mathrm{R}_i^j)_{i,j\in\mathcal{AG}}, \mathbb{L})$ be the resulting structure of the algorithm. The formula $\varphi \in \mathcal{L}$ is satisfiable iff $\varphi \in s$, for some $s \in S$.

Proof. In order to show the part right-to-left of the above property, note that either the resulting structure is a pseudo-model for φ, or $S = \emptyset$ (this can be shown inductively on the structure of the algorithm). So, if $\varphi \in s$ for some $s \in S$, φ is satisfiable by Theorem 1.

Conversely, if φ is satisfiable, then there exists a model M^* such that $M^* \models \varphi$. Let $M^*_{\leftrightarrow_{FL(\varphi)}} = M' = (S', T', (\mathrm{R}_i'^K)_{i\in\mathcal{AG}}, (\mathrm{R}_i'^O)_{i\in\mathcal{AG}}, (\mathrm{R}_i'^j)_{i,j\in\mathcal{AG}}, \mathbb{L}')$ be the quotient structure of M^* by $\leftrightarrow_{FL(\varphi)}$. By Theorem 1 we have that M' is a pseudo-model for φ. So, $\mathbb{L}'$ satisfies all the propositional consistency rules, the local consistency rules, and properties $H11$ and $H'12$. Moreover, by the definition of $\mathbb{L}'$ in the quotient structure, $\mathbb{L}'(s)$ is maximal with respect to $FL(\varphi)$ for all $s \in S'$. Now, let $M'' = (S'', T'', (\mathrm{R}_i''^K)_{i\in\mathcal{AG}}, (\mathrm{R}_i''^O)_{i\in\mathcal{AG}}, (\mathrm{R}_i''^j)_{i,j\in\mathcal{AG}}, \mathbb{L}'')$ be a structure defined by step 1 of the decidability algorithm, and $f : S' \to S''$ a function defined by $f(s) = \mathbb{L}'(s)$. The following conditions hold:

1. If $(s, t) \in T'$, then $(f(s), f(t)) \in T''$;
Proof (via contradiction): Let $(s, t) \in T'$ and $(f(s), f(t)) \notin T''$. By the definition of T'' we have that $\neg \mathrm{EX}\alpha \in f(s)$ and $\alpha \in f(t)$. Then, by the definition of f, we have that $\neg \mathrm{EX}\alpha \in \mathbb{L}'(s)$ and $\alpha \in \mathbb{L}'(t)$. So, by the definition of $\mathbb{L}'$

in the quotient structure we have that $M^*, s \models \neg\mathrm{EX}\alpha$ and $M^*, t \models \alpha$, which contradict the fact that $(s,t) \in T'$.

2. If $(s,t) \in \mathrm{R}_i'^K$, then $(f(s), f(t)) \in \mathrm{R}_i''^K$;
Proof (via contradiction): Let $(s,t) \in \mathrm{R}_i'^K$ and $(f(s), f(t)) \notin \mathrm{R}_i''^K$. By the definition of $\mathrm{R}_i''^K$ we have that $\mathrm{K}_i\alpha \in f(s)$ and $\alpha \notin f(t)$. Then, by the definition of f, we have that $\mathrm{K}_i\alpha \in \mathbb{L}'(s)$ and $\alpha \notin \mathbb{L}'(t)$. So, by the definition of $\mathbb{L}'$ in the quotient structure we have that $M^*, s \models \mathrm{K}_i\alpha$ and $M^*, t \models \neg\alpha$, which contradict the fact that $(s,t) \in \mathrm{R}_i'^K$.
3. If $(s,t) \in \mathrm{R}_i'^O$, then $(f(s), f(t)) \in \mathrm{R}_i''^O$;
Proof (via contradiction): Let $(s,t) \in \mathrm{R}_i'^O$ and $(f(s), f(t)) \notin \mathrm{R}_i''^O$. By the definition of $\mathrm{R}_i''^O$ we have that $\mathrm{O}_i\alpha \in f(s)$ and $\alpha \notin f(t)$. Then, by the definition of f, we have that $\mathrm{O}_i\alpha \in \mathbb{L}'(s)$ and $\alpha \notin \mathbb{L}'(t)$. So, by the definition of $\mathbb{L}'$ in the quotient structure we have that $M^*, s \models \mathrm{O}_i\alpha$ and $M^*, t \models \neg\alpha$, which contradict the fact that $(s,t) \in \mathrm{R}_i'^O$.
4. If $(s,t) \in \mathrm{R}_i'^j$, then $(f(s), f(t)) \in \mathrm{R}_i''^j$;
Proof (via contradiction): Let $(s,t) \in \mathrm{R}_i'^j$ and $(f(s), f(t)) \notin \mathrm{R}_i''^j$. By the definition of $\mathrm{R}_i''^j$ we have that $\widehat{\mathrm{K}}_i^j\alpha \in f(s)$ and $\alpha \notin f(t)$. Then, by the definition of f, we have that $\widehat{\mathrm{K}}_i^j\alpha \in \mathbb{L}'(s)$ and $\alpha \notin \mathbb{L}'(t)$. So, by the definition of $\mathbb{L}'$ in the quotient structure we have that $M^*, s \models \widehat{\mathrm{K}}_i^j\alpha$ and $M^*, t \models \neg\alpha$, which contradict the fact that $(s,t) \in \mathrm{R}_i'^j$.

Thus, the image of M' under f is contained in M'', i.e., $M' \subseteq M''$. It remains to show that if $s \in S'$, then $f(s) \in S''$ will not be eliminated in step 2 of the decidability algorithm. This can be checked by induction on the order in which states of S'' are eliminated. For instance, assume that $s \in S'$, and $\mathrm{A}(\alpha \mathrm{U} \beta) \in f(s)$. By the definition of f, we have that $\mathrm{A}(\alpha \mathrm{U} \beta) \in \mathbb{L}'(s)$. Now, since M' is a pseudo-model, by Definition 9 we have that there exists a fragment rooted at s that is contained in M' and it satisfies property $H'12$. Thus, since f preserves the above condition (a), we have that there exists a fragment rooted at $f(s)$ that is contained in M'' and it satisfies property $H'12$. This implies that $f(s) \in S''$ will not be eliminated in step 2b of the decidability algorithm. Other cases can be proven similarly. Therefore, it follows that for some $s \in S$ we have $\varphi \in \mathbb{L}(s)$. □

5 A Complete Axiomatic System for $\mathcal{L}$

An *axiomatic system* consists of a collection of *axioms* and *inference rules*. An axiom is a formula, and an inference rule has the form "from formulas $\varphi_1, \ldots, \varphi_m$ infer formula φ". We say that φ is *provable* (written $\vdash \varphi$) if there is a sequence of formulas ending with φ, such that each formula is either an instance of an axiom, or follows from other provable formulas by applying an inference rule. We say that a formula φ is *consistent* if $\neg\varphi$ is not provable. A finite set $\{\varphi_1, \ldots, \varphi_m\}$ of formulas is *consistent* if and only if the conjunction $\varphi_1 \wedge \ldots \wedge \varphi_m$ of its members is consistent, and an infinite set of formulas is *consistent* if all of its finite subsets are consistent. A set F of formulas is a *maximal consistent set* if it is consistent and for all $\varphi \notin F$, the set $F \cup \{\varphi\}$ is inconsistent. An axiom system is *sound* (resp. *complete*) with respect to the class of models, if $\vdash \varphi$ implies $\models \varphi$ (resp. if $\models \varphi$ implies $\vdash \varphi$).

Definition 10 (Axiomatisation of deontic interpreted systems). *Let $i \in \{1, \ldots, n\}$. Consider the following axiomatic system for $\mathcal{L}$:*

PC.	*All substitution instances of classical tautologies.*		
$\mathbf{X_1}$.	$\mathrm{EX}\top$	$\mathbf{U_1}$.	$\mathrm{E}(\alpha \mathrm{U} \beta) \Leftrightarrow \beta \vee (\alpha \wedge \mathrm{EXE}(\alpha \mathrm{U} \beta))$
$\mathbf{X_2}$.	$\mathrm{EX}(\alpha \vee \beta) \Leftrightarrow \mathrm{EX}\alpha \vee \mathrm{EX}\beta$	$\mathbf{U_2}$.	$\mathrm{A}(\alpha \mathrm{U} \beta) \Leftrightarrow \beta \vee (\alpha \wedge \mathrm{AXA}(\alpha \mathrm{U} \beta))$
$\mathbf{K_{K_i}}$.	$(\mathrm{K}_i\alpha \wedge \mathrm{K}_i(\alpha \Rightarrow \beta)) \Rightarrow \mathrm{K}_i\beta$	$\mathbf{T_{K_i}}$.	$\mathrm{K}_i\alpha \Rightarrow \alpha$
$\mathbf{5_{K_i}}$.	$\neg\mathrm{K}_i\alpha \Rightarrow \mathrm{K}_i\neg\mathrm{K}_i\alpha$	$\mathbf{K_{O_i}}$.	$(\mathrm{O}_i\alpha \wedge \mathrm{O}_i(\alpha \Rightarrow \beta)) \Rightarrow \mathrm{O}_i\beta$
$\mathbf{4_{O_i}}$.	$\mathrm{O}_i\alpha \Rightarrow \mathrm{O}_i\mathrm{O}_i\alpha$	$\mathbf{D_{O_i}}$.	$\mathrm{O}_i\alpha \Rightarrow \neg\mathrm{O}_i\neg\alpha$
$\mathbf{5^{i-j}_{O_i}}$.	$\neg\mathrm{O}_i\alpha \Rightarrow \mathrm{O}_j\neg\mathrm{O}_i\alpha$	$\mathrm{O} - \widehat{\mathrm{K}}^j_i$.	$\mathrm{O}_j\alpha \Rightarrow \widehat{\mathrm{K}}^j_i\alpha$
$\mathrm{K} - \widehat{\mathrm{K}}^j_i$.	$\mathrm{K}_i\alpha \Rightarrow \widehat{\mathrm{K}}^j_i\alpha$	$\mathbf{K_{\widehat{K}^j_i}}$.	$(\widehat{\mathrm{K}}^j_i\alpha \wedge \widehat{\mathrm{K}}^j_i(\alpha \Rightarrow \beta)) \Rightarrow \widehat{\mathrm{K}}^j_i\beta$
$\mathbf{4_{\widehat{K}^j_i}}$.	$\widehat{\mathrm{K}}^j_i\alpha \Rightarrow \widehat{\mathrm{K}}^j_i\widehat{\mathrm{K}}^j_i\alpha$	$\mathbf{5_{\widehat{K}^j_i}}$.	$\neg\widehat{\mathrm{K}}^j_i\alpha \Rightarrow \widehat{\mathrm{K}}^j_i\neg\widehat{\mathrm{K}}^j_i\alpha$
MP.	*From* α *and* $\alpha \Rightarrow \beta$ *infer* β	$\mathbf{Nec_{K_i}}$.	*From* α *infer* $\mathrm{K}_i\alpha$
$\mathbf{Nec_{O_i}}$.	*From* α *infer* $\mathrm{O}_i\alpha$	$\mathbf{R1_X}$.	*From* $\alpha \Rightarrow \beta$ *infer* $\mathrm{EX}\alpha \Rightarrow \mathrm{EX}\beta$
$\mathbf{R2_X}$.	*From* $\gamma \Rightarrow (\neg\beta \wedge \mathrm{EX}\gamma)$ *infer* $\gamma \Rightarrow \neg\mathrm{A}(\alpha \mathrm{U} \beta)$		
$\mathbf{R3_X}$.	*From* $\gamma \Rightarrow (\neg\beta \wedge \mathrm{AX}(\gamma \vee \neg\mathrm{E}(\alpha \mathrm{U} \beta)))$ *infer* $\gamma \Rightarrow \neg\mathrm{E}(\alpha \mathrm{U} \beta)$		

We note that the system above includes the axiomatisation for CTL [2], S5 [3] for K_i and $\mathrm{KD45}^{i-j}$ [5] for O_i. The fragment for the operators $\widehat{\mathrm{K}}^j_i$, previously not explored, is K45. In line with the traditional interpretation of these axioms in an epistemic setting these are to be interpreted from the point of view of an external observer ascribing properties to the system. They both seem in line with the interpretation of the modality of knowledge under the assumption of correct behaviour. Further note that axioms $4_{\widehat{\mathrm{K}}^j_i}$, and $5_{\widehat{\mathrm{K}}^j_i}$ are to be expected given that both the underlying relations are transitive and Euclidean.

The interaction axioms $\mathrm{O}_i - \widehat{\mathrm{K}}^j_i$ and $\mathrm{K}_i - \widehat{\mathrm{K}}^j_i$ regulate the relationship between $\mathrm{O}_i, \mathrm{K}_i$ and $\widehat{\mathrm{K}}^j_i$. They were both discussed in [5] and correspond to our intuition regarding the meaning of the modalities. Note also that they closely match the interaction axioms for distributed versus standard knowledge, which again confirms our intuition given that distributed knowledge is defined on the intersection of the relations for standard knowledge.

The inference rules for all the components are also entirely expected — note that while Necessitation for $\widehat{\mathrm{K}}^j_i$ is not explicitly listed, it may easily be deduced from Nec_{K_i} or Nec_{O_i}.

Theorem 3. *The axiomatic system for $\mathcal{L}$ is sound and complete with respect to the deontic interpreted systems, i.e. $\models \varphi$ iff $\vdash \varphi$, for any formula $\varphi \in \mathcal{L}$.*

Proof. Soundness can be checked inductively as standard. For completeness, we show that any consistent formula φ is satisfiable. To do this, we first consider the structure $M = (S, T, (\mathrm{R}^K_i)_{i \in \mathcal{AG}}, (\mathrm{R}^O_i)_{i \in \mathcal{AG}}, (\mathrm{R}^j_i)_{i,j \in \mathcal{AG}}, \mathbb{L})$ for φ as defined in step 1 of the decidability algorithm. We then execute step 2 of the algorithm, obtaining a pseudo-model for φ. Crucially we show below that if a state $s \in S$ is eliminated at step 2 of the algorithm, then the formula $\psi_s = \bigwedge_{\alpha \in s} \alpha$ is inconsistent. Observe now that that for any $\alpha \in FL(\varphi)$ we have $\vdash \alpha \Leftrightarrow \bigvee_{\{s \mid \alpha \in s \text{ and } \psi_s \text{ is consistent}\}} \psi_s$. In particular, $\vdash \varphi \Leftrightarrow \bigvee_{\{s \mid \varphi \in s \text{ and } \psi_s \text{ is consistent}\}} \psi_s$. Thus, if φ is consistent, then ψ_s is consistent as well for some $s \in S$. It follows by Claim 2 of Theorem 2 that this particular s is present in the pseudo-model resulting from the execution of the algorithm. So, by Theorem 1, φ is satisfiable. Note that pseudo-models share the structural properties of models, i.e., their underlying frames have the same properties.

It remains to show that if a state $s \in S$ is eliminated at step 2 of the algorithm then the formula ψ_s is inconsistent. Before we do it, we need some auxiliary claims.

Claim (3). Let $s \in S$ and $\alpha \in FL(\varphi)$. Then, $\alpha \in s$ iff $\vdash \psi_s \Rightarrow \alpha$.

Proof. ('if'). Let $\alpha \in s$. Then, by PC we have that $\vdash \psi_s \Rightarrow \alpha$. ('only if'). Let $\vdash \psi_s \Rightarrow \alpha$. Then, since s is maximal and propositionally consitent, we have that $\alpha \in s$. □

Claim (4). Let $s, t \in S$, both of them be maximal and propositionally consistent, $s\mathrm{R}_i^K t$ (respectively $s\mathrm{R}_i^O t$ and $s\mathrm{R}_i^j t$), and $\alpha \in FL(\varphi)$. If $\alpha \in t$, then $\neg K_i \neg \alpha \in s$ (respectively $\neg O_i \neg \alpha \in s$ and $\neg \widehat{\mathrm{K}}_i^j \neg \alpha \in s$).

Proof.[By contraposition] Let $\alpha \in t$ and $\neg K_i \neg \alpha \notin s$. Then, since s is maximal we have that $K_i \neg \alpha \in s$. Thus, since $s\mathrm{R}_i^K t$, we have that $\neg \alpha \in t$. This contradicts the fact that $\alpha \in t$, since t is propositionally consistent.

The same proof applies to O_i and $\widehat{\mathrm{K}}_i^j$. □

Claim (5). Let $s \in S$ be a maximal and consistent set of formulas, $\alpha \in FL(\varphi)$ and α such that $\vdash \alpha$. Then $\alpha \in s$.

Proof. Suppose $\alpha \notin s$ and $\vdash \alpha$. Since s is maximal, $\neg \alpha \in s$. Thus $\neg \alpha \wedge \psi_s$ is consistent where ψ_s where $\psi_s \in s$. So by definition of consistency we have that $\nvdash \neg(\neg \alpha \wedge \psi_s)$, so $\nvdash \alpha \vee \neg \psi_s$. But we have $\vdash \alpha \vee \psi_s$, so this is a contradiction. □

We now show, by induction on the structure of the decidability algorithm for $\mathcal{L}$, that if a state $s \in S$ is eliminated at step 2 of the decidability algorithm, then $\vdash \neg \psi_s$.

Claim (6). If ψ_s is consistent, then s is not eliminated at step 2 of the decidability algorithm for $\mathcal{L}$.

Proof.

(a). Let $\mathrm{EX}\alpha \in s$ and ψ_s be consistent. By the same reasoning as in the proof of Claim 4(a) in [2], we conclude that s satisfies $H9$. So s is not eliminated.

(b). Let $\mathrm{E}(\alpha \mathrm{U} \beta) \in s$ (respectively $\mathrm{A}(\alpha \mathrm{U} \beta) \in s$) and suppose s is eliminated at step 2 because $H11$ (respectively $H'12$) is not satisfied. Then ψ_s is inconsistent. The proof showing that fact is the same as the proof of Claim 4(c) (respectively Claim 4(d)) in [2].

(c). Let $\neg \mathrm{K}_i \alpha \in s$ and ψ_s be consistent. Consider the set $S_{\neg \alpha} = \{\neg \alpha\} \cup \{\beta \mid \mathrm{K}_i \beta \in s\}$. We will show that $S_{\neg \alpha}$ is consistent. Suppose that $S_{\neg \alpha}$ is inconsistent. Then, $\vdash \beta_1 \wedge \ldots \wedge \beta_m \Rightarrow \alpha$, where $\beta_j \in \{\beta \mid \mathrm{K}_i \beta \in s\}$ for

$j \in \{1,\ldots,m\}$. By rule Nec_{K_i} we have $\vdash \mathrm{K}_i((\beta_1 \wedge \ldots \wedge \beta_m) \Rightarrow \alpha)$. By axioms K_{K_i} and PC we have $\vdash (\mathrm{K}_i\beta_1 \wedge \ldots \wedge \mathrm{K}_i\beta_m) \Rightarrow \mathrm{K}_i\alpha$. Thus, since each $\mathrm{K}_i\beta_j \in s$ for $j \in \{1,\ldots,m\}$ and s is maximal and propositionally consistent, we have $\mathrm{K}_i\alpha \in s$. This contradicts the fact that ψ_s is consistent. So, $S_{\neg\alpha}$ is consistent. Now, since each set of formulas can be extended to a maximal one, we have that $S_{\neg\alpha}$ is contained in some maximal set t. Thus $\neg\alpha \in t$, and moreover, by the definition of R_i^K in M and the definition of $S_{\neg\alpha}$ we have that $s\mathrm{R}_i^K t$. Thus, s satisfies $H14$, and it is not eliminated by step (c) of the decidability algorithm.

(d). Let $\neg\mathrm{O}_i\alpha \in s$ and ψ_s be consistent. Consider the set $S_{\neg\alpha} = \{\neg\alpha\} \cup \{\beta \mid \mathrm{O}_i\beta \in s\}$. We will show that $S_{\neg\alpha}$ is consistent. Suppose that $S_{\neg\alpha}$ is inconsistent. Then, $\vdash \beta_1 \wedge \ldots \wedge \beta_m \Rightarrow \alpha$, where $\beta_j \in \{\beta \mid \mathrm{O}_i\beta \in s\}$ for $j \in \{1,\ldots,m\}$. By rule Nec_{O_i} we have $\vdash \mathrm{O}_i((\beta_1 \wedge \ldots \wedge \beta_m) \Rightarrow \alpha)$. By axioms K_{O_i} and PC we have $\vdash (\mathrm{O}_i\beta_1 \wedge \ldots \wedge \mathrm{O}_i\beta_m) \Rightarrow \mathrm{O}_i\alpha$. Since each $\mathrm{O}_i\beta_j \in s$ for $j \in \{1,\ldots,m\}$ and s is maximal and propositionally consistent, we have $\mathrm{O}_i\alpha \in s$. This contradicts the fact that ψ_s is consistent. So, $S_{\neg\alpha}$ is consistent. Now, since each set of formulas can be extended to a maximal one, we have that $S_{\neg\alpha}$ is contained in some maximal set t. Thus $\neg\alpha \in t$, and moreover, by the definition of R_i^O in M and the definition of $S_{\neg\alpha}$ we have that $s\mathrm{R}_i^O t$. Thus, s satisfies $H16$, and it is not eliminated by step (d) of the decidability algorithm.

(e). Let $\neg\widehat{\mathrm{K}}_i^j\alpha \in s$ and ψ_s be consistent. Consider the set $S_{\neg\alpha} = \{\neg\alpha\} \cup \{\beta \mid \widehat{\mathrm{K}}_i^j\beta \in s\}$. We will show that $S_{\neg\alpha}$ is consistent. Suppose that $S_{\neg\alpha}$ is inconsistent. Then, $\vdash \beta_1 \wedge \ldots \wedge \beta_m \Rightarrow \alpha$, where $\beta_j \in \{\beta \mid \widehat{\mathrm{K}}_i^j\beta \in s\}$ for $j \in \{1,\ldots,m\}$. By rule Nec_{K_i} we have $\vdash \mathrm{K}_i((\beta_1 \wedge \ldots \wedge \beta_m) \Rightarrow \alpha)$. By axiom $(K - \widehat{\mathrm{K}}_j^j)$ we have $\vdash \widehat{\mathrm{K}}_i^j((\beta_1 \wedge \ldots \wedge \beta_m) \Rightarrow \alpha)$. By axioms $K_{\widehat{\mathrm{K}}_i^j}$ and PC we have $\vdash (\widehat{\mathrm{K}}_i^j\beta_1 \wedge \ldots \wedge \widehat{\mathrm{K}}_i^j\beta_m) \Rightarrow \widehat{\mathrm{K}}_i^j\alpha$. Since each $\widehat{\mathrm{K}}_i^j\beta_j \in s$ for $j \in \{1,\ldots,m\}$ and s is maximal and propositionally consistent, we have $\widehat{\mathrm{K}}_i^j\alpha \in s$. This contradicts the fact that ψ_s is consistent. So, $S_{\neg\alpha}$ is consistent. Now, since each set of formulas can be extended to a maximal one, we have that $S_{\neg\alpha}$ is contained in some maximal set t. Thus $\neg\alpha \in t$, and moreover, by the definition of R_i^j in M and the definition of $S_{\neg\alpha}$ we have that $sR_i^j t$. Thus, s satisfies $H18$, and it is not eliminated by step (e) of the decidability algorithm.

(f). Suppose that s is consistent and it is eliminated at step (f) of the decidability algorithm. Thus, we have that $s\mathrm{R}_i^K t$, $s\mathrm{R}_i^K u$, $\alpha \in t$, and $\mathrm{K}_i\neg\alpha \in u$. So, since $s\mathrm{R}_i^K t$, $\alpha \in t$, s and t are maximal and propositionally consistent, by Claim 4 we have that $\neg K_i\neg\alpha \in s$. Since s is maximal and consistent, by axiom 5_{K_i} and Claim 5, we have that $\neg\mathrm{K}_i\neg\alpha \Rightarrow \mathrm{K}_i\neg\mathrm{K}_i\neg\alpha \in s$. Therefore, we have that $\mathrm{K}_i\neg\mathrm{K}_i\neg\alpha \in s$. Thus, since $s\mathrm{R}_i^K u$, we have that $\neg\mathrm{K}_i\neg\alpha \in u$. But this is a contradictions given that $\mathrm{K}_i\neg\alpha \in u$ an u is propositionally consistent. So s is inconsistent. Therefore s cannot be eliminated at step (f) of the decidability algorithm.

(g). Suppose that ψ_s is consistent and s is eliminated at step (g) of the decidability algorithm. Then, we have that $sR_i^O t$, $O_i\alpha \in s$ and $\neg O_i\alpha \in t$. Thus, since s and t are maximal and propositionally consistent, by Claim 4 we have that $\neg O_i O_i\alpha \in s$. By axiom 4_{O_i} and Claim 5 we have that $O_i\alpha \Rightarrow O_i O_i\alpha \in s$. So, since $O_i\alpha \in s$ we have that $O_i O_i\alpha \in s$. So s is inconsistent. Therefore s cannot be eliminated at step (g) of the decidability algorithm.

(h). If ψ_s is consistent, s cannot be eliminated at step (h) (respectively (i) and (j)) of the decidability algorithm. The proof can be done similarly to the one in (f) (respectively (g) and (f)) by using axiom $5_{O_i}^{i-j}$ (respectively $4_{\widehat{K}_i^j}$ and $5_{\widehat{K}_i^j}$).

(i). Suppose that s is consistent and s is eliminated at step (k) of the decidability algorithm. Thus, we have that $K_i\alpha \in s$ and $\widehat{K}_i^j\alpha \notin s$. Since s is maximal we have that $\neg\widehat{K}_i^j\alpha \in s$. Since s is consistent, by axiom $K - \widehat{K}_i^j$ and Claim 5 we have that $K_i\alpha \Rightarrow \widehat{K}_i^j\alpha \in s$. So, since $K_i\alpha \in s$ we have that $\widehat{K}_i^j\alpha \in s$. So s is inconsistent. Therefore s cannot be eliminated at step (k) of the decidability algorithm.

(j). Suppose that s is consistent and s is eliminated at step (l) of the decidability algorithm. Thus, we have that $O_j\alpha \in s$ and $\widehat{K}_i^j\alpha \notin s$. Since s is maximal we have that $\neg\widehat{K}_i^j\alpha \in s$. Since s is consistent, by axiom $O - \widehat{K}_i^j$ and Claim 5 we have that $O_j\alpha \Rightarrow \widehat{K}_i^j\alpha \in s$. So, since $O_j\alpha \in s$ we have that $\widehat{K}_i^j\alpha \in s$. So s is inconsistent. Therefore s cannot be eliminated at step (l) of the decidability algorithm. □

We have now shown that only states s with ψ_s inconsistent are eliminated. This ends the completeness proof. □

6 Conclusion

We have given a complete axiomatisation of deontic interpreted systems on a language that includes full CTL as well as the the K_i, O_i and $\widehat{K}_i^j$ modalities. Thereby, we have solved the problem left open in [5]. Further, we have shown that the language considered here has the finite model property, so it is decidable.

The $\widehat{K}_i^j$ modality can be straightforwardly extended to $\widehat{K}_i^X$ [5] representing knowledge of i under the assumption of correctness of all agents in X. We believe that the technique of this paper can be extended to prove completeness for axiomatisation for $\widehat{K}_i^X$ without difficulty. For clarity this is not presented in this paper.

References

1. E. Clarke and E. Emerson. Design and synthesis of synchronization skeletons for branching-time temporal logic. In *Proceedings of Workshop on Logic of Programs*, volume 131 of *LNCS*, pages 52–71. Springer-Verlag, 1981.
2. E. A. Emerson and J. Y. Halpern. Decision procedures and expressiveness in the temporal logic of branching time. *Journal of Computer and System Sciences*, 30(1):1–24, 1985.

3. R. Fagin, J. Y. Halpern, Y. Moses, and M. Y. Vardi. *Reasoning about Knowledge*. MIT Press, Cambridge, 1995.
4. A. Lomuscio and F. Raimondi. MCMAS: A model checker for multi-agent systems. In H. Hermanns and J. Palsberg, editors, *Proceedings of TACAS 2006, Vienna*, volume 3920, pages 450–454. Springer Verlag, March 2006.
5. A. Lomuscio and M. Sergot. Deontic interpreted systems. *Studia Logica*, 75(1):63–92, 2003.
6. A. Lomuscio and M. Sergot. A formalisation of violation, error recovery, and enforcement in the bit transmission problem. *Journal of Applied Logic*, 2(1):93–116, March 2004.
7. W. Nabialek, A. Niewiadomski, W. Penczek, A. Pólrola, and M. Szreter. VerICS 2004: A model checker for real time and multi-agent systems. In *Proceedings of the International Workshop on Concurrency, Specification and Programming (CS&P'04)*, volume 170 of *Informatik-Berichte*, pages 88–99. Humboldt University, 2004.
8. F. Raimondi and A. Lomuscio. Automatic verification of multi-agent systems by model checking via OBDDs. *Journal of Applied Logic*, 2005. To appear in Special issue on Logic-based agent verification.
9. M. Wooldridge and A. Lomuscio. A computationally grounded logic of visibility, perception, and knowledge. *Logic Journal of the IGPL*, 9(2):273–288, 2001.
10. B. Woźna, A. Lomuscio, and W. Penczek. Bounded model checking for deontic interpreted systems. In *Proc. of the 2nd Workshop on Logic and Communication in Multi-Agent Systems (LCMAS'04)*, volume 126 of *ENTCS*, pages 93–114. Elsevier, 2004.

Sequences, Obligations, and the Contrary-to-Duty Paradox*

Adam Zachary Wyner

King's College London
London, UK
adam@wyner.info

Abstract. In order to provide an implemented language of deontic concepts on complex actions for the purposes of social simulation, we consider the logical representation of obligations, sequences of actions, and the *Contrary to Duty* (CTD) Paradox. We show that approaches which follow Standard Deontic Logic (Carmo and Jones (2002)) or Dynamic Deontic Logic (Khosla and Maibaum (1987) and Meyer (1988)) encounter problems with obligations, sequences, and CTDs. In particular, it is crucial to differentiate sequences of obligations from obligations on sequences and to consider contract change over time. Contra Meyer (1988), we argue that the CTD problem cannot be reduced to a a sequence of obligations. Contra Carmo and Jones (2002), the analysis of CTDs needs explicit state change and does not need a concept of ideality. We discuss *Pörn's Criterion*, which states that it is critical to a *comprehensive* theory of deontic reasoning to take dynamic aspects into account (Pörn (1977:ix-x)); in our view, this ought to encompass *Contract State Change*. In a theory of deontic specifications on actions, we show that articulated, compositional, and productive markers for violation and fulfillment are key to address the problems identified. The theorical arguments inform the *Abstract Contract Calculator*, a prototype implementation in Haskell of a language for reasoning with and simulating the results of deontically specified actions (Wyner (2006a) and Wyner (2006b)). With the language, one can represent and study the outcomes of multi-agent artificial normative systems as agents execute actions over time.

1 Introduction

We consider the logical representation of obligations, sequences of actions, and the *Contrary to Duty* (CTD) Paradox. We agree with Carmo and Jones (2002) that the CTD problem is the key defining problem for deontic reasoning. Though Carmo and Jones (2002), which develops the Standard Deontic Logic (SDL)

 This work was prepared while the author was a postgraduate student at King's College London under the supervision of Tom Maibaum and Andrew Jones, which was funded by a studentship from Hewlett-Packard Research Labs, Bristol, UK. The author thanks Tom, Andrew, HP, and anonymous reviewers for their support and comments. Errors rest with the author.

L. Goble and J.-J.C. Meyer (Eds.): DEON 2006, LNAI 4048, pp. 255–271, 2006.

analysis, as well as Meyer (1988), which presents one version of the Dynamic Deontic Logic (DDL) analysis, claim to have solved the CTD problem, we argue that there are problems with each solution. We show that a better solution to the CTD problem must also resolve other interlocking issues in deontic reasoning – the relationship of obligations and sequences, the negation of actions, and the expression of violations and fulfillments. While each of these subtopics has been discussed in the literature (cf. Carmo and Jones (2002), Meyer (1988), Royakkers (1996), van der Meyden (1996), Khosla and Maibaum (1987), and Kent, Maibaum, and Quirk (1993)), the problems in the relationship between them have not been shown, an integrated solution has not been provided, nor has an implementation been given. In a companion paper (Wyner (2006b)) and related work (Wyner (2006a), we provide an integrated solution in an implementation – the *Abstract Action Calculator* (ACC). The ACC is a prototype language in which one can express deontic specifications on complex actions, show the results of executing actions relative to deontic specifications, as well as express inferential relationships between deontic specifications. With the ACC, one has a prototype program in which one can simulate executions of actions in multi-agent artificial normative systems. Our primary objective in this paper is to show the problems and sketch their solution.

We develop arguments to show the following. It is crucial to differentiate sequences of obligations from obligations on sequences, a distinction mentioned in Khosla and Maibaum (1987); we argue that one cannot be reduced to the other (and conflated in Meyer (1988)). The CTD problem *cannot* be reduced to an obligation on a sequence, and a sequence of obligations does *not* account for the CTD problem, contra Meyer (1988). Contra Carmo and Jones (2002) but agreeing with Meyer (1988) and Khosla and Maibaum (1987), the analysis of CTDs needs explicit state change and the concept of ideality is problematic. In contrast to Meyer (1988) and Khosla and Maibaum (1987), we claim that to be of use in practical deontic reasoning, negation of an action *cannot* be the complement set of actions from the domain of action, but is like the notion of *antonym* in natural language lexical semantics. Antonyms are opposites of one another, but otherwise undefined in opposition to other actions. In addition, though markers for violation and fulfillment have been proposed (Anderson and Moore (1957), Meyer (1988), Khosla and Maibaum (1987, van der Meyden (1996), and Carmo and Jones (2002)), we show that not only do they have a central role in guiding the process of deontic reasoning, but that we must have articulated, compositional, and productive markers. This later point has not, to our knowledge, previously been made in the literature.

To show how these issues relate, we introduce and discuss a problem related to what we call *Pörn's Criterion* (Pörn (1977:ix-x)), which states that it is critical to a *comprehensive* theory of deontic reasoning to take dynamic aspects into account. While sequences of obligations *do* involve a dynamic aspect, the more problematic cases are inferences relative to *contract state change*, where the set of deontic expressions change from context to context. We show that previous theories make the wrong inferences under these circumstances.

The layout of the paper is as follows. In the next section, narrow the scope of the presentation. Then we briefly outline elements of SDL and DDL. In the subsequent section, we present several problems, comparing the informal cases to formal analyses and a more desireable solution. In the final section, we very briefly sketch our implementation in the ACC, which addresses the problems raised in the paper.

2 Scope of Discussion

Having outlined our presentation, we should point out its scope. While we do discuss the CTD problem, we do not discuss the whole range of paradoxes of deontic logic, but see Wyner (2006a) where we argue that many of the paradoxes are not problems when given linguistically well-motivated semantic representations. In addition, our goal is a logic-based implemented prototype language which expresses deontic concepts as applied to complex actions. The role of logical analysis is to clarify issues and problems in the design of the language. Moreover, the langugage can be used to express *alternative* definitions of the concepts. With the implementation, we can simulate the execution of agentive actions relative to a contract. Thus, we are not giving a logic, and we have not addressed formal properties of the language such as completeness or decidability for two reasons. First, the choice of definitions in the language is still under discussion. Second, the formal properties which a *simulation* ought to satisfy is under intensive discussion (cf. Dignum, Edmonds, and Sonenberg (2004), Dignum and Sonenberg (2002), Edmonds (2002), and Fasli (2004)). While this paper expresses a view on the relation of logic, language, and simulation, it is beyond the scope of this paper to explicitly present it.

3 SDL and DDL

In this section, we very briefly review some basic elements of SDL and DDL with the goal to make clear the problems discussed in the following section.

3.1 SDL

Standard deontic logic (SDL) is the weakest normal modal system of type KD (in the Chellas classification). The theorems of KD are characterized by the smallest set of formulas of the propositional calculus together with propositional operators O and P and including the axioms K and D in (1). The formulas are closed under the rules of O-necessitation and Modus Ponens in (2). We suppose that O stands for the *obligation* operator and P for the *permission* operator; as we are not concerned with permission in this paper, we do not discuss it further.

Definition 1. *a.* ***PC****: All instances of tautologies of Propositional Calculus*
b. ***K****:* $O(A \rightarrow B) \rightarrow (OA \rightarrow OB)$
c. ***D****:* $(OA \rightarrow PA)$

Definition 2. *a.* ***O-necessitation:*** *$A \vdash OA$,*
where A is an axiom, theorem, or logical tautology
b. ***Modus Ponens:*** *$A, A \rightarrow B \vdash B$*

The models for the semantics of SDL are standard modal logic models. A model is *M = (W, R, V)*, where W is a non-empty set of worlds, R is a binary relation on worlds (the accessibility relation), and V is the valuation function which assigns sets of worlds to atomic sentences. V(p) denotes the set of worlds where p is true. The D schema is valid where the accessibility relation R is serial, where wRv is read as w is in the accessibility relation to v: ∀w ∃v wRv.

For deontic logic, we assume that the accessibility relation wRv means that v is a deontic alternative to w. Another way to say it is that wRv means that v is an ideal version of w; it is an *ideal* world. We have the definition of a formula A true in a world w of a model *M*, written $M \models_w$ A. For the deontic expressions, this means the following:

Definition 3. *Models for Obligated(A)*
$M \models_w$ Obligated(A) ***if and only if*** *∀v (**if** wRv, **then** $M \models_v$ A)*

This is to be understood informally as the formula Obligated(A) is true in world w if and only if A is true in all of the ideal versions v which are accessible from w. We also can understand that all those worlds in which A is false and are accessible from w are *subideal* worlds. Thus, relative to an obligation, a proposition, and a world, the accessible the worlds are *partitioned* between the ideal and subideal.

It is worth emphasizing that in SDL, there is no *context change*, even though there is evaluation of the truth of a proposition relative to alternative worlds. By the same token, there are no *actions* in the sense of Dynamic Logic (cf. Harel (2000)) which change contexts.

3.2 DDL

Meyer (1988) expresses the logic of obligation, permission, and prohibition on actions in dynamic logic. Dynamic logic is a very weak modal logic like K, but with extra axioms for actions. One of the key aspects of a dynamic logical system is that actions and assertions are strictly separated, which avoids paradoxes and counterintuitive propositions which appear in SDL (Meyer 1988p.109). In DDL, we have action names such as α and β, which are syntactic entities that we use to define atomic actions. The action names denote abstract semantic actions α' and β'. In DDL, there is no way to further specify properties of atomic actions or relations among them in terms of more basic attributes; there is no fine-grained structure to them. Complex actions are constructed from atomic (or complex) actions by action combinators. For example, given α and β are actions, then (α;β) is the sequence formed by first executing α and then executing β. Given an action name α, we may form the action $[\alpha]\phi$. We suppose that the action $[\alpha]\phi$ applies in a context in which the weakest preconditions defined by the action hold and results in a context in which the postconditions, here ϕ, hold. Action negation, the negation of α indicated by $\overline{\alpha}$, is largely given axiomatically; the negation of an action denotes the set of actions of the domain of actions other than the action.

The deontic notions apply directly to action names, but are *reduced to* an action and a violation marker. In Meyer (1988) this is the special propositional letter V (which first appeared in Anderson and Moore (1957) and is related to the the normative proposition of Khosla and Maibaum (1987)). Thus, given an arbitrary action α and a world where σ holds is, Obligation O, Prohibition F, and Permission P, appear as:

Definition 4. *a.* $\sigma \models F\alpha$ *iff* $\sigma \models [\alpha](V)$
b. $\sigma \models O\alpha$ *iff* $\sigma \models F\overline{\alpha}$ *iff* $[\overline{\alpha}](V)$
c. $\sigma \models P\alpha$ *iff* $\sigma \models \neg F\alpha$ *iff* $\neg[\alpha](V)$

Given the single marker of violation, V does not differentiate among who executed which action with respect to which deontic specification; that is, there are no distinctions among what follows should a violation hold.

It is a *theorem* of DDL that sequences of obligations are equivalent to obligations on sequences.

Theorem 1. $O(\alpha_1;\alpha_2) \equiv O(\alpha_1) \wedge [\alpha_1](O(\alpha_2))$

3.3 Comparisons

Carmo and Jones (2002) and Meyer (1988) both find fault with the approach of the other. Carmo and Jones (2002) claim that DDL does not handle deontic specifications on static expressions (but see d'Altan, Meyer, and Wieringa (1996) and Wyner (2004) for solutions). Meyer (1988) argues that SDL uses the problematic concept of *ideality*, incurs a host of paradoxes, and cannot accommodate context change. We tend to agree with Meyer (1988). However, we believe that SDL has maintained a key insight that is obscured in DDL, namely a reference to the context-sensitivity of secondary obligations. Moreover, we claim that SDL and DDL share similar problems, though to save space, we represent the issues in DDL.

4 The Problems

We focus the discussion on CTDs, which cover a range of different cases and issues. We consider three issues: changing deontic specifications of contexts; partitioning the action space; and the relationship between sequences and obligations. Following this discussion, we touch on a range of tangential issues before the formal issues. We discuss examples and relate them to formal theories to show the problems and indicate the issues which drive our analysis and implementation.

4.1 Changing Deontic Contexts

A Basic CTD Case. Suppose we have the following set of statements, which all hold consistently in one context (cf. Carmo and Jones (2002) for discussion of a broad range of issues that arise for CTD arguments). We comment below on our choice of CTD case.

Example 1. a. It is obligatory that Bill walk through Hyde Park.
b. If Bill does walk through Hyde Park,
then it is obligatory that Bill walk up Primrose Hill.
c. If Bill does not walk through Hyde Park,
then it is obligatory that Bill walk along
the South Bank Promenade.
d. Bill does not walk through Hyde Park.

The places mentioned here are various parks in London. We make the following assumptions to facilitate the discussion. First, let us assume that there are only four available actions in this model, one of which is not explicit in the examples: walking through Hyde Park, walking up Primrose Hill, walking along the South Bank Promenade, and walking through Finsbury Park. The parks are all distinct locations, none a part of the other. All of the locations are of finite extent, and execution of the action implies covering the space from end to end. Finally, we allow modest violations of space and time such that having walked through one park or the other, one is in a position to carry on walking in one of the other locations, as the case may be.

There is a clear intuition that from (1d) and (1c), we can infer:

Example 2. It is obligatory that Bill walk along the South Bank Promenade.

Additional intuitions are associated with the argument in (1). From (1a) and (1d), we may say that Bill has not done what he has been obligated to do. Let us assume that instead of walking through Hyde Park, he has walked through Finsbury Park. Thus, he has *violated* his obligation to walk through Hyde Park by walking through Finsbury Park. Thus, (2) is said to be a contrary-to-duty obligation, for it is an obligation which is implied where someone has done something which has violated another obligation. We can say that (1a) is the primary obligation and (2) is a secondary obligation, for it arises relative to the primary obligation. From the violation, other consequences may follow.

We may consider, counterfactually, had Bill instead have walked through Hyde Park, then it would follow, in such a circumstance, that he would be obligated to walk up Primrose Hill. In addition, in this counterfactual context, we would say that Bill had *fulfilled* his obligation. Consequences may follow from Bill's having fulfilled his obligation such as he is endowed with a novel permission. In the counterfactual context, we have a different secondary obligation. In Carmo and Jones (2002), this is called an *ideal obligation*, for it is the obligation which arises in the ideal circumstance; they devote considerable effort to providing an analysis which implies the ideal obligation in the subideal context. For our purposes, we do not discuss it further except to point out that different secondary obligations arise relative to the primary obligation. Furthermore, the different contexts are associated with fulfillment or violation of the primary obligation.

Carmo and Jones (2002), Meyer, Wieringa, and Dignum (1998), and Dignum (2004) discuss a range of CTD cases. The one we consider here is called the *forward* version of the Chisholm set by Meyer, Wieringa, and Dignum (1998). What they call the *parallel* version incorporates Forrester's Paradox (the gentle murderer) and the *backwards* version includes a secondary obligation which is temporally prior to the action which violates the primary obligation. We have not used the original Chisholm set of arguments, where we find a wide-scope deontic operator in the second argument (scope over the conditional) and a narrow-scope deontic operator in the third argument (scope over the consequent only)). In Wyner (2006a), we argue that when informed linguistic considerations are brought to bear on the basic expressions, we can provide well-motivated alternative logical forms which do not introduce paradoxes. In general, following the formal semantic analysis of Montague (1974), we should provide independent analyses for components of the CTD puzzles so as not to conflate issues. In other words, we should provide a semantics for adverbial modification, sequence of tense, and interaction of the conditional with modal operators. With these in place, we can provide a compositional and well-motivated analysis. However, we do not discuss these alternatives further here.

An Analysis. Consider an analysis of (1). Let us assume four actions α, β, γ, and δ, which we associate with the four actions of Bill's walking through Hyde Park, up Primrose Hill, along the South Bank Promenade, and through Finsbury Park, respectively. In DDL, we can represent the CTD problem as:

Example 3. a. $O(\alpha)$
b. $[\alpha](O(\beta))$
c. $[\overline{\alpha}](O(\gamma))$
d. Execute an action which is an element of $\overline{\alpha}$.

We assume, for the moment, that $\overline{\alpha}$ denotes a set of actions, namely the set of actions of the domain of actions other than α. Thus, executing an action that is an element of $\overline{\alpha}$ is well-formed. Furthermore, the execution of an action in $\overline{\alpha}$ means that in the subsequent context, V holds (since the obligation on α has been violated) and also that $O(\gamma)$ holds. Let us assume that, certeris paribus, what held in the precondition context holds in the postcondition context as well, unless this results in inconsistency. We assume, then, that $O(\alpha)$ and $[\alpha](O(\beta))$ hold in the postcondition context as well. In addition, given the equivalence in (1), we have an obligation on a sequence: $O(\alpha;\beta)$. So far as we are aware, Meyer (1988) is not concerned with *ideal obligations*.

Changing the Primary Obligation. Let us consider what we can infer should we have the following case, where we have an obligation on Bill's walking through Finsbury Park, which is another park in London. This is part of *Pörn's Criterion* in that we consider what follows from a changed set of deontic specifications.

Example 4. a. It is obligatory that Bill walk through Finsbury Park.
b. If Bill does walk through Hyde Park,
then it is obligatory that Bill walk up Primrose Hill.
c. If Bill does not walk through Hyde Park,
then it is obligatory that Bill walk along
the South Bank Promenade.
d. Bill does not walk through Hyde Park.

We can represent this set of arguments as follows.

Example 5. a. $O(\delta)$
b. $[\alpha](O(\beta))$
c. $[\overline{\alpha}](O(\gamma))$
d. Execute an action which is an element of $\overline{\alpha}$.

After execution of the action and from this set of arguments, we may infer that $(O(\gamma))$ holds. As δ is an element of $\overline{\alpha}$, the action executed could be δ, in which case, we infer that V does *not* hold; but, since $\overline{\delta}$ and $\overline{\alpha}$ intersect, some action *other* than δ could be executed, in which case V *does* hold. Morever, since δ is an element of $\overline{\alpha}$, we can infer that $O(\delta;\gamma)$ holds as well.

Whatever the logical consequences, it seems intuitively unreasonable to infer that Bill's not walking through Hyde Park should be considered a violation or fulfillment relative to his obligation to walk through Finsbury Park. Rather, we are *indeterminate* as to whether we are in a context where Bill has violated his primary obligation or fulfilled it. Along the same lines, it seems unreasonable to say that the secondary obligation which is introduced, Bill's walking along the South Bank Promenade, is a contrary-to-duty obligation, which is essentially an obligation which is inferred in a context where a primary obligation has been violated. By the same token, we should not say that it follows from fulfillment of a primary duty, where Bill's not walking through Hyde Park was Bill's walking through Finsbury Park.

Finally, it seems intuitively odd to infer that we have an obligation on a sequence in this example:

Example 6. It is obligatory that Bill walk through Finsbury Park, and then walk along the South Bank Promenade.

Our problem is not with the logic *per se*, but that the logic does not correlate with our intuitions; the logic allows us to draw *definite* inferences which do not seem intuitively plausible. As an abstraction, the logic may serve a purpose, but as a model of natural legal reasoning, it seems to be overdetermined. It does not seem that (1) and (4) should follow the same reasoning patterns. It appears that the *link* between the primary and secondary obligations as well as the violation marker in (1) has been broken in (4). We see that the issue arises where we change primary deontic specifications, a problem neither SDL nor DDL seem to have previously accounted for (cf. Khosla and Maibaum (1987) who mention it, but do not provide an account).

4.2 Partitioning the Action Space

The reason the cases are treated alike is related to a second issue. Suppose one obligation and our four actions.

Example 7. It is obligatory that Bill walk up Primrose Hill.

Given definitions of obligation and action negation, it seems clear that Bill's walking up Primrose Hill *fulfills* the obligation, while *any other action violates it.* In other words, no matter what Bill does, he either violates or fulfills his obligation. By the same token, as soon as Bill *incurs* his obligation, and supposing he must do something, then he immediately induces either a violation or a fulfillment. Similarly, were Bill to bear two obligations which could not both be simultaneously satisfied, then he is sure to violate one or the other. While in an abstract and theoretical domain, these conclusions might not seem unreasonable, in any application or for real world reasoning, they are untenable.

The underlying reason for the problem is that, relative to a particular obligation, one can consider the domain of actions *partitioned* between those actions which fulfill the obligation (the explicitly given action) and those which violate it (the complement set of the given action) (cf. Meyer (1988) and Royakkers (1996)). There are no actions which are underspecified with respect to the obligation such that executing that action induces neither a violation, nor a fulfillment. But consider a case where one were obligated to deliver pizzas for an hour. It is reasonable that delivering 4 pizzas, one every 15 minutes, counts as fulfilling the obligation, while delivering no (prepared) pizzas in the hour counts as violating it. However, we do not want just *any* action to count as *not delivering pizzas for an hour.* For example, eating an apple or many other actions would seem to be deontically underspecified. We want a more refined abstract analysis.[1]

While it is relatively elementary to determine which actions count towards violation or fulfillment where we have deontic specifications on *atomic* actions, the issues are more significant where we consider deontic specification on *complex* actions such as obligations on sequences. Given an obligation on a sequence, exactly which actions count as *not* executing the sequence such that a violation arises? As complex actions are *compositional* (the meaning of the action arising from the meanings of the component actions and the mode of combination) and *productive* (the processes apply to *any appropriate actions*), then negation on an action must also be sensitive to compositionality and productivity.

The objective is to be able to *calculate* the denotation of $\overline{\alpha}$ for any action α. For our purposes here, we can say that $\overline{\alpha}$ denotes a proper subset of the domain of actions minus α. Thus, there may be actions which are not deontically specified relative; that is, in $\mathrm{O}(\alpha)$, there are actions which lead to fulfillment, others which lead to violation, and others which are underspecified either way. One way to calculate such a denotation is given in the implementation (cf. Wyner (2006a) for the specifics).

4.3 Obligations on Sequences and Sequences of Obligations

Our third consideration is the relationship between an obligation on a sequence (OOS) as in (8a) and a sequence of obligations (SOO) as in (8b).

[1] One could take the tack that *every* deontic specification has *some* temporal specification such that other actions can be executed without inducing violation or fulfillment (cf. Dignum (2004)). However, in Wyner (2006a), we argue that *aspectual* distinctions (as in the previous example) rather than temporal extents are key.

Example 8. a. It is obligatory that Bill walk through Hyde Park, then walk up Primrose Hill.
b. It is obligatory that Bill walk through Hyde Park. After having walked up Hyde Park, it is obligatory that Bill walk up Primrose Hill.

Meyer (1988) and Royakkers (1996) prove that (8a) and (8b) are *equivalent* in DDL. In contrast, Khosla and Maibaum (1987) claim they are distinct – an obligation on a sequence is an obligation on an action which is *distinct* from its components. They introduce distinct deontic operators for complex expressions which are not equivalent to deontic operators on the parts. However, they do not discuss the problem it introduces for Meyer's analysis. We are unaware of any subsequent discussion of the distinction. There are several ways to implement this idea, and we give one example, though the implementation allows a range of alternatives.

One way to see that an OOS and a SOO are not equivalent is to consider *consequences* which follow from violation or fulfillment, what we refer to as the *violation conditions*: if OOSs and SOOs are equivalent, then they must have the same violation conditions. We can show that they do not, so they are not equivalent.[2]

Suppose the following, where we have co-indexed expressions of the form *It is obligatory P*, for P a proposition, and *this obligation* to make it clear exactly which obligation we are referring to. We suppose that the statements in (9) hold in one context and (10) hold in another.

Example 9. a. [It is obligatory that Bill walk through Hyde Park.]$_i$
If he fulfills [this obligation]$_i$, then he gets paid £3.
If he violates [this obligation]$_i$, then he owes £3.
b. [It is obligatory that Bill walk up Primrose Hill.]$_j$
If he fulfills [this obligation]$_j$, then he gets paid £2.
If he violates [this obligation]$_j$, then he owes £2.

Example 10. [It is obligatory that Bill walk through Hyde Park, then walk up Primrose Hill.]$_k$
If he fulfills [this obligation]$_k$, then he gets paid £10.
If he violates [this obligation]$_k$, then he owes £10.

We assume the middle two arguments of the CTD set in (1) which introduce secondary obligations. Suppose (9). Bill first walks through Hyde Park, for which he gets paid £3 as he has fulfilled his obligation to walk through Hyde Park. Given the CTD set, he incurs a secondary obligation to walk up Primrose Hill.

---

[2] An anonymous reviewer provided another case where an obligation on a sequence and a sequence of obligations are not equivalent. Suppose $O(\alpha;\beta)$. If the execution of α is such that the execution of β is no longer obligated, then the obligation on the sequence cannot be equivalent to the sequence of obligations given by $O(\alpha) \wedge [\alpha](O(\beta))$, where the obligation on β *does follow* the execution of α. This is entirely within the spirit of our analysis.

He does not walk up Primrose Hill, so he incurs a penalty, and loses £2. In the end, he has £1. Alternatively, consider that Bill executes the same sequence of actions relative to (10). In this case, it is plausible that Bill ends up with *nothing* because it is clear that he has not fulfilled his obligation on the sequence *per se.* One could jigger the cases to suit, but the basic point is clear: there is no *logical* relationship between OOS and SOO in terms of what follows from violations and fulfillments. Meyer (1988) assumes that there is; as this is a *proven* equivalence, it indicates a deeper problem with his analysis. What Meyer (1988) misses is that the deontic specifier can apply to the *compositional meaning of sequences*, which are not necessarily reducible to their parts.

Notice that the key *tool* we have used to make the argument are *articulated* violation and fulfillment markers. They are, in our view, the key for deontic reasoning. This is in contrast to Meyer (1988) and Khosla and Maibaum (1987), where there is only one *atomic* violation and fulfillment proposition. In these systems, the same consequences follow from a violation no matter who is the agent, what is the action, or what is the deontic operator. In an application, this is infeasible. Others have suggested articulating the markers, though not with the same motivations or of the same form as in our implementation (cf. Kent, Maibaum, and Quirk (1993), van den Meyden (1996), Dignum (2004), among others).

To formalize the analysis, let us suppose that rather than one violation proposition, we have as many violations as there are basic and complex actions, indicating each with a subscript as in (Vio_{O_α}), which is read *The obligation to execute α has been violated.* We could further articulate this with respect to the agent (cf. Wyner (2006a) along the lines of *Bill has violated his obligation to execute α.* For clarity, we also assume there are fulfillment markers, (Ful_{O_α}), which are not equivalent to the negation of the violation markers. Fulfillent markers allow us to explicitly reason with what follows from satisfaction of a deontic specification such as rewards. In addition, we explicitly introduce the conditions under which an obligation is fulfilled. We have the conditions for obligations on basic actions as follows.

Definition 5. *$O(\alpha) =_{def} [\alpha](Ful_{O_\alpha}) \wedge [\overline{\alpha}](Vio_{O_\alpha})$, where α is a basic action.*

We read this as saying that an obligation on an action α holds where execution of α leads to a context where the proposition *The obligation on the execution of α has been fulfilled* holds, and where execution of an action among $\overline{\alpha}$ leads to a context where the proposition *The obligation on the execution of α has been violated* holds. Where we understand $\overline{\alpha}$ to be some specified proper subset of actions from the domain of actions, then (5) allows for the execution of actions which neither induce fulfillment or violation of the obligation.

Next, we assume that we can make our deontic operators *sensitive to the compositional structure of the complex action.* As we discuss in Wyner (2006a), there are many possible alternative analyses of the operators. We present two versions of obligations on sequences. The first is along the lines of Meyer, where an obligation operator on a sequence *distributes* the basic operator to each of the component actions; in this case, the violation conditions are per component action. We give this as an *distributed obligation on a sequence* – O_{dis}. This is the version closest to Meyer's (1988) definition.

Definition 6. $O_{dis}(\alpha;\beta) =_{def}$
$O(\alpha) \wedge [\alpha](O(\beta))$

The second version allows the sequence to be broken up into parts (so is distinct from Khosla and Maibaum's (1987) obligation on sequences operator), yet there are violation flags relative to the sequence per se. We can indicate at which points of the sequence violation or fulfillment flags are introduced. We give this as an *obligation on an interruptable sequence* – O_{int}.

Definition 7. $O_{int}(\alpha;\beta) =_{def}$
$[\overline{\alpha}](V_{O(\alpha;\beta)}) \wedge [\alpha]([\overline{\beta}](V_{O(\alpha;\beta)}) \wedge [\beta](F_{O(\alpha;\beta)}))$

$O_{dis}(\alpha;\beta)$ is not equivalent to $O_{int}(\alpha;\beta)$, for different violation conditions arise in each; nor does one imply the other, by the same token. A sequence of obligations and an obligation on a sequence are not necessarily equivalent.

Once one allows complex violation and fulfillment markers and variants of the deontic operators, we can define the deontic operators in a variety of ways.

4.4 Inference

Let us consider inference with respect to definitions along the lines of (5) and (7). One advantage of the analysis is that it allows us to provide rules about what follows from *particular* violations (or fulfillments). For example, what follows from (Vio_{O_α}) may be distinct from what follows from (Vio_{O_β}). Were we to add additional distinctions in terms of agents and in terms of the particular deontic operator, as we do in the implementation, we would make fine-grained inferences based on who violated what deontic specification on which action; legal systems have this flavor.

However, such an analysis does not, in and of itself, allow for inferential relationships *between* deontically specified actions exactly because the reductions to actions and violation flags is so specific. For instance, where $O(\alpha)$ holds, then we would like to infer that it is prohibited to not do α, which we can represent as $Pr(\overline{\alpha})$. However, the reduction of $O(\alpha)$ does not itself allow this inference. Nor, by the same token, does the analysis account for consistency of deontic specifications. Indeed, there is nothing inconsistent with $O(\alpha)$ and $Pr(\alpha)$ since they are reduced to distinct expressions.

The solution, given fuller expression in the implementation (Wyner (2006a)), is to provide an additional, finite set of *lexical axioms* to ensure such inferences. These are in the spirit of the *Meaning Postulates* of Montague (1974), which are introduced to restrict the class of admissible models. In addition, we discuss negation of deontic specifications and consistency so as to provide consistent sets of deontic specifications. The resultant language may be less general and abstract that deontic logics, but it also avoids the overgeneration of problematic inferences (i.e. the paradoxes). However, it is beyond the scope of this paper to discuss this further.

4.5 A Summary

By way of summarizing our observations, we outline of our approach to CTDs, using enriched markers for violation conditions. First, in contrast to Carmo and Jones (2002) and Meyer (1988), secondary obligations do *not* follow from actions (or propositions) directly, but rather from the articulated violation or fulfillment markers themselves; we make the markers syntactically active in the language.

Example 11. a. It is obligatory that Bill walks through Hyde Park
b. If Bill has fulfilled his obligation to walk through Hyde Park, then it is obligatory that Bill walk up Primrose Hill.
c. If Bill has violated his obligation to walk through Hyde Park, then it is obligatory that Bill walk along the South Bank Promenade.
d. Bill does not walk through Hyde Park.

And in our logical language, this appears as follows.

Example 12. a. $O(\alpha)$
b. $Ful_{O_\alpha} \rightarrow (O(\beta))$
c. $Vio_{O_\alpha} \rightarrow (O(\gamma))$
d. Execute an action which is an element of $\overline{\alpha}$.

Bill's not walking through Hyde Park implies that Bill has violated his obligation to walk through Hyde Park (from (5)). Furthermore, this violation implies that Bill is obligated to walk along the South Bank Promenade, which is the secondary obligation. We can similarly calculate a secondary obligation in a counterfactual context where Bill does walk through Hyde Park. In contrast to Meyer (1988), we cannot infer from (12) that *any* sequence of obligations or obligation on a sequence holds; we view this as an important result.

Let us then consider the case where we *change the primary obligation*, which we represent as follows.

Example 13. a. It is obligatory that Bill walks through Finsbury Park
b. If Bill has fulfilled his obligation to walk through Hyde Park, then it is obligatory that Bill walk up Primrose Hill.
c. If Bill has violated his obligation to walk through Hyde Park, then it is obligatory that Bill walk along the South Bank Promenade.
d. Bill does not walk through Hyde Park.

And in our logical language, this appears as follows.

Example 14. a. $O(\delta)$
b. $Ful_{O_\alpha} \rightarrow (O(\beta))$
c. $Vio_{O_\alpha} \rightarrow (O(\gamma))$
d. Execute an action which is an element of $\overline{\alpha}$.

If Bill's not walking through Finsbury Park is δ, then Bill fulfills his obligation on δ; if Bill's not walking through Finsbury Park is some action from $\overline{\delta}$, then Bill violates his obligation on δ. However, given the articulated violation and fulfillment markers, these violation and fulfillment markers are *distinct* from those for an obligation on α. Thus, in the argument in (14), we *cannot infer any secondary obligations*. This analysis comports much better with our intuitions about (4). We can refine the analysis further if we suppose that $\overline{\alpha}$ is a functionally specified particular action rather than any action other than α; suppose it is β. Similarly, let us suppose that $\overline{\delta}$ is γ. If this is the case, then from (14) we cannot make *any* further inferences. As we argued earlier, this seems to be intuitively the case in (4).

The articulated violation and fulfillment markers plus a lexical semantic approach to action negation allow an accurate analysis of CTD cases where we change the primary obligation. In our view, we have made productive and overt use of metatheoretical goal of *ideality*, though we have not used ideality anywhere in the analysis. In our view, the goal is to make inferences to secondary obligations *depend* on properties of the context which are ascribed relative to the *primary obligation*. Informally, it is not an action itself which induces the secondary obligations, but the action *in relation to its deontic specification*.

4.6 Additional Issues

We have not discussed a host of issues relating to deontic reasoning in general or CTDs in particular (cf. Wyner (2006a)). We mention some of these in order to put them aside. We have focussed on one construction using *It is obligatory that*; there are other constructions or lexical items. For example, one can consider the relative scope of the deontic operator and elements of the conditional. One can consider *ought* rather than *obligatory*. There are many other CTD cases to consider such as the *Considerate Assassin* and *Reykjavik*. One can employ alternative axioms to account for implications. We have argued (op. cit.) that much of this is not relevant for our purposes or have provided other (linguistic) reasons to set the issues aside.

5 A Reference to the Implementation

In Wyner (2006a) and Wyner (2006b), we implement our language of deontic specifications on complex actions so as to take into account the problems and analyses outlined above. Here we mention a few of the aspects of the program so as to relate it to this paper. The implementation, *Abstract Contract Calculator* (ACC), has been written in Haskell, which is a functional programming language, (cf. Wyner (2006a) for the code and documentation; on Haskell and Computational Semantics, see Doets and van Eijck (2004) and van Eijck (2004)). The ACC processes the deontic notions of *prohibition*, *permission*, and *obligation* applied to complex actions. The ACC is a prototype language and not a deontic logic. Alternative notions of actions, complex actions, and deontic specification can be systematically examined and animated. The tool enables us to abstractly simulate environments in which agents behave relative to actions, contract states, and contract changes. It is intended to be used for simulation and modelling of Multi-Agent systems where deontic specifications govern the behavior of individual or collectives of agents (see Gilbert and Troitzsch (2005) for a discussion of social science simulations).

The ACC has the following modules. We define *States of Affairs* (SOAs), which are consistent lists of propositions along with indices for worlds and times. *Basic Actions* are functions from SOAs to SOAs. An action is executed where the preconditions of the action are satisfied and the postconditions do not induce inconsistency. Given abstract actions, we may define exactly which propositions are changed from context to context, and otherwise leave propositions unaffected (inertia). We have *Lexical Semantic Functions* to allow us to *calculate* actions in the lexical semantic relation of *opposition*. In particular, we observe that for the purposes of deontic specification, actions in opposition must be executable *in the same context*. The analogy to natural language is that if one is obligated to leave a room, then leaving the room fulfills the obligation, while remaining in the room violates it; leaving the room and remaining in the room have the same preconditions, but different postconditions. Thus, we can calculate the denotation of a proper subset of actions which are reasonably construed to be in opposition. Alternative formulations are possible. *Deontic Operators* apply to actions so as to specify what actions lead to contexts where fulfillment or violation is marked relative to the action and agent. We call such a specification a *Contract Flag State*. When an action is executed, we check whether the action is deontically specified in the contract flag state. If it is, then we record in a history not only that the action has been executed, but the *value* of the execution of the action relative to the deontic specification. For example, if it is prohibited to execute α, then executing α implies that a violation of that prohibition holds in the history after the execution. We implement reasoning for *Contrary-to-Duty Obligations* by modifying contract flag states relative to violation or fulfillment *flags*. For example, if we find in the history that the prohibition to execute α has been violated, then that may *trigger* contract state modification, for example by adding an obligation to the contract flag state. In the course of the discussion, we introduce consistency constraints on contracts. To provide fine-grained analyses of deontic specifications on complex actions, we provide a structure in which the input actions and output actions are available for deontic specification. With this, we may distinguish between deontic specification on the *parts* of a complex action from deontic specification on the *whole*. Thus, we can implement the alternative definitions of obligations on sequences or sequences of obligations as above. In Wyner (2006b), we discuss related proposals in the literature.

6 Conclusion

We have argued for an alternative approach to dynamic deontic logic. It allows articulated violation and fulfillment markers as well as a spectrum of deontic specifications on complex actions. We have also argued for an alternative approach to action negation, drawing on analogies to natural language antonyms.

We have shown how our analysis provides a better approach to CTDs, particularly where we consider alternative primary obligations, which is crucial given *Pörn's Criterion*; when primary obligations change and all else remains the same, we want to ensure that we make intuitively plausible inferences. Finally, we have given some indication of how our analysis is implemented in Haskell so as to allow simulations of agentive behavior relative to contract specifications.

References

Anderson, A., Moore, O.: The Formal Analysis of Normative Concepts. The American Sociological Review. **22** (1957) 9-17

Carmo, J., Jones, A.: Deontic Logic and Contrary-to-duties. In D. Gabbay and Franz Guenthner (eds.) Handbook of Philosophical Logic, Dordrecht: Kluwer Academic Publishers, (2002)

d'Altan, P., Meyer, J.-J.Ch., Wieringa, M.: An integrated framework for ought–to–be and ought–to–do constraints. Artificial Intelligence and Law. **4** (1996) 77–111

Dignum, F., Edmonds, B., and Sonenberg, L.: Editorial: The Use of Logic in Agent-Based Social Simulation. Journal of Artificial Societies and Social Simulation. **7.4** (2004). http://jasss.soc.surrey.ac.uk/7/4/8.html

Dignum, F. and Sonenberg, L.: A Dialogical Argument for the Usefulness of Logic in MAS. RASTA'02 (Proceedings of the International Workshop on Regulated Agent-Based Social Systems: Theories and Applications). See http://jasss.soc.surrey.ac.uk/7/4/8.html

Dignum, V.: A Model for Organizational Interaction: Based on Agents, Founded in Logic. Ph.D. Thesis, Utrecht University, Netherlands.

Dowty, D.: Word Meaning and Montague Grammar. Dordrecht, Holldand: Reidel Publishing Company (1979)

Edmonds, B.: Comments on A Dialogical Argument for the Usefulness of Logic in MAS. RASTA'02 (Proceedings of the International Workshop on Regulated Agent-Based Social Systems: Theories and Applications). See http://jasss.soc.surrey.ac.uk/7/4/8.html

Fasli, M.: Formal Systems $\wedge$ Agent-Based Social Simulation = $\perp$?. Journal of Artificial Societies and Social Simulation. **7.4** (2004). http://jasss.soc.surrey.ac.uk/7/4/8.html

Harel, D., Kozen, D., and Tiuryn, J.: Dynamic Logic. Cambridge, MA: The MIT Press (2000)

Doets, K., van Eijck, J.: The Haskell Road to Logic, Maths and Programming, London: King's College Publications, (2004)

van Eijck, J.: Computational Semantics and Type Theory. Website download – http://homepages.cwi.nl/ jve/cs/, (2004)

Gilbert, N., Troitzsch, K.: Simulation for the Social Scientist, London, UK: Open University Press, (2005)

Jones, A., Sergot, M.: On the Characterisation of Law and Computer Systems: the Normative Systems Perspective. In J.-J.Ch Meyer and R.J. Wieringa (eds.) Deontic Logic in Computer Science – Normative System Specification. Wiley (1993) 275-307

Kent, S., Maibaum, T., and Quirk, W.: Formally Specifying Temporal Contraints and Error Recovery. In Proceedings of the IEEE International Symposium on Requirements Engineering, IEEE C.S. Press, 208-215

Khosla, S., Maibaum, T.: The Prescription and Description of State-Based Systems. In B. Banieqbal, H. Barringer, and A. Pneuli (eds.) Temporal Logic in Specification. Springer-Verlag (1987) 243-294

Meyden, R. v. d.: The Dynamic Logic of Permission. Journal of Logic and Computation. **6** (1996) 465-479

Meyer, J.-J.Ch.: A Different Approach to Deontic Logic: Deontic Logic Viewed as a Variant of Dynamic Logic. Notre Dame Journal of Formal Logic. **1** (1988) 109-136

Meyer, J.-J.Ch.: Dynamic Logic for Reasoning about Actions and Agents. In J. Minker (ed.) Workshop on Logic-Based Artificial Intelligence. Washington, DC. (1999)

Meyer, J.-J.Ch., Wieringa, R.J.: Actors, Actions, and Initiative in Normative System Specification. Annals of Mathematics and Artificial Intelligence. **7** (1993) 289-346

Meyer, J.-J.Ch., Wieringa, R.J., and Dignum, F.: The Role of Deontic Logic in the Specification of Information Systems. In *Logics for Databases and Information Systems.* (1998) 71-115.

Montague, R.: Formal Philosophy: Selected Papers of Richard Montague. R. Thomason (ed.), New Haven, Yale University Press, (1974)

Penner, J., Schiff, D., Nobles, R. (eds.): Introduction to Legal Theory and Jurisprudence: Commentary and Materials. London, Buttersworth Law (2002)

Pörn, P.: Action Theory and Social Science: Some Formal Models, Dordrecht: Reidel (1977)

Royakkers, L.: Representing Legal Rules in Deontic Logic. Ph.D. Thesis, Katholieke Universiteit Brabant, Tilburg (1996)

Sergot, M.: A Computational Theory of Normative Positions. ACM Transactions on Computational Logic. **2** (2001) 581-622

Wieringa, R.J., Meyer, J.: Deontic Logic in Computer Science: Normative System Specification. John Wiley and Sons (1993)

Wyner, A.Z.: Violations and Fulfillments in the Formal Representation of Contracts. ms King's College London, Department of Computer Science, submitted for the Ph.D. in Computer Science, (2006a)

Wyner, A.Z.: A Functional Program for Agents, Actions, and Deontic Specifications. In Ulle Endriss (ed.) Proceedings of the Workshop on Declarative Agent Language Technologies. AAMAS'06, Hakodate, Japan, May 7-12, 2006. (2006b)

Wyner, A.Z.: Maintaining Obligations on Stative Expressions in a Deontic Action Logic. In A. Lomuscio and D. Nute (eds.) Deontic Logic in Computer Science. Springer (2004), 258-274

Author Index

Vol. 3835: G. Sutcliffe, A. Voronkov (Eds.), Logic for Programming, Artificial Intelligence, and Reasoning. XIV, 744 pages. 2005.

Vol. 3830: D. Weyns, H. V.D. Parunak, F. Michel (Eds.), Environments for Multi-Agent Systems II. VIII, 291 pages. 2006.

Vol. 3817: M. Faundez-Zanuy, L. Janer, A. Esposito, A. Satue-Villar, J. Roure, V. Espinosa-Duro (Eds.), Nonlinear Analyses and Algorithms for Speech Processing. XII, 380 pages. 2006.

Vol. 3814: M. Maybury, O. Stock, W. Wahlster (Eds.), Intelligent Technologies for Interactive Entertainment. XV, 342 pages. 2005.

Vol. 3809: S. Zhang, R. Jarvis (Eds.), AI 2005: Advances in Artificial Intelligence. XXVII, 1344 pages. 2005.

Vol. 3808: C. Bento, A. Cardoso, G. Dias (Eds.), Progress in Artificial Intelligence. XVIII, 704 pages. 2005.

Vol. 3802: Y. Hao, J. Liu, Y.-P. Wang, Y.-m. Cheung, H. Yin, L. Jiao, J. Ma, Y.-C. Jiao (Eds.), Computational Intelligence and Security, Part II. XLII, 1166 pages. 2005.

Vol. 3801: Y. Hao, J. Liu, Y.-P. Wang, Y.-m. Cheung, H. Yin, L. Jiao, J. Ma, Y.-C. Jiao (Eds.), Computational Intelligence and Security, Part I. XLI, 1122 pages. 2005.

Vol. 3789: A. Gelbukh, Á. de Albornoz, H. Terashima-Marín (Eds.), MICAI 2005: Advances in Artificial Intelligence. XXVI, 1198 pages. 2005.

Vol. 3782: K.-D. Althoff, A. Dengel, R. Bergmann, M. Nick, T.R. Roth-Berghofer (Eds.), Professional Knowledge Management. XXIII, 739 pages. 2005.

Vol. 3763: H. Hong, D. Wang (Eds.), Automated Deduction in Geometry. X, 213 pages. 2006.

Vol. 3755: G.J. Williams, S.J. Simoff (Eds.), Data Mining. XI, 331 pages. 2006.

Vol. 3735: A. Hoffmann, H. Motoda, T. Scheffer (Eds.), Discovery Science. XVI, 400 pages. 2005.

Vol. 3734: S. Jain, H.U. Simon, E. Tomita (Eds.), Algorithmic Learning Theory. XII, 490 pages. 2005.

Vol. 3721: A.M. Jorge, L. Torgo, P.B. Brazdil, R. Camacho, J. Gama (Eds.), Knowledge Discovery in Databases: PKDD 2005. XXIII, 719 pages. 2005.

Vol. 3720: J. Gama, R. Camacho, P.B. Brazdil, A.M. Jorge, L. Torgo (Eds.), Machine Learning: ECML 2005. XXIII, 769 pages. 2005.

Vol. 3717: B. Gramlich (Ed.), Frontiers of Combining Systems. X, 321 pages. 2005.

Vol. 3702: B. Beckert (Ed.), Automated Reasoning with Analytic Tableaux and Related Methods. XIII, 343 pages. 2005.

Vol. 3698: U. Furbach (Ed.), KI 2005: Advances in Artificial Intelligence. XIII, 409 pages. 2005.

Vol. 3690: M. Pěchouček, P. Petta, L.Z. Varga (Eds.), Multi-Agent Systems and Applications IV. XVII, 667 pages. 2005.

Vol. 3684: R. Khosla, R.J. Howlett, L.C. Jain (Eds.), Knowledge-Based Intelligent Information and Engineering Systems, Part IV. LXXIX, 933 pages. 2005.

Vol. 3683: R. Khosla, R.J. Howlett, L.C. Jain (Eds.), Knowledge-Based Intelligent Information and Engineering Systems, Part III. LXXX, 1397 pages. 2005.

Vol. 3682: R. Khosla, R.J. Howlett, L.C. Jain (Eds.), Knowledge-Based Intelligent Information and Engineering Systems, Part II. LXXIX, 1371 pages. 2005.

Vol. 3681: R. Khosla, R.J. Howlett, L.C. Jain (Eds.), Knowledge-Based Intelligent Information and Engineering Systems, Part I. LXXX, 1319 pages. 2005.

Vol. 3673: S. Bandini, S. Manzoni (Eds.), AI*IA 2005: Advances in Artificial Intelligence. XIV, 614 pages. 2005.

Vol. 3662: C. Baral, G. Greco, N. Leone, G. Terracina (Eds.), Logic Programming and Nonmonotonic Reasoning. XIII, 454 pages. 2005.

Vol. 3661: T. Panayiotopoulos, J. Gratch, R. Aylett, D. Ballin, P. Olivier, T. Rist (Eds.), Intelligent Virtual Agents. XIII, 506 pages. 2005.

Vol. 3658: V. Matoušek, P. Mautner, T. Pavelka (Eds.), Text, Speech and Dialogue. XV, 460 pages. 2005.

Vol. 3651: R. Dale, K.-F. Wong, J. Su, O.Y. Kwong (Eds.), Natural Language Processing – IJCNLP 2005. XXI, 1031 pages. 2005.

Vol. 3642: D. Ślęzak, J. Yao, J.F. Peters, W. Ziarko, X. Hu (Eds.), Rough Sets, Fuzzy Sets, Data Mining, and Granular Computing, Part II. XXIII, 738 pages. 2005.

Vol. 3641: D. Ślęzak, G. Wang, M. Szczuka, I. Düntsch, Y. Yao (Eds.), Rough Sets, Fuzzy Sets, Data Mining, and Granular Computing, Part I. XXIV, 742 pages. 2005.

Vol. 3635: J.R. Winkler, M. Niranjan, N.D. Lawrence (Eds.), Deterministic and Statistical Methods in Machine Learning. VIII, 341 pages. 2005.

Vol. 3632: R. Nieuwenhuis (Ed.), Automated Deduction – CADE-20. XIII, 459 pages. 2005.

Vol. 3630: M.S. Capcarrère, A.A. Freitas, P.J. Bentley, C.G. Johnson, J. Timmis (Eds.), Advances in Artificial Life. XIX, 949 pages. 2005.

Vol. 3626: B. Ganter, G. Stumme, R. Wille (Eds.), Formal Concept Analysis. X, 349 pages. 2005.

Vol. 3625: S. Kramer, B. Pfahringer (Eds.), Inductive Logic Programming. XIII, 427 pages. 2005.

Vol. 3620: H. Muñoz-Ávila, F. Ricci (Eds.), Case-Based Reasoning Research and Development. XV, 654 pages. 2005.

Vol. 3614: L. Wang, Y. Jin (Eds.), Fuzzy Systems and Knowledge Discovery, Part II. XLI, 1314 pages. 2005.

Vol. 3613: L. Wang, Y. Jin (Eds.), Fuzzy Systems and Knowledge Discovery, Part I. XLI, 1334 pages. 2005.

Vol. 3607: J.-D. Zucker, L. Saitta (Eds.), Abstraction, Reformulation and Approximation. XII, 376 pages. 2005.

Vol. 3601: G. Moro, S. Bergamaschi, K. Aberer (Eds.), Agents and Peer-to-Peer Computing. XII, 245 pages. 2005.

Vol. 3600: F. Wiedijk (Ed.), The Seventeen Provers of the World. XVI, 159 pages. 2006.

Vol. 3596: F. Dau, M.-L. Mugnier, G. Stumme (Eds.), Conceptual Structures: Common Semantics for Sharing Knowledge. XI, 467 pages. 2005.

Lecture Notes in Artificial Intelligence (LNAI)

Vol. 4048: L. Goble, J.-J.C. Meyer (Eds.), Deontic Logic and Artificial Normative Systems. X, 273 pages. 2006.

Vol. 4045: D. Barker-Plummer, R. Cox, N. Swoboda (Eds.), Diagrammatic Representation and Inference. XII, 301 pages. 2006.

Vol. 4031: M. Ali, R. Dapoigny (Eds.), Innovations in Applied Artificial Intelligence. XXIII, 1353 pages. 2006.

Vol. 4027: H.L. Larsen, G. Pasi, D. Ortiz-Arroyo, T. Andreasen, H. Christiansen (Eds.), Flexible Query Answering Systems. XVIII, 714 pages. 2006.

Vol. 4021: E. André, L. Dybkjær, W. Minker, H. Neumann, M. Weber (Eds.), Perception and Interactive Technologies. XI, 217 pages. 2006.

Vol. 4020: A. Bredenfeld, A. Jacoff, I. Noda, Y. Takahashi (Eds.), RoboCup 2005: Robot Soccer World Cup IX. XVII, 727 pages. 2006.

Vol. 4013: L. Lamontagne, M. Marchand (Eds.), Advances in Artificial Intelligence. XIII, 564 pages. 2006.

Vol. 4012: T. Washio, A. Sakurai, K. Nakajima, H. Takeda, S. Tojo, M. Yokoo (Eds.), New Frontiers in Artificial Intelligence. XIII, 484 pages. 2006.

Vol. 4005: G. Lugosi, H.U. Simon (Eds.), Learning Theory. XI, 656 pages. 2006.

Vol. 3978: B. Hnich, M. Carlsson, F. Fages, F. Rossi (Eds.), Recent Advances in Constraints. VIII, 179 pages. 2006.

Vol. 3963: O. Dikenelli, M.-P. Gleizes, A. Ricci (Eds.), Engineering Societies in the Agents World VI. XII, 303 pages. 2006.

Vol. 3960: R. Vieira, P. Quaresma, M.d.G.V. Nunes, N.J. Mamede, C. Oliveira, M.C. Dias (Eds.), Computational Processing of the Portuguese Language. XII, 274 pages. 2006.

Vol. 3955: G. Antoniou, G. Potamias, C. Spyropoulos, D. Plexousakis (Eds.), Advances in Artificial Intelligence. XVII, 611 pages. 2006.

Vol. 3946: T.R. Roth-Berghofer, S. Schulz, D.B. Leake (Eds.), Modeling and Retrieval of Context. XI, 149 pages. 2006.

Vol. 3944: J. Quiñonero-Candela, I. Dagan, B. Magnini, F. d'Alché-Buc (Eds.), Machine Learning Challenges. XIII, 462 pages. 2006.

Vol. 3930: D.S. Yeung, Z.-Q. Liu, X.-Z. Wang, H. Yan (Eds.), Advances in Machine Learning and Cybernetics. XXI, 1110 pages. 2006.

Vol. 3918: W.K. Ng, M. Kitsuregawa, J. Li, K. Chang (Eds.), Advances in Knowledge Discovery and Data Mining. XXIV, 879 pages. 2006.

Vol. 3913: O. Boissier, J. Padget, V. Dignum, G. Lindemann, E. Matson, S. Ossowski, J.S. Sichman, J. Vázquez-Salceda (Eds.), Coordination, Organizations, Institutions, and Norms in Multi-Agent Systems. XII, 259 pages. 2006.

Vol. 3910: S.A. Brueckner, G.D.M. Serugendo, D. Hales, F. Zambonelli (Eds.), Engineering Self-Organising Systems. XII, 245 pages. 2006.

Vol. 3904: M. Baldoni, U. Endriss, A. Omicini, P. Torroni (Eds.), Declarative Agent Languages and Technologies III. XII, 245 pages. 2006.

Vol. 3900: F. Toni, P. Torroni (Eds.), Computational Logic in Multi-Agent Systems. XVII, 427 pages. 2006.

Vol. 3899: S. Frintrop, VOCUS: A Visual Attention System for Object Detection and Goal-Directed Search. XIV, 216 pages. 2006.

Vol. 3898: K. Tuyls, P.J. 't Hoen, K. Verbeeck, S. Sen (Eds.), Learning and Adaption in Multi-Agent Systems. X, 217 pages. 2006.

Vol. 3891: J.S. Sichman, L. Antunes (Eds.), Multi-Agent-Based Simulation VI. X, 191 pages. 2006.

Vol. 3890: S.G. Thompson, R. Ghanea-Hercock (Eds.), Defence Applications of Multi-Agent Systems. XII, 141 pages. 2006.

Vol. 3885: V. Torra, Y. Narukawa, A. Valls, J. Domingo-Ferrer (Eds.), Modeling Decisions for Artificial Intelligence. XII, 374 pages. 2006.

Vol. 3881: S. Gibet, N. Courty, J.-F. Kamp (Eds.), Gesture in Human-Computer Interaction and Simulation. XIII, 344 pages. 2006.

Vol. 3874: R. Missaoui, J. Schmidt (Eds.), Formal Concept Analysis. X, 309 pages. 2006.

Vol. 3873: L. Maicher, J. Park (Eds.), Charting the Topic Maps Research and Applications Landscape. VIII, 281 pages. 2006.

Vol. 3863: M. Kohlhase (Ed.), Mathematical Knowledge Management. XI, 405 pages. 2006.

Vol. 3862: R.H. Bordini, M. Dastani, J. Dix, A.E.F. Seghrouchni (Eds.), Programming Multi-Agent Systems. XIV, 267 pages. 2006.

Vol. 3849: I. Bloch, A. Petrosino, A.G.B. Tettamanzi (Eds.), Fuzzy Logic and Applications. XIV, 438 pages. 2006.

Vol. 3848: J.-F. Boulicaut, L. De Raedt, H. Mannila (Eds.), Constraint-Based Mining and Inductive Databases. X, 401 pages. 2006.

Vol. 3847: K.P. Jantke, A. Lunzer, N. Spyratos, Y. Tanaka (Eds.), Federation over the Web. X, 215 pages. 2006.